Main Currents of Marxism

'the discussion as a whole is unique and unparalleled in the entire literature on Marx and Marxism, and should dominate the field for decades to come'

Tom Rockmore, *Man and World*

'a deeply impressive examination of the entire Marxist tradition, marked by lucid and accurate exposition, sustained and high-level analysis, and a passionately committed point of view which ... distorts neither exposition nor analysis'

Steven Lukes, *New York Review of Books*

MAIN CURRENTS OF MARXISM is a handbook and a thorough survey of the varieties of Marxism. Leszek Kolakowski describes the development of Marx's own thought and the contributions of his best-known followers. No survey of the doctrines of the Marxist tradition could fail to be controversial, but Kolakowski's treatment is detached and pluralistic, and he does not attempt to identify a pure or essentially Marxist strand in the tradition as a whole. There is no better example of the variety of Marxism than the diversity which results from the tension between the Utopian and fatalist impulses in Marx's thought. In the author's own words, 'The surprising diversity of views expressed by Marxists in regard to Marx's so-called historical determinism is a factor which makes it possible to present and schematize with precision the trends of twentieth-century Marxism. It is also clear that one's answer to the question concerning the place of human consciousness and will in the historical process goes far towards determining the sense one ascribes to socialist ideals and is directly linked with the theory of revolutions and of crises.'

LESZEK KOLAKOWSKI was for many years, until March 1968 when he was expelled for political reasons, Professor of the History of Philosophy at the University of Warsaw. He is now a Fellow of All Souls College, Oxford, and has been a visiting Professor at the universities of Montreal, Yale, and California, Berkeley.

Main Currents of Marxism

ITS ORIGINS, GROWTH AND DISSOLUTION

LESZEK KOLAKOWSKI

*

I. THE FOUNDERS

Translated from the Polish by P. S. Falla

Oxford New York

OXFORD UNIVERSITY PRESS

Oxford University Press, Walton Street, Oxford OX2 6DP

London   Glasgow   New York   Toronto
Delhi   Bombay   Calcutta   Madras   Karachi
Kuala Lumpur   Singapore   Hong Kong   Tokyo
Nairobi   Dar es Salaam   Cape Town
Melbourne   Auckland
and associates in
Beirut   Berlin   Ibadan   Mexico City   Nicosia

First published 1978 by Oxford University Press
First published as an Oxford University Press paperback 1981
Reprinted 1982

British Library Cataloguing in Publication Data
Kołakowski, Leszek
Main currents of Marxism.
Vol. I: The founders
I. Communism — History
I. Title II. Falla, Paul Stephen
335.4'09'034   HX21   78-40247
ISBN 0-19-285107-1

Printed in the
United States of America

PREFACE

THE present work is intended to serve as a handbook. In saying this I am not putting forward the absurd claim to have succeeded in presenting the history of Marxism in a non-controversial manner, eliminating my own opinions, preferences, and principles of interpretation. All I mean is that I have endeavoured to present that history not in the form of a loose essay but rather so as to include the principal facts that are likely to be of use to anyone seeking an introduction to the subject, whether or not he agrees with my assessment of them. I have also done my best not to merge comment with exposition, but to present my own views in separate, clearly defined sections.

Naturally an author's opinions and preferences are bound to be reflected in his presentation of the material, his selection of themes, and the relative importance he attaches to different ideas, events, writings, and individuals. But it would be impossible to compile a historical manual of any kind—whether of political history, the history of ideas, or the history of art—if we were to suppose that every presentation of the facts is equally distorted by the author's personal views and is in fact a more or less arbitrary construction, so that there is no such thing as a historical account but only a series of historical assessments.

This book is an attempt at a history of Marxism, i.e. the history of a doctrine. It is not a history of socialist ideas, nor of the parties or political movements that have adopted one or another version of the doctrine as their own ideology. I need not emphasize that this distinction is a difficult one to observe, especially in the case of Marxism where there is manifestly a close link between theory and ideology on the one hand and political contests on the other. However, a writer on any subject is bound to extract from the 'living whole' separate portions which, as he is well aware, are not wholly self-contained or independent. If this were not permitted we should have to confine ourselves to writing histories of the world, since all things are interconnected in one way or another.

Another feature that gives the work the character of a hand-book is that I have indicated, though as briefly as possible, the basic facts showing the connection between the development of the doctrine and its function as a political ideology. The whole is a narrative strewn with my glosses.

There is scarcely any question relating to the interpretation of Marxism that is not a matter of dispute. I have tried to record the principal controversies, but it would altogether exceed the scope of this book to enter into a detailed analysis of the views of all historians and critics whose works I have studied, but whose opinions or interpretations I do not share. The book does not pretend to propose a particularly original interpretation of Marx. And it is easy to see that my reading of Marx was influenced more by Lukács than by other commentators, though I am far from sharing his attitude to the doctrine.

It will be observed that the book is not subdivided according to a single principle. It proved impossible to adhere to a purely chronological arrangement, as I found it necessary to present certain individuals or tendencies as part of a self-contained whole. The division into volumes is essentially chronological, but here too I had to permit myself some inconsistency in order, as far as possible, to treat different trends in Marxism as separate themes.

The first volume was originally drafted in 1968, during the leisure time at my disposal after dismissal from my professorship at Warsaw University. Within a year or two it became clear that the draft required a good deal of supplementing, amendment, and alteration. The second and third volumes were written in 1970–6, during my Fellowship at All Souls College, Oxford, and I am almost certain that it could not have been written but for the privileges I enjoyed as a Fellow.

The book does not contain an exhaustive bibliography, but only indications for the reader who wishes to refer to the sources and principal commentaries. In the works I have mentioned it will be easy for anyone to find references to literature which today, unfortunately, is altogether too extensive for a single reader to master.

The second volume has been read in typescript by two of my Warsaw friends, Dr. Andrzej Walicki and Dr. Ryszard Herczyński. The former is a historian of ideas, the latter a mathematician; both have made many valuable critical remarks and suggestions. The whole work has been read, prior to translation, only by myself and my wife, Dr. Tamara Kołakowska, who is a psychiatrist by profession; like all my other writings, it owes much to her good sense and critical comments.

Oxford   LESZEK KOŁAKOWSKI

CONTENTS

BIBLIOGRAPHICAL NOTE

Sources of quotations used in the text:

ST. ANSELM, *Proslogium* etc., trans. S. N. Deane, Kegan Paul, London, 1903.

ENGELS, F., *Herr E. Dühring's Revolution in Science*, Martin Lawrence, London, 1935.

— — *Dialectics of Nature*, Progress Publishers, Moscow, 1954.

HEGEL, G. W. F., *Phenomenology of Mind*, trans. Baillie, Allen & Unwin, London, 1910.

— — *Philosophy of Right*, trans. T. M. Knox, Oxford, 1942.

— — *Science of Logic*, trans. A. V. Miller, Allen & Unwin, London, 1969.

— — *Lectures on the Philosophy of History*, trans, J. Sibree, Dover, New York, 1956.

MCLELLAN, D., *Karl Marx: his Life and Thought*, Macmillan, London, 1973.

— — (ed.), *Karl Marx: Early Texts*, Blackwell, Oxford, 1971.

MARX, K., *Selected Works*, 2 vols., Lawrence & Wishart, London, 1942.

— — *Capital*, trans. in 3 vols., Lawrence & Wishart and Progress Publishers, Moscow, 1970.

— — *Grundrisse*, trans. M. Nicolaus, Penguin, London, 1973.

— — *Surveys from Exile*, ed. Fernbach, Penguin, London, 1973.

— — and F. Engels, *Selected Works*, Lawrence & Wishart, London, 1968.

PLOTINUS, *Enneads*, trans. S. McKenna, Faber, London, 1962.

Introduction

KARL MARX was a German philosopher. This does not sound a particularly enlightening statement, yet it is not so commonplace as it may at first appear. Jules Michelet, it will be recalled, used to begin his lectures on British history with the words: 'Messieurs, l'Angleterre est une île.' It makes a good deal of difference whether we simply know that Britain is an island, or whether we interpret its history in the light of that fact, which thus takes on a significance of its own. Similarly, the statement that Marx was a German philosopher may imply a certain interpretation of his thought and of its philosophical or historical importance, as a system unfolded in terms of economic analysis and political doctrine. A presentation of this kind is neither self-evident nor uncontroversial. Moreover, although it is clear to us that Marx was a German philosopher, half a century ago things were somewhat different. In the days of the Second International the majority of Marxists considered him rather as the author of a certain economic and social theory which, according to some, was compatible with various types of metaphysical or epistemological outlook; while others took the view that it had been furnished with a philosophical basis by Engels, so that Marxism in the proper sense was a body of theory compounded of two or three parts elaborated by Marx and Engels respectively.

We are all familiar with the political background to the present-day interest in Marxism, regarded as the ideological tradition on which Communism is based. Those who consider themselves Marxists, and also their opponents, are concerned with the question whether modern Communism, in its ideology and institutions, is the legitimate heir of Marxian doctrine. The three commonest answers to this question may be expressed in simplified terms as follows: (1) Yes, modern Communism is the perfect embodiment of Marxism, which proves that the latter is a doctrine leading to enslavement, tyranny, and crime; (2) Yes, modern Communism is the perfect embodiment of Marxism, which therefore signifies a hope of liberation and happiness for mankind; (3) No, Communism as we know it is a profound deformation of Marx's gospel and a betrayal of the fundamentals of Marxian socialism. The first answer corresponds to traditional anti-Communist orthodoxy, the second to traditional Communist orthodoxy, and the third to various forms of critical, revisionistic, or 'open' Marxism. The argument of the present work, however, is that the question is wrongly formulated and that attempts to answer it are not worth while. More precisely, it is impossible to answer the questions 'How can the various problems of the modern world be solved in accordance with Marxism?', or 'What would Marx say if he could see what his followers have done?' Both these are sterile questions and there is no rational way of seeking an answer to them. Marxism does not provide any specific method of solving questions that Marx did not put to himself or that did not exist in his time. If his life had been prolonged for ninety years he would have had to alter his views in ways that we have no means of conjecturing.

Those who hold that Communism is a 'betrayal' or 'distortion' of Marxism are seeking, as it were, to absolve Marx of responsibility for the actions of those who call themselves his spiritual posterity. In the same way, heretics and schismatics of the sixteenth and seventeenth centuries accused the Roman Church of betraying its mission and sought to vindicate St. Paul from the association with Roman corruption. In the same way, too, admirers of Nietzsche sought to clear his name from responsibility for the ideology and practice of Nazism. The ideological motivation of such attempts is clear enough, but their informative value is next to nothing. There is abundant evidence that all social movements are to be explained by a variety of circumstances and that the ideological sources to which they appeal, and to which they seek to remain faithful, are only one of the factors determining the form they assume and their patterns of thought and action. We may therefore be certain in advance that no political or religious movement is a perfect expression of that movement's 'essence' as laid down in its sacred writings;

on the other hand, these writings are not merely passive, but exercise an influence of their own on the course of the movement. What normally happens is that the social forces which make themselves the representatives of a given ideology are stronger than that ideology, but are to some extent dependent on its own tradition.

The problem facing the historian of ideas, therefore, does not consist in comparing the 'essence' of a particular idea with its practical 'existence' in terms of social movements. The question is rather how, and as a result of what circumstances, the original idea came to serve as a rallying-point for so many different and mutually hostile forces; or what were the ambiguities and conflicting tendencies in the idea itself which led to its developing as it did? It is a well-known fact, to which the history of civilization records no exception, that all important ideas are subject to division and differentiation as their influence continues to spread. So there is no point in asking who is a 'true' Marxist in the modern world, as such questions can only arise within an ideological perspective which assumes that the canonical writings are the authentic source of truth, and that whoever interprets them rightly must therefore be possessed of the truth. There is no reason, in fact, why we should not acknowledge that different movements and ideologies, however antagonistic to one another, are equally entitled to invoke the name of Marx —except for some extreme cases with which this work is not concerned. In the same way, it is sterile to inquire 'Who was a true Aristotelian—Averroës, Thomas Aquinas, or Pompon-azzi?', or 'Who was the truest Christian—Calvin, Erasmus, Bellarmine, or Loyola?' The latter question may have a meaning for Christian believers, but it has no relevance to the history of ideas. The historian may, however, be concerned to inquire what it was in primitive Christianity that made it possible for men so unlike as Calvin, Erasmus, Bellarmine, and Loyola to appeal to the same source. In other words, the historian treats ideas seriously and does not regard them as completely subservient to events and possessing no life of their own (for in that case there would be no point in studying them), but he does not believe that they can endure from one generation to another without some change of meaning.

The relationship between the Marxism of Marx and that of

the Marxists is a legitimate field of inquiry, but it does not enable us to decide who are the 'truest' Marxists.

If, as historians of ideas, we place ourselves outside ideology, this does not mean placing ourselves outside the culture within which we live. On the contrary, the history of ideas, and especially those which have been and continue to be the most influential, is to some extent an exercise in cultural self-criticism. I propose in this work to study Marxism from a point of view similar to that which Thomas Mann adopted in *Doktor Faustus* *vis-à-vis* Nazism and its relation to German culture. Thomas Mann was entitled to say that Nazism had nothing to do with German culture or was a gross denial and travesty of it. In fact, however, he did not say this: instead, he inquired how such phenomena as the Hitler movement and Nazi ideology could have come about in Germany, and what were the elements in German culture that made this possible. Every German, he maintained, would recognize with horror, in the bestialities of Nazism, the distortion of features which could be discerned even in the noblest representatives (this is the important point) of the national culture. Mann was not content to pass over the question of the birth of Nazism in the usual manner, or to contend that it had no legitimate claim to any part of the German inheritance. Instead, he frankly criticized that culture of which he was himself a part and a creative element. It is indeed not enough to say that Nazi ideology was a 'caricature' of Nietzsche, since the essence of a caricature is that it helps us to recognize the original. The Nazis told their supermen to read *The Will to Power*, and it is no good saying that this was a mere chance and that they might equally well have chosen the *Critique of Practical Reason*. It is not a question of establishing the 'guilt' of Nietzsche, who as an individual was not responsible for the use made of his writings; nevertheless, the fact that they were so used is bound to cause alarm and cannot be dismissed as irrelevant to the understanding of what was in his mind. St. Paul was not personally responsible for the Inquisition and for the Roman Church at the end of the fifteenth century, but the inquirer, whether Christian or not, cannot be content to observe that Christianity was depraved or distorted by the conduct of unworthy popes and bishops; he must rather seek to discover what it was in the Pauline epistles that gave rise, in the fullness of time, to unworthy and criminal actions. Our attitude to the problem of Marx and Marxism should be the same, and in this sense the present study is not only a historical account but an attempt to analyse the strange fate of an idea which began in Promethean humanism and culminated in the monstrous tyranny of Stalin.

The chronology of Marxism is complicated for the chief reason that many of what are now considered Marx's most important writings were not printed until the twenties and thirties of the present century or even later. This applies, for example, to the full text of *The German Ideology*; the full text of the doctoral thesis on the *Difference between the Democritean and Epicurean Philosophy of Nature*; *Contribution to the Critique of the Hegelian Philosophy of Law*; the *Economic and Philosophic Manuscripts* of 1844; *Foundations of the Critique of Political Economy (Grundrisse)*; and also Engels's *Dialectic of Nature*. These works could not affect the epoch in which they were written, but today they are regarded as important not only from the biographical point of view but also as integral components of a doctrine which cannot be understood without them. It is still disputed whether, and how far, what are considered to be Marx's mature ideas, as reflected especially in *Capital*, are a natural development of his philosophy as a young man, or whether, as some critics hold, they represent a radical intellectual change: did Marx, in other words, abandon in the fifties and sixties a mode of thought and inquiry bounded by the horizon of Hegelian and Young Hegelian philosophy? Some believe that the social philosophy of *Capital* is, as it were, prefigured by the earlier writings and is a development or particularization of them, while others maintain that the analysis of capitalist society denotes a breakaway from the utopian and normative rhetoric of the early period; and the two conflicting views are correlated with opposing interpretations of the whole body of Marx's thought.

It is a premiss of this work that, logically as well as chronologically, the starting-point of Marxism is to be found in philosophic anthropology. At the same time, it is virtually impossible to isolate the philosophic content from the main body of Marx's thought. Marx was not an academic writer but a humanist in the Renaissance sense of the term: his mind was concerned with the totality of human affairs, and his vision of social liberation embraced, as an interdependent whole, all the major problems with which humanity is faced. It has become customary to divide Marxism into three fields of speculation— basic philosophic anthropology, socialist doctrine, and economic analysis—and to point to three corresponding sources in German dialectics, French socialist thought, and British political economy. Many are of the opinion, however, that this clear-cut division is contrary to Marx's own purpose, which was to provide a global interpretation of human behaviour and history and to reconstruct an integral theory of mankind in which particular questions are only significant in relation to the whole. As to the manner in which the elements of Marxism are interrelated, and the nature of its internal coherence, this is not something which can be defined in a single sentence. It would seem, however, that Marx endeavoured to discern those aspects of the historical process which confer a common significance on epistemological and economic questions and social ideals; or, to put it another way, he sought to create instruments of thought or categories of knowledge that were sufficiently general to make all human phenomena intelligible. If, however, we attempt to reconstruct these categories and display Marx's thought in accordance with them, we run the risk of neglecting his evolution as a thinker and of treating the whole of his work as a single homogeneous block. It seems better, therefore, to pursue the development of his thought in its main lines and only afterwards to consider which of its elements were present from the outset, albeit implicitly, and which may be regarded as transient and accidental.

The present conspectus of the history of Marxism will be focused on the question which appears at all times to have occupied a central place in Marx's independent thinking: viz. how is it possible to avoid the dilemma of utopianism versus historical fatalism? In other words, how can one articulate and defend a viewpoint which is neither the arbitrary proclamation of imagined ideals, nor resigned acceptance of the proposition that human affairs are subject to an anonymous historical process in which all participate but which no one is able to control? The surprising diversity of views expressed by Marxists in regard to Marx's so-called historical determinism is a factor which makes it possible to present and schematize with precision the trends of twentieth-century Marxism. It is also clear that one's answer to the question concerning the place of human consciousness and will in the historical process goes far towards determining the sense one ascribes to socialist ideals, and is directly linked with the theory of revolutions and crises.

The starting-point of Marx's thinking, however, was provided by the philosophic questions comprised in the Hegelian inheritance, and the break-up of that inheritance is the natural background to any attempt at expounding his ideas.

CHAPTER I

The Origins of Dialectic

ALL living trends of modern philosophy have their own pre-history, which can be traced back almost to the beginnings of recorded philosophical thought. They have, in consequence, a history which is older than their names and clearly distinguishable forms: it is meaningful to speak of positivism before Comte, or existential philosophy before Jaspers. At first sight it may appear that Marxism is in a different position, since it derives its name from that of its founder: 'Marxism before Marx' would seem to be as much of a paradox as 'Cartesianism before Descartes' or 'Christianity before Christ'. Yet even intellectual trends that originate with a given person have a prehistory of their own, embodied in a range of questions that have come to the fore, or a series of isolated answers that are knitted into a single whole by some outstanding mind and are thus transformed into a new cultural phenomenon. 'Christianity before Christ' may, of course, be a mere play on words, using 'Christianity' in a sense different from that which it normally bears; there is general agreement, after all, that the history of early Christianity cannot be understood without the knowledge that scholars have been at pains to acquire concerning the spiritual life of Judaea immediately before the advent of Christ. Something analogous may be predicated of Marxism. The phrase 'Marxism before Marx' has no meaning, but Marx's thought would be emptied of its content if it were not considered in the setting of European cultural history as a whole, as an answer to certain fundamental questions that philosophers have posed for centuries in one form or another. It is only in relation to these questions, to their evolution and the different ways in which they have been formulated, that Marx's philosophy can be understood in its historical uniqueness and the permanency of its values.

In the last quarter of a century many historians of Marxism have done valuable work in studying the questions that classical German philosophy presented to Marx and to which he offered new answers. But that philosophy itself, from Kant to Hegel, was an attempt to devise new conceptual forms for basic, immemorial questions. It makes no sense except in terms of such questions, though certainly it is not exhausted by them—if such a simplification were possible the history of philosophy would cease to exist, as every philosophical development would be deprived of its unique relation to its own time. In general the history of philosophy is subject to two principles that limit each other. On the one hand, the questions of basic interest to each philosopher must be regarded as aspects of the same curiosity of the human mind in the face of the unaltering conditions with which life confronts it; on the other hand, it behoves us to bring to light the historical uniqueness of every intellectual trend or observable fact and relate it as closely as possible to the epoch that gave birth to the philosopher in question and that he himself helped to form. It is difficult to observe both these rules at once, since, although we know they are bound to limit each other, we do not know precisely in what manner and are therefore thrown back on fallible intuition. The two principles are thus far from being so reliable or unequivocal as the method of setting up a scientific experiment or identifying documents, but they are none the less useful as guidelines and as a means of avoiding two extreme forms of historical nihilism. One is based on the systematic reduction of every philosophical effort to a set of eternally repeated questions, thus ignoring the panorama of the cultural evolution of mankind and, in general, disparaging that evolution. The second form of nihilism consists in that we are satisfied with grasping the specific quality of every pheno-menon or cultural epoch, on the premiss, expressed or implied, that the only factor of importance is that which constitutes the uniqueness of a particular historical complex, every detail of which—although it may be indisputably a repetition of former ideas—acquires a new meaning in its relationship to that com-plex and is no longer significant in any other way. This hermeneutic assumption clearly leads to a historical nihilism of its own, since by insisting on the exclusive relationship of every detail to a synchronic whole (whether the whole be an individual mind or an entire cultural epoch) it rules out all continuity of interpretation, obliging us to treat the mind or the epoch as one of a series of closed, monadic entities. It lays down in advance that there is no possibility of communication among such entities and no language capable of describing them collectively: every concept takes on a different meaning according to the complex to which it is applied, and the construction of superior or non-historical categories is ruled out as contrary to the basic principle of investigation.

In seeking to avoid both these nihilistic extremes, it is the purpose of this inquiry to understand Marx's basic thoughts as answers to questions that have long exercised the minds of philosophers, but at the same time to comprehend them in their uniqueness both as emanations of Marx's genius and as phenomena of a particular age. It is easier to formulate such a directive than to apply it successfully; to do so to perfection, one would have to write a complete history of philosophy, or indeed of human civilization. As a modest substitute for that impossible task, we propose to give a brief account of the questions in regard to which Marxism can be described as constituting a new step in the development of European philosophy.

1. The contingency of human existence

If the aspiration of philosophy was and is to comprehend intellectually the whole of Being its initial stimulus came from awareness of human imperfection. Both this awareness and the resolve to overcome man's imperfection by means of understanding the Whole were inherited by philosophy from the realm of myth.

Philosophical interest centred on the limitations and misery of the human condition—not in its obvious, tangible, and remediable forms, but the fundamental impoverishment which cannot be cured by technical devices and which, when once apprehended, was felt to be the cause of man's more obvious, empirical deficiencies, the latter being mere secondary phenomena. The fundamental, innate deficiency was given various names: medieval Christian philosophy spoke of the 'contingency' of human existence, as of all other created beings. The term 'contingency' was derived from the Aristotelian tradition (the *De interpretatione* refers to contingent judgements as those which predicate of an object something which may or may not apply to it without altering its nature) and denoted the state of a finite being that might or might not exist but was not necessary, i.e. its essence did not involve existence. Every created thing has a beginning in time: there was a time when it did not exist, and consequently it does not exist of necessity. For the Scholastics, following Aristotle, the distinction between essence and existence served to distinguish created beings from the Creator who existed necessarily (God's essence and existence were one and the same), and was the most evident proof of the transitoriness of creation; but it was not regarded as a misfortune or a manifestation of decay. The fact that man was a contingent or accidental being was a cause for humility and worship of the Creator; it was an inevitable, ineradicable aspect of his being, but it did not denote a fall from a higher state. Man's bodily and temporal existence was not the result of any degradation, but were the natural characteristics of the human species within the hierarchy of created beings.

In the Platonic tradition, on the other hand, the term 'contingent' was seldom or never used, and the fact that man as a finite, temporal being was different from the essence of humanity signified that 'man was other than himself', i.e. his empirical, temporal, factual existence was not identical with the ideal, perfect, extra-temporal Being of humanity as such. But to be 'other than oneself' is to suffer from an unbearable disjunction, to live in awareness of one's own decline and in perpetual longing for the perfect identification from which we are debarred by our existence in time and in a physical body subject to corruption. The world in which we live as finite individuals, conscious of our own transience, is a place of exile.

2. *The soteriology of Plotinus*

Plato and the Platonists formulated in philosophical language the question, originating in religious tradition and pervading the whole history of European civilization: is there a remedy for the contingent state of man? Is his life incurably accidental, as Lucretius thought and as the existentialists maintain today, or has man, despite his duality, preserved some discoverable link with non-accidental and non-contingent Being, so that he may entertain a hope of self-identification? Or, in other terms, is he summoned or destined to return to a state of completeness and non-contingency?

For the Platonists, especially Plotinus, and also for St. Augustine, the deficiency of human existence is most evident in its temporal character—not only in the fact that man has a beginning in time, but in his being subject to the time-process at all. Plotinus follows up a line of thought originated by Parmenides, and although his intellectual construction culminates in a stage higher than Being as Parmenides conceived it (which Plotinus treats as secondary to the One or the Absolute), his basic philosophical outlook is nevertheless the same as his predecessor's. Plotinus does not argue, like the Aristotelians, *ex contingente ad necessarium*; that is, he does not try to show that the reality of the One can be conclusively deduced from observation of finite beings, as a logical presupposition of their existence. The reality of the One is inexpressible yet self-evident, since 'to be' in its most basic sense signifies to be immutably and absolutely, to be undifferentiated and outside time. That which truly *is* cannot be subject to time, to the distinction of past and future. Finite and conditional beings, on the other hand, are constantly moving from a past that has ceased to be into a future that does not yet exist; they are obliged to see themselves in terms of memory or anticipation; their self-knowledge is not direct, but mediated by the distinction of what was and what will be. They are not self-identical or 'all of a piece': they live in a present that vanishes even as it comes to be, and can then only be revealed through memory. The One is truly self-identical and for that reason cannot be properly understood in opposition to the transitory world, but only in and through itself. ('We cannot think of the First as moving towards any other; He holds his own manner of being before any other was; even Being we withhold, and therefore all relation to beings': *Enneads*, VI. 8. 8. 'Certainly that which has never passed outside its own orbit, unbendingly what it is, its own unchangeably, is that which may most strictly be said to possess its own being' (ibid. 9).) Composite beings, on the other hand, are not self-identical, inasmuch as it is one thing to say that they exist and another to say that they are such and such. That which is what it is is not amenable to language: even 'the

One', even 'Being' are clumsy attempts to express the inexpressible; those who have experienced this Being know what they are speaking of, but can never communicate their experience. The *Enneads* revolve with infinite persistence around this fundamental intuition which perpetually eludes the grasp of speech. It cannot truly be said of finite creatures that they 'are', since they fade away at each moment of their duration and cannot perceive themselves as identities, but are obliged to go forwards or backwards beyond themselves to achieve self-understanding. But the term 'existence' is inappropriate to the Absolute also, since in ordinary language it is applied to that which can be grasped by means of concepts, whereas the One is not a conceptual entity. Our reason approaches the One by way of negation, and with our insufficient minds we apprehend it as that which is radically other than the world of limitations —not merely the sensual one, but even the world of eternal rational ideas. But this negative approach is only an unfortunate necessity, and in reality things are the other way about: it is the world of transitory objects which is negative, characterized by limitation and by participation in non-being. The One is not 'something', for to be something is merely not to be some other thing: it is to be definable by qualities that the object possesses and that are opposed to those it does not possess. To be something is to be limited, in other words to be in some measure nothing.

The hypostases of reality are so many stages of degradation. Being or intellect, as a secondary hypostasis, represents the One degraded by multiplicity, since it involves taking cognizance of oneself and hence a kind of duality between that which apprehends and that which is apprehended. (We cannot speak of knowledge in the One, since the act of cognition distinguishes a subject and an object: *Enn.* v. 3. 12–13 and v. 6. 2–4.) The soul, which is the third hypostasis, consists of mind degraded by contact with the physical world, that is to say with evil or non-existence. Matter, and the bodies which are its qualitative manifestation, are the last stage of degradation, representing radical passivity and non-self-sufficiency: incomplete, divested of harmony, little more than shadows, their Being signifies virtually no more than non-being (*Enn.* i. 8. 3–5). The attenuation of existence is measured by the descent from unity to multiplicity, from immobility to motion and from eternity to time. Movement is a degradation of quiescence, activity is enfeebled contemplation (*Enn.* III. 8. 4), time is a corruption of eternity. The human mind can only conceive eternity as non-time, but in fact time is non-eternity, the negation or dilution of Being. To be in time signifies not to be all at once. Strictly speaking, according to the intricate exposition of *Enn.* III. 7, it is not souls which are in time but time which is in them, for they have created time by concerning themselves with objects of sense. Plotinus's definition of eternity, fore-shadowing the celebrated language of the *Consolations* of Boethius, is that it is 'the Life—instantaneously entire, complete ... which belongs to the Authentic Existent by its very existence' (*Enn.* III. 7. 3). It knows no distinction between what was and what is not yet, and is therefore identical with true Being: for 'to be in reality' signifies 'never not to be' and 'never to be different', i.e. to be self-identical and unchangeable (III. 7. 6).

But the soul imprisoned in transience, impelled without ceasing from the nothingness of what was to the nothingness of what is to be, is not after all condemned to endless exile. The sixth *Ennead* is not only a description of the infinite distance between supreme reality and the life of our minds and senses, speech and concepts: it also points the way by which we may return from exile to union with the Absolute. This return, however, is not an exaltation of man above his natural state (the conception of the supernatural cannot in general be read into Plotinus's thought), but is a reversion of the soul to its own self. 'It is not in the soul's nature to touch utter nothingness; the lowest descent is into evil and, so far, into non-being; but into utter nothing, never. When the soul begins again to mount, it comes not to something alien but to its very self; thus detached, it is in nothing but in itself' (*Enn.* VI. 9. 11). Even at its lowest descent, the soul is not cut off from its source and is always free to return. 'We have not been cut away; we are not separate, though the body-nature had closed about us to press us to itself; we breathe and hold our ground because the Supreme does not give and pass but gives of its bounty for ever, so long as it remains what it is' (VI. 9. 9). The road to unity does not mean a quest for something outside the seeker: it involves, on the contrary, casting off all ties with external reality, first with the physical world and then with that of ideas, so that the soul may commune with that which constitutes its inmost being. Plotinus's work is not a metaphysical system, since language cannot express the most important truths; it is not a theory but a work of spiritual counsel, a guide for the use of those who wish to set about liberating themselves from temporal being.

Plotinus, Iamblichus, and other Platonists exerted influence both directly and to the extent that their ideas were accepted by early Christian thinkers. The conception which they thus disseminated, and which has never disappeared from our culture, was a philosophical articulation of the mythopoeic longing for a lost paradise and faith in One who Is—who presents himself to man not only as a creator or a self-sufficient being but also as the supreme good, as the fulfilment of man's fine purpose and as a voice summoning him to itself. The Platonists familiarized philosophy with categories intended to express the difference between empirical, factual, and finite existence and man's true Being, self-identical and free from the shackles of time: they pointed to a 'native home' beyond everyday reality, a place where man might be what he truly was. They explained the process of decline and reascent, the difference between man's contingent and his authentic Being, and sought to show how he might overcome this duality by an effort of self-deification. They refused to accept contingency as the human lot, and believed that the way lay open to the Absolute.

At the same time Plotinus pointed out the link between the duality of man's nature and the limitation which obliges us to regard the known world as essentially different from ourselves, so that our thoughts and perceptions move in an alien universe. If we can overcome the soul's alienation from itself (an alienation forced upon it by time, since we can only know ourselves as we are no longer, or as we have yet to be), then we shall also have overcome the alienation between the soul and everything it knows, loves, or desires. Plato wrote in one of his *Letters* that 'he who is not linked by a tie of kinship with the object will not acquire insight through ease of apprehension or a good memory; for basically he does not accept the object, as its nature is foreign to him' (*Letter* 7, 344a). In true knowledge, by contrast, the subject is not merely an absorber of information about realities that are completely external to him: he enters into an intimate contact with the object, and this cognition is his mode of becoming better than he was before. For Plato and the Platonists, therefore, the soul's urge to liberate itself from contingency involves overcoming the alienation between the soul and its object. Whatever makes the world alien and essentially different from me is, by the same token, a cause of my own limitation, insufficiency, and imperfection. To rediscover oneself is to make the world one's own again, to come to terms with reality. My own unity signifies my unity with the world, and my ascent to knowledge is identical with the aspiration of the universe to a lost unity. As the human mind is the guiding light of creation, logic, that is to say the movement of my thoughts about reality, is the process by which reality seeks its own reintegration. This may sound like a piece of Hegelian exposition, but it is quite in accordance with Plotinus's thinking: 'Dialectic does not consist of bare theories and rules: it deals with verities; Existences are, as it were, Matter to it, or at least it proceeds methodically towards Existences, and possesses itself, at the one step, of the notions and of the realities' (*Enn.* 1. 3. 5). Since the cosmic odyssey is the history of the soul, and the soul's activity is logical thought, ideas and reality tend to converge in their evolution and there is no longer any distinction between dialectics and metaphysics. Thought, in the true sense, is and must be self-directed. 'If the intellectual act is directed upon something outside, then it is deficient and the intellection faulty' (*Enn.* v. 3. 13).

To resume: the only single reality, for Plotinus, is that which is absolute, non-contingent, and identical with its own existence. The contingency of man's Being lies in the fact that his true essence is outside himself and differs from his empirical life, as is most evident in the latter's subjection to time. A return to non-contingency signifies a return to unity with the Absolute, in a way which cannot be more closely defined and is indeed inexpressible. The return involves liberation from time, so that memory ceases to exist (*Enn.* iv. 4. 1). The process by which the soul frees itself from time is likewise an evolution of the whole of reality from a conditional to an absolute state. The effect of the process is to obliterate the distinction between knower and known; the subject and object are once more unified, and the world ceases to be a foreign realm into which the soul makes its entrance from outside.

3. Plotinus and Christian Platonism. The search for the reason of creation
The Christian version of Platonism, i.e. the philosophy of
St. Augustine, differs fundamentally from that of Plotinus in that
it is based on the Incarnation and Redemption and the idea
of a personal God who calls the world into being by his own
free choice. But for Augustine, too, the contingency of man is
most evident in his temporal being. Book XI of the *Confessions*,
which is doubtless influenced by Plotinus, reflects the overwhelm-
ing experience of becoming aware of one's existence between
the unreal past and the unreal future. Time must be subjective,
an attribute of the soul experiencing its own Being, since that
which was and that which will be have no existence except what
is apprehended by human thought. It is only in relation to the
soul, therefore, that we can speak intelligently of a distinction
between past and future reality. But this distinction itself betrays
the contingency of a being who is aware that his own life is
a perpetual evanescence, represented at any given time by a
point which has no extension and is isolated between two
stretches of nothingness. Augustine, like Plotinus, depicts the
insufficiency of man, but the idea of Providence alters the
picture. Since the one basic dichotomy is that between the
personal God and the created world, and since that world is
encompassed by God's providence; since, moreover, the earth
is a place of exile to which we are relegated on account of sin
and not by an ineluctable process of emanation, while our
liberation from sin is the work of the incarnate Redeemer, it
is not surprising that Augustine's writings are a cry for help rather
than, like Plotinus's, a summons to effort. Augustine's thought,
profoundly influenced by the controversy with Manichaeism,
lays paramount weight on the omnipotence of God watching
over his creation, whereas for Plotinus reality is, first and
foremost, 'a road leading both upwards and downwards'.
Plotinus's Absolute is, in the sense we have indicated, 'human
nature': man discovers it within him as his true self and
recognizes that Eternity is his native home, whereas Augustinian
man identifies himself as a helpless, miserable being incapable
of self-liberation. As we have seen, a division between the natural
and the supernatural would be meaningless in Plotinus's system,
whereas in Augustine's it is the basic framework of metaphysics.

God is not the essence of man, but a Ruler and a source of help. Temporal existence is the visible token of man's insignificance, through which he becomes aware of his need for protection and support.

In short, the return to a lost paradise means different things to the two philosophers, and they give different accounts of how it is to be achieved. For Plotinus it signifies identification with the Absolute, and can be attained by the unaided effort of any man who can free himself from the bonds of corporeal and intellectual being; in principle, the Absolute is 'within us'. For Augustine the return is only possible with the help of grace, and the exertion of the individual will plays a secondary part or no part at all. Moreover, it does not do away with the difference between the Creator and creation, nor is there any question of recovering a lost identity between them; on the contrary, the first step towards returning is the soul's awareness of the gulf which separates fallen man from God.

Both systems, however, the emanational and that of Christianity, leave unanswered the question which they regard as beyond the power of human mind to solve, although they make some attempt to do so: namely, how did the degradation of Being take place? The way this question is formulated varies according to the conception of the Absolute: in the first case it is 'Why did the One give rise to the manifold?', in the second 'Why did God create the world?' The One in Plotinus's thinking, like the Creator in Augustine's, is characterized by absolute self-sufficiency, and it would be blasphemous to suppose that they needed other beings or lacked anything that could be supplied by the created world. Nor can the question 'Why?' be asked in the sense of discovering an external cause that could influence the will of God or the emanational activity of the Absolute. A being which is completely self-sufficient, lacking and needing nothing, unable to be more perfect than it is, cannot display to the human mind any 'reason' prompting the act of creation. The very notion of an Absolute Creator contains within itself a kind of contradiction: if absolute, why does he or it create human kind? If created reality includes evil—even though we regard this evil as mere negation, defect, or insufficiency—how can we explain its presence in a world brought into being by an Absolute which is itself supreme Power and supreme Goodness?

Plotinus and Augustine give essentially the same answer to this question, by which they are both equally baffled. According to Plotinus, everything depends on the Good and aspires to it, as all things need it while it needs nothing (*Enn*. I. 8. 2). Since there exists not only the Good but that which radiates from it, the limit of that radiation must of necessity be Evil, that is to say pure deficiency, which is matter. ('As necessarily as there is something after the First, so necessarily there is a Last; this Last is Matter, the thing which has no residue of good in it': *Enn*. I. 8. 7). The road leading downwards from the One to lower and lower hypostases has a kind of inevitability about it, entailing successive degrees of deficiency or evil. But as to why the Supreme Good had to go outside itself in order to produce a reality that it does not need and, in so doing, to introduce the disturbance of evil into the closed autarky of the Absolute—of this Plotinus has nothing to say except for a laconic remark about 'superfluity' or 'superabundance' (*hyperpleres*). ('Seeking nothing, possessing nothing, lacking nothing, the One is perfect and, in our metaphor, has overflowed, and its superfluity has produced the new: this product has turned again to its begetter and been filled and has become its contemplator and so an intellectual principle [*nous*]': *Enn*. V. 2. 1).

This enigmatic notion of a 'superfluity' of existence or of goodness has continued to serve Christian philosophy as a solution to the awkward problem, although its inadequacy is obvious enough—we may ask, for instance, 'superfluity' in relation to what? Augustine himself does not seem to be bothered by the problem or to see that there is anything to answer, and he expresses astonishment at what he calls Origen's errors on the subject. God, he declares, does not experience any lack; the creation is the effect of his goodness; he did not create the world from necessity or from any need of his own, but because he is good and because it is fitting for the Supreme Good to create good things. (*Conf*. XIII. 2. 2; *Civitas Dei*, XI. 21–3).

This motif recurs, almost without change, throughout the course of Christian philosophy in so far as it is free from suspicion of unorthodoxy. As St. Thomas Aquinas puts it, '... excessus autem divinae bonitatis supra creaturam per hoc maxime exprimitur quod creaturae non semper fuerunt' (*Summa contra Gentiles*, II. 35). This indeed is all that can be said, given the premiss of

God's perfect self-sufficiency, but the flimsiness of the explanation could not go altogether unperceived. What exactly can be the *excessus bonitatis* which creates a universe that nobody needs? Kindness, or bounty, is a relative quality, at any rate to the human mind; it is impossible for us to comprehend the goodness of a self-sufficient God without any creature to which it can be extended, and we are thus led to conclude that the goodness of God without the universe is a virtual and not an actual goodness—but this conflicts with the principle that there is no potentiality in God. We might suppose that the act of creation was necessary to God in order that his goodness might manifest itself, so that in creation God attains to a higher perfection than before; but this in turn conflicts with the principle that God's perfection is absolute and cannot be increased. Theology, of course, has answers to these objections, pointing out that it is meaningless to speak of God 'before' the act of creation, because time itself is part of the created universe and God is not subject to temporality; as Augustine says, he does not precede his creation. In any case, the theologians continue, our minds are not capable of fathoming the depths of God's nature, but can only understand him in relation to the work of his hands—as Creator, as almighty, kind, and merciful; while, on the other hand, it is certain that no relative attributes can pertain to God, that he exists in and by himself and that the universe cannot modify his Being. These answers, however, amount merely to an admission that no answer can be given. For if we are only able to know the divine nature in relation to ourselves, and if we know that this relativity is not a reality in God himself, then it follows that the question we are seeking to answer concerning the essence of God in himself and its relation to his essence 'after' the creation is not a question that can properly be asked, and that we must fall back on the sacred formulas without attempting to probe their meaning.

But there is yet another difficulty in explaining creation by the 'excess of divine goodness', namely the existence of evil. True, the whole of Christian theology since the controversies with Gnosticism and Manichaeism agrees in holding that evil is not a reality in itself but is pure negativity, deficiency, the absence of good. Evil is the lack of what ought to be, and the notion of evil thus introduces a normative idea to which reality itself is inadequate. The inequality of created things is not an evil, but a matter of order and degree. Evil in the strictest sense, i.e. moral evil, proceeds solely from beings endowed with reason and is caused by the sin of disobedience. Such beings are able to exert their own will against that of the Creator, and thus evil is not God's work. The various passages of Holy Writ, debated over the centuries, which suggest plainly and disquietingly that God is the author of evil as well as of good (e.g. Isa. 45: 7; Eccles. 7: 14 (in the Vulgate); Ecclus. 33: 12; Amos 3: 6) can of course be reconciled with orthodoxy by skilful exegesis (God permits evil but does not perform it), but this does not explain the creation of a world which itself produces evil. Christian theodicy wavers between two basic solutions. The first argues that evil is an indispensable component of the cosmos as a whole; this amounts to suggesting that in fact there is no such thing as evil, or that it only appears to exist from a partial point of view and vanishes when the universe is contemplated in its entirety—a standpoint characteristic of doctrines that verge upon pantheism. The second solution argues that although evil is mere negation, *privatio* or *carentia*, its source is the corruption of the will that disobeys the divine commands. (Both forms of theodicy, as Bréhier has shown, are present, unreconciled, in the philosophy of Plotinus.) The second version, which denies the responsibility of God for evil, suggests by the same token that man is endowed with a spontaneous creative initiative, albeit restricted to evil, so that the freedom whereby he disobeys God is complete and equal to the freedom of God himself, though naturally man does not share God's goodness and omnipotence. The effect is to regard man as a source of completely independent initiative, an Absolute on a par with the Deity. This ultimate conclusion is first explicitly brought out in the Cartesian theory of freedom.

The first version, according to which good presupposes evil, is hard to accept in so far as it implies that there is no such thing as evil pure and simple. It can only be sustained on the basis of a dynamic view of the universe, i.e. on the premiss that evil is an essential condition of the final efflorescence and complete realization of good. The answer to the problem of evil, and also to that of contingency, thus leads to a dialectic of negation, i.e. to the idea that evil and contingency must exist if all the possibilities of being are to be realized. This dialectic provides an answer to the questions why the world was created, why evil exists, and why human beings are imperfect, but it does so in a way that places it outside the bounds of orthodox Christianity. It involves believing that God needed the world, that he only achieves fulfilment in creation and that his perfection depends on giving life to imperfect reality. This again is contrary to what Scripture tells us of God's self-sufficiency (Acts 17: 25). It means introducing the divine principle into history and subjecting it to the process of self-multiplication through creation.

4. Eriugena and Christian theogony

This idea probably found expression for the first time, though incompletely, in the work of Eriugena. It has since been essential to all Northern mysticism of the pantheistic kind, and we can trace it in different variations almost from one generation to another, from the Carolingian renaissance to Hegel. Speaking in the most general terms, it is the idea of the potential Absolute (a semi-Absolute, if this expression can be permitted) which attains to full actuality by evolving out of itself a non-absolute reality characterized by transience, contingency, and evil; such non-absolute realities are a necessary phase of the Absolute's growth towards self-realization, and this function of theirs justifies the course of world history. In and through them, and above all in and through mankind, the Deity attains to itself: having created a finite spirit it liberates that spirit from its finitude and receives it back into itself, and by so doing it enriches its own Being. The human soul is the instrument whereby God achieves maturity and thereby infinitude; at the same time, by this process the soul itself becomes infinite, ceases to be alien to the world, and liberates itself from contingency and from the opposition between subject and object. Deity and humanity are thus alike fulfilled in the cosmic drama; the problem of the Absolute and that of creation are resolved at a single stroke. The prospect of the final consummation of the unity of Being gives a meaning to human existence from the point of view of the evolution of God, and also from the point of view of man himself as he attains to the realization of his own humanity or divinity.

This, of course, is a simplified schema, expressed in terms that cannot be found in the actual writings of Eriugena, Eckhart, Nicholas of Cusa, Böhme, or Silesius—to mention the principal philosophers and mystics who come into question here. Nevertheless, despite the differences of exposition, their works can be seen as formulations of the same basic intuition which constitutes the historical background of the Hegelian dialectic and therefore of Marxian historiosophy. Naturally we cannot describe the history of that dialectic, in all its variations, in an account confined to the antecedents of Marx's thought, but some aspects of its history should be briefly noted here.

Eriugena's principal work, *De divisione naturae*, by its initial distinction of four natures in effect introduces the concept of a historical God, a God who comes into existence in and through the world. God as Creator (*natura naturans non naturata*), and God as the location of the ultimate unity of creation (*natura non naturata non naturans*), is not presented under a twofold guise for didactic reasons or because the infirmity of our understanding requires it thus: the juxtaposition of the two names signifies the actual evolution of God, who is not the same at the end of all things as he was at the beginning.

Eriugena makes frequent appeals to tradition: to the Cappadocian Fathers, St. Augustine and St. Ambrose, sometimes Origen, but most often the Pseudo-Dionysus and Maximus the Confessor. The most important borrowing from the Pseudo-Dionysus is the whole idea of negative theology as expressed in *De nominibus Dei* (the royal road to knowledge of God consisting in knowing what he is not). But from all these sources Eriugena constructs an original theogony of a neo-Platonist kind, which he attempts, despite immense difficulty and incessant contradictions, to reconcile with the truths of faith.

De divisione naturae is in fact a prototype of Hegel's *Phenomenology of Mind*, which it precedes by almost a thousand years—a dramatic history of the return of the Spirit to itself through the created world; a history of the Absolute recognizing itself in its works and drawing them into unity with itself, to the point at which all difference, all alienation, and all contingency is removed, yet the wealth of creation is not simply annihilated but incorporated in a higher form of existence, a higher form which entails a previous decline.

Eriugena accepts the premiss, common to all Platonists and all Christian theologians, that God is not prior to the world in time, since time itself is part of creation: God exists in a *nunc stans* where there is no distinction of past and future (*De divisione naturae*, III. 6, 8). God is unchangeable and the act of creation does not make any alteration in him, nor is it accidental in relation to his Being (v. 24). However, though Eriugena pays lip-service to God's unchangeability, it is called into question as soon as we come to consider the reason for creation: for at this stage it appears that

God, in a marvellous and inexpressible way, is created in creation in so far as he manifests himself and becomes visible instead of invisible, comprehensible instead of incomprehensible, revealed instead of hidden, known instead of unknown; when instead of being without form or shape he becomes beautiful and attractive; from super-essential he becomes essential, from supernatural natural, from uncompounded compound, from non-contingent contingent and accidental, from infinite finite, from limitless limited, from timeless temporal, from spaceless located in space, from the creator of all things to that which is created in all things. (III. 17)

Eriugena makes clear that he is not speaking merely of the Incarnate Word but of the whole manifestation of Deity in the created universe, This indeed is intelligible on the premiss that God alone truly *is*, that he is 'the being of all things' (I. 2), the form of all things (I. 56), so that everything that exists is God as far as its Being is concerned. On the other hand, the statement that 'God is' is itself misleading in so far as it suggests that he is some one thing and not another (III. 19). If, however, the Being is divinity itself, it will be true to say that

the divine nature both creates and is created. For it is created by itself in the primordial causes and thereby creates itself, that is to say it begins to manifest itself in its own theophanies, desiring to pass beyond the most secret boundaries of its nature, in which it is as yet unknown to itself and recognizes itself in nothing, inasmuch as it is unlimited, supernatural and supereternal and is above all things that can and cannot be understood. (III. 23)

We can thus comprehend the reasons for creation as it relates to God himself: God enters into nature in order to manifest himself, to become 'all in all', and thereafter, having called everything back into himself, to return into his own Being.

But not all created things participate in this process directly and on an equal footing. The whole visible world has been called into existence for man's sake, in order that he may rule over it; consequently, human nature is present in the whole of created nature, all creation is comprised in human nature and is destined to achieve its freedom through man (IV. 4). Man, as a microcosm of creation, contains in himself all the attributes of the visible and invisible world (V. 20). Mankind is, as it were, the leader of the cosmos, which follows it into the depths and back into union with the divine source of all Being.

It is clear that Eriugena sees the creative act of God as a satisfaction of the creator's own need, and that he regards the circuit whereby creation returns to the creator as a process which restores God's nature to him in a form other than its original one. In one passage he actually puts the question: why was everything created from nothingness in order to return to its first beginning? Having stated that the answer to this question is beyond human understanding, he at once proceeds to offer an answer: everything was created in order that the fullness and immensity of God's goodness should be manifested and adored in his works. If the divine goodness had remained inactive and at rest there would have been no occasion to glorify it, but, as it overflows into the wealth of the visible and invisible world and makes itself known to the rational creation, the whole of creation sings its praises. Moreover, the Good which exists in and by itself had to create another Good that only participates in the original bounty, otherwise God would not be lord and creator, judge and fountain of all benefits (V. 33).

Thus the Absolute had to exceed its own boundaries and create a contingent, finite, and transient world in which it could contemplate itself as in a mirror, so that, having reabsorbed this exteriorization of itself, it might become other than it originally was, richer by the totality of its relationship with the world: instead of a closed self-sufficient system it becomes an Absolute known and loved by its own creation. We have here a complete schema of 'enriching alienation', serving to explain the whole history of being; the vision of a Deity who develops by a process of decline and reascent.

The term 'decline' must, however, be used with reservation. For the Deity to enter into the world of creation is in itself, of course, a descent into a lower form of existence; but are we to understand that evil, or non-being, is also part of the universal cycle? Does it too perform a necessary part in the process of emanation and return? Eriugena nowhere says expressly that it does. The fall of man cannot, of course, be ascribed to his nature, which is good; nor can it be the effect of free will, which is also good (v. 36), even though it belongs to man's animal nature (iv. 4). It is the result of evil desires, which are harmless in animals but contrary to man's true nature (v. 7). Eriugena does not explain specifically how the fall was possible, but concentrates on man's return to his lost perfection. 'Paradise' signifies no more or less than human nature as it was created by God and destined to immortal life; death and all the consequences of our exile are the effect of sin, but the exile itself is a manifestation of the mercy of God, whose desire is not to condemn but to renew and sanctify mankind and enable us to eat of the tree of life (v. 2). Eriugena repeats time and again that when fallen man returns to God he will recover his original greatness and dignity, and he explains that this return consists of five stages: corporal death, resurrection, the transformation of the body into spirit, the return of the spirit and of man's whole nature to its 'primordial causes', and finally the return of these 'causes' (principles, ideas), and all else with them, to God (v. 7). The 'causes' or essential forms have nothing about them that is contingent, changeable, or composite; every species comes into being by participation in its form, which is one and one only and is fully present in every individual of the species; every human being contains in himself one and the same form of humanity (iii. 27). It might seem, therefore, that the unity of mankind with God signifies a loss of individuality, the whole species being identified with its universal, which belongs to the divine essence. This 'monopsychism' is suggested by various remarks on the unity and simplicity of the 'first principles', which are not creatures and are not bounded in space and time (v. 15, 16), and by the statement that a thing which begins to be what it was not ceases to be what is was—including presumably human individuals as differentiated by the contingent attributes of each (v. 19). We are also told expressly (v. 23, 27) that in heaven there will be no contingent differences. However, different human beings will enjoy different stations in heaven according to the degrees of their love of God, even though they will all be saved and evil will cease to exist. Evidently the nature of our unification with the Deity is not clear to Eriugena, and he is unable to say whether or to what extent human individuality will survive in that final unity. It is certain, however, that whatever has been created by God cannot cease to be, though it may change its character. The lower will be absorbed by the higher, but will not be destroyed; the corporal will become spiritual, not losing its nature but ennobling it, and in the same way the soul will be united with God (v. 8). There will thus be a complete resorption of each lower level of Being by the next higher or more perfect, but nothing that has been created will be lost: this is the pattern of Hegel's *Aufhebung* or 'sublation'.

The whole process of return, in which man takes the lead, is in no way enforced by God upon nature: on the contrary, it is engrafted upon humanity, as the philosopher shows by a fanciful piece of medieval etymology, equating *anthropia* with *anotropia* or ascent (v. 31). Resurrection is a natural phenomenon (Eriugena disclaims his previous opinion that it was the effect of grace alone: v. 23), and so is our return to the house of God, in which there is room for all. The supernatural gift of grace will consist merely in the fact that the elect, sanctified in Christ, will be received into the very heart of paradise and will undergo deification.

But in the same way as God, when the cosmic epic is concluded, will find himself in a different state than before, enriched by his own creatures' knowledge of him, so also man, though he returns to his 'first beginnings', will not be purely and simply in his original state: for he will be in a condition in which a second fall is impossible and his unity with God is eternal and indissoluble (*theosis* being, however, reserved for the elect). It also appears that the work of the Incarnate Word does not consist, for Eriugena, simply in restoring men to the bliss of paradise by erasing the consequence of sin. Christ's incarnation has effects over and above redemption: Christ has set all men free, but while some will merely be restored to their primal state, others will be deified, thus raising humanity to the dignity of Godhead (v. 25).

It thus appears that the degradation of Being was not in vain. In the final result, the duality of man (who, as a composite being, cannot be a true image of God, who is uncompounded: v. 35) is a condition of his return to God and thereby to his own self. Humanity recovers its lost nature and even surpasses it by being deified. The drama ends with the attainment of godlike existence, self-identification, the abolition of the division between forms of Being, and hence the coincidence, once again, of the soul with its object. In accordance with the spirit of his dialectic, Eriugena finally declares that evil is only apparent when we view things in part; when we consider the whole there is no such thing as evil, since it plays its part in the divine plan and enables the good to shine with brighter radiance (v. 35). In this theodicy everything finds its justification and, in the eschatological perspective, the history of the cosmos is ultimately the history of God's growth in the human spirit and man's ripening into divinity—in short, a history of the salvation of Being by negation. If creation is the negation of divinity by reason of its finitude, differentiation, and lack of unity, divinity as the point of return—*natura non naturans non naturata*—may be called the 'negation of a negation'. Eriugena himself does not use this expression, which probably appears for the first time in Eckhart.

There are many hesitations and contradictions in Eriugena's work: we read, for instance, that evil has no cause, and also that it is caused by the corruption of will; that no one is condemned, and that some will suffer eternal sadness; that all will be one in God, and that there will be a system of ranks in heaven; that eternal ideas are part of creation, and that they are infinite, and so forth. Nevertheless, he was the first Latin philosopher to propound a system of categories, based on the Greek patristic tradition, which made it possible to link the history of mankind with that of a self-creating God, thus justifying the miseries of life by the hope of deification and offering man the prospect of final reconciliation with himself through reconciliation with Absolute Being.

No such majestic theogony as the *De divisione naturae* was composed in the Christian world between Eriugena's time and that of Teilhard de Chardin. None the less, its main themes can be traced again and again—not always in the same arrangement —throughout Christian philosophy, theology, and theosophy in so far as these are influenced directly or indirectly by the ideas of Plotinus, Iamblichus, and Proclus, or, in later centuries, by Arabic and Jewish thought which was inspired by those ideas. Among the themes in question are these:

Only the Absolute is perfectly identical with itself; man suffers from disjunction and, as a temporal being, cannot achieve self-identification.

Man's essence lies outside him or, which is the same thing, is present in him as an Absolute that is unrealized and aspires to realization.

It is possible for man to escape from the contingency of existence by union with the Absolute.

This escape, to which man is called, signifies a return to his own Being; and it is also the way in which the Absolute attains fulfilment, which was not possible without the defective world of creation.

Thus the process whereby conditional existence evolves from the Absolute is, for the Absolute, a loss of itself in order to achieve self-enrichment; and degradation is a condition of the furtherance of the highest mode of Being.

Hence the history of the world is also the history of unconditional Being, which attains to its final perfection as a result of being reflected in the mirror of the finite spirit.

In this final phase the difference between finite and infinite disappears, as the Absolute reassimilates its own works and they are incorporated in the divine Being.

Consequently the difference between subject and object also disappears, as does the estrangement between the soul that knows and loves and the rest of Being; the soul puts on infinitude, and ceases to be 'something' in opposition to something else which it is not.

All these thoughts recur persistently in Christian philosophy, despite various criticisms and condemnations, and were taken up in due course by the dissidents of the Reformation.

'Thou alone, O Lord,' says St. Anselm, 'art what thou art; and thou art he who thou art. For that which is one thing in the whole and another in the parts, and in which there is any mutable element, is not altogether what it is. And that which begins from non-existence, and can be conceived not to exist, and unless it subsists through something else, returns to non-existence; and which has a past existence that is no longer, or a future existence that is not yet,—this does not properly and absolutely exist.' (*Proslogium*, XXII.)

However, although this opposition between God and man is

strictly orthodox, it at once raises the question: Can man be saved unless he is delivered from contingency, yet is not contingency a necessary correlative of his particular mode of existence? In other words, can man achieve self-identification without losing that which makes him a separate entity and becoming passively transformed into the divine Being?

5. *Eckhart and the dialectic of deification*

This consequence was accepted by Northern mysticism, which freed itself in this respect from a certain ambiguity of Eriugena's Platonism. To Eckhart the maxim of 'co-operation with one's own God' signifies the same as self-annihilation—a mystic kenosis that is not a mere moral precept but an ontological transformation. 'He who desires to possess everything must renounce everything'; to possess everything means to possess God, and to renounce everything involves renouncing oneself. God himself desires only to belong to me, but to belong wholly. When the soul achieves the fullness of inner poverty or denudation it makes God wholly its own, and he belongs to it in exactly the same way as to himself. There is then nothing in the soul that is not God. But the soul also achieves release from itself as a creature, that is to say from nothingness: for all creation (according to the well-known formula from Eckhart's sermon on James 1: 17, quoted in the bull of John XXII) is pure nothingness, not in the sense of being insignificant but in the literal sense of non-being. Thus the self-annihilation of the mystic is, paradoxically, the destruction of nothingness or, if we may so express it, the overcoming of the resistance that the void opposes to Being. When the soul is completely emptied of its particular nature, God gives himself to it in the fullness of his Being and belongs to it as to himself. But by this self-destruction the soul attains to that which it truly is: for there is within it a latent spark of divinity, obscured by the link with created things and by attachment to its individual, limited form. There is present in the soul that which is not created, namely the Son of God; and therefore any man can, like Christ himself, be united with the Father. Thus for a man to be one with himself is the same as to be one with God. In this way man's will is identified with God's and shares in his omnipotence. For the soul which has thus found itself, or found God in itself, there ceases to be any problem of the relationship between its will and that of the Absolute—both are fundamentally the same, and the question of obedience or disobedience does not arise. Eckhart distinguishes the particular, contingent self-will that tries to maintain a partial, isolated, subjective existence from the real will which is identical with that of universal Being, the only Being that truly deserves to be so called (although, unlike Aquinas, Eckhart regards Being as secondary in relation to the mind of God).

Eckhart's thought is dominated by the intense, unremitting conviction that Being and God are one and the same. The multitude of individual existence is nothing in so far as each of them is limited and partial; in so far as each one is possessed of Being, it is identical with God. Hence the question as to the reason of creation does not figure, properly speaking, in his sermons and writings. At the same time, he makes a distinction between the Godhead or indescribable Absolute—the One of Plotinus—and the personal Absolute which is God. This God—corresponding to Plotinus's second hypostasis, Being or Mind—realizes himself as God in creation; or, more precisely, it is only in the human soul that God, as its hidden nature, becomes what he is. In this sense we can speak of the meaning of creation from the standpoint of God himself. But the final aim of human endeavour is not to discover God in oneself but to destroy him, i.e. to destroy the last barrier that separates the soul from Godhead and prevents it returning into the inexpressible unity of the Absolute. This return takes place in the form of cognition and is consummated in a state in which all difference between the knower and known is obliterated.

In this way Eckhart's pantheistic mysticism embodies some of the basic ideas we have been considering. The contingency of human existence is only apparent ('Man is essentially a heavenly being'—sermon on Hebrews 11: 37), but this appearance must be overcome by the soul exercising its faculty of knowledge, and only in this way can the soul discover itself. In so doing it loses itself as a partial being and enters into possession of itself as an entirety, as divinity, as the Absolute. The particularization of Being pertains to the history of God realizing his own Being, which he can only do in and through the soul, but it does not pertain to the history of the Godhead or first hypostasis, which is not subject to any process of becoming.

6. *Nicholas of Cusa. The contradictions of Absolute Being*

North European spiritual writing in the fourteenth century preserves much of the Eckhart tradition, but in terms of practical devotion rather than speculation. We cannot deal here with the social and ecclesiastical conditions which encouraged this type of mystic piety in the late Middle Ages. A substantial attempt at a new, speculative neo-Platonic theogony is found in the fifteenth century in the work of Nicholas Cusanus, who speaks, more clearly perhaps than his predecessors, of God's need for creation. God desired to manifest his glory, and for this he needed rational beings to know and worship him. 'Nihil enim movit creatorem ut hoc universum conderet pulcherrimum opus nisi laus et gloria sua, quam ostendere voluit; finis igitur creatoris ipse est, qui et principium. Et quia omnis rex incognitus est sine laude et gloria, cognosci voluit omnium creator, ut gloriam suam ostendere posset. Immo qui voluit cognosci, creavit intellectualem naturam cognitionis capacem' (Letters written in 1463 to a monk at Monte Oliveto; published by W. Rubczyński in *Przegląd Filozoficzny*, v. 2).

This, however, suggests too strongly the idea of a God who needs something other than himself, which is contrary to the principle of divine self-sufficiency. In his chief work, *De docta ignorantia*, when discussing the question of God's relationship to created beings, Cusanus avows himself defeated by the mystery of contradictions in the divine essence. The absolute unity of God is all that there can be, i.e. it is complete actuality, and is therefore not multipliable. ('Haec unitas, cum maxima sit, non est multiplicabilis, quoniam est omne id quod esse potest' —*Doct. ign.* I. 6). On the other hand, God (as *rerum entitas, forma essendi, actus omnium, quidditas absoluta mundi*, etc.) descends into the manifold, differentiated world and creates the whole of its existential reality. Creation in itself is nothing; in so far as it exists, it is God; one cannot speak of it as a combination of Being and non-being. As the *esse Dei* it is eternal; as something temporal, it is not of God (II. 2). It is, as it were, a finite infinitude or a created God: 'Ac si dixisset creator "Fiat," et quia Deus fieri non potuit, qui est ipsa aeternitas, hoc factum est, quod fieri potuit Deo similius ... Communicat enim piissimus Deus esse omnibus eo modo quo percipi potest' (ibid.). God is the *complicatio* (wrapping-up, involution) of all things, as unity is that of number, rest of motion, presentness of time, identity of diversity, equality of inequality, simplicity of divisibility. In God, however, unity and identity are not opposed to the multiplicity of the world that is 'involved' in him. The converse relationship is *explicatio* or unfolding: thus the world is the *explicatio* of God, multiplicity is that of unity, motion of rest, etc. But, Cusanus explains, the character of this mutual relationship is beyond our comprehension; for, as understanding and Being are in God one and the same, then in understanding multiplicity he should himself be multiplied, which is impossible. ('... videtur quasi Deus, qui est unitas, sit in rebus multiplicatus, postquam intelligere eius est esse; et tamen intelligis non esse possibile illam unitatem, quae est infinita et maxima, multiplicari'—II. 3). It would seem that God cannot 'unfold' himself in multiplicity without violating his absolute unity or his complete actuality, or the exclusiveness of his Being; yet one of these attributes, and hence all of them, must be forfeited if we accept that the development from unity to multiplicity, or, more simply, the process of creation, involves turning potential into actual Being. Yet the actual fact is that amid the multiplicity of things all that has Being consists of God alone. Thus we know only that everything is in God, since he is the *complicatio* of all things, and God is in everything, since creation is the *explicatio* of God; but we cannot fathom how this is so. The universe considered as an intermediary between God and the manifold or *unitas contracta*—the undifferentiated Being that is no particular thing but is, in everything, that thing itself ('universum, licet non sit nec sol nec luna, est tamen in sole sol et in luna luna'—ibid.)—does not resolve the contradiction, since the Being of all things is God and nothing else.

The difficulty felt by Cusanus is that of all monism. He seeks in vain for a formula that will make it possible to regard the development from unity into multiplicity as a real development, but not as a change from potential to actual Being, which would imply the ascription of potentiality to God himself. Cusanus's thought is in a state of tension between two extremes, neither of which can be reconciled with even the loosest form of orthodoxy. On the one hand is the eternal temptation to regard the whole manifold universe as an illusion and a mere semblance of being, while the only reality is the unity of the Absolute. Alternatively, the world must be regarded as God in a state of evolution, from which it follows that God is not fully actual, nor is he the Absolute, but that he merely becomes so at the end of the history of creation and by virtue of that history. Pantheists often waver between these opposing views, which represent the dilemma of all monistic thought. The first alternative leads to the contemplative morality of self-annihilation; the second to religious Prometheanism, animated by the hope of achieving deification by one's own efforts.

There is no doubt that Cusanus was more attracted by the idea (though he does not express it outright) of God realizing himself in his creation than by that of the created world as an illusion. Like all 'emanationists' he regards the human spirit as the medium through which the Deity achieves actuality, which means that the Absolute is at the same time the true fulfilment of humanity. The soul returns into the actualized Absolute by means of knowledge, specifically knowledge of the whole and its relation to the parts: a paradoxical form of knowledge, discarding the principle of contradiction in favour of the *coincidentia oppositorum* which finds its prototype in the mathematics of infinitely large or limiting quantities. With the aid of knowledge the soul discovers itself as divinity and adopts the infinite object of its knowledge as its own self.

Cusanus found an ineradicable contradiction in the divine nature, but it was, to use Hegelian language, an immobile contradiction, i.e. the result of a speculation leading to an antinomy. Reflection on the divine nature leads to the conclusion that it must contain within itself qualities that are incompatible in finite beings: for, as God is pure actuality and at the same time embraces the whole of reality, there can be nothing in that reality that is not actualized, in some unfathomable way, in the divine unity. Cusanus's thought thus led him to the antinomy that results from simply developing the notion of the Absolute. The contradiction appeared under a logical, not a dynamic form: it was not a collision of real forces whose antagonism gives rise to something new. Nor was it an explanation of God's creation, but rather a recognition of the absurdity in which the finite mind becomes involved when it seeks to probe the infinite.

7. *Böhme and the duality of Being*

Contradiction, or rather antagonism, conceived as an ontological category makes its first appearance in the works of Böhme, which resemble a dense, swirling cloud of vapours, yet which open a new chapter in the history of dialectic. The picture of the world as a scene of cosmic conflict between hostile forces was of course a traditional one, and recurred from time to time in different versions of Manichaean theology. But it is one thing to regard the whole of reality as a battlefield between rivals, and another to ascribe the conflict to a rift within a single Absolute.

Böhme's visionary writings are a continuation of the Platonism which was active among the pantheistic dissidents of the Reformation and which, as in the cases of Franck and Weigel, repeated in different language many ideas found in Eckhart and the *Theologia germanica*. Within this school of thought Böhme was something of an innovator. Following the tradition of the alchemists, he regarded the visible world as a collection of sensory and legible signs revealing invisible realities; but this revelation was in his view a necessary means whereby the Deity exteriorized and displayed itself. The 'eternal self-seeker and self-discoverer' duplicates himself, as it were, and emerges from a state of undifferentiated immobility to become truly God. Thus we find in Böhme's notion of divinity the same ambiguity as in Eckhart's works, an echo of the two first hypostases of Plotinus. God as revealed is the God who transmutes himself into creation, but he can only do so in such a way that what is actually unity in him appears in the guise of opposing forces of light and darkness. 'In the light, this power is the fire of divine love; in the darkness it is the fire of God's wrath, and yet there is only one fire. It divides itself into two principles, so that one should become manifest in the other. For the fire of wrath is a manifestation of great love: we perceive light in darkness, otherwise it could not be seen' (*Mysterium magnum*, VIII. 27). By emerging from his solitude and overstepping his own boundaries in search of himself, God inevitably creates a divided world in which qualities can only be recognized thanks to their opposites. Böhme has chiefly in mind the internal antagonism aroused in the human soul by conflicting desires. The essential drama of creation is played out within the individual torn by opposing forces. The soul's true home is in God, who has sown in it the seed of grace, but at the same time it longs to assert its own will. Thus there is no return to God without an internal conflict in which, by means of self-denial, the desire for harmony finally conquers the urge towards self-affirmation.

Böhme's theosophy is, as it were, the obscure self-knowledge of a central antinomy inherent in the idea of an unconditional being creating a finite world: the latter is both a manifestation and a denial of its creator, and cannot be one without the other. In so far as the absolute spirit chooses to become manifest, it is bound to contradict itself. The world of finite beings, inspired by the unity of its source, cannot altogether resist the force that bids it return to its origin; but, as it has come into existence, it also cannot escape the urge to assert itself in its own finitude. In Böhme's theosophy this conflict is for the first time clearly presented as the antagonism of two cosmic energies arising from a cleavage in the primal impulse of creation.

8. *Angelus Silesius and Fénelon: salvation through annihilation*

The dialectic of God's self-limitation and the idea of the non-self-identity of man's Being recur throughout the seventeenth and eighteenth centuries, chiefly in Northern mysticism; they can be traced without difficulty in Benedictus de Canfield and in Angelus Silesius. However, while the former emphasizes the 'nothingness' of all created beings and the exclusive reality of God, Silesius in *Der cherubinische Wandersmann*, written no doubt before his conversion to Catholicism, is not content with this and returns to Eckhart's theme of the divinity as man's true essence and final home. The call of eternity is constantly present within each of us; in answering it we become 'essential' instead of 'contingent', putting off the particularity of individual existence and becoming absorbed in absolute being. The contrast between 'essential' and 'contingent' is clearly expressed: 'Mensch, werde wesentlich! denn wenn die Welt vergeht—So fällt der Zufall weg, das Wesen, das besteht' (*Cher. Wand.* II. 30). But whereas in some of Silesius's epigrams the contingency of individual Being appears simply as an evil whose presence is incomprehensible and which must be cured by voluntary renouncing all attachment to one's 'selfhood' (*Selbstheit, Seinheit*), in others we find Eriugena's notion of a cycle wherein creation restores to God his own Being in an altered form. Only in me can God find his 'double', equal to him for all eternity (I. 278); I alone am the image in which he can contemplate himself (I. 105); only in me does God become something (I. 200). It may be said that God descends into the world of chance and wretchedness in order that man in his turn may achieve divinity (III. 20). We thus have the same model of the Absolute exteriorizing itself into finitude so as ultimately to put an end to that finitude and return to unity with itself, but a unity enriched by all the effects of the polarization of the spirit, and hence, we may suppose, a reflective unity of self-contemplation. Contingency, evil, finitude—which all mean the same thing—are not a gratuitous, inexplicable decline on the part of the Deity, nor are they the work of any rival or adversary of his. They belong to the last phase of the circular dialectical movement, involving first a negation of divinity and then a counter-negation by the finite soul which wills its own annihilation. Once again, therefore, the return of God to himself is also a return of the human spirit to itself, to that eternity which is its true nature and resting-place but which is eclipsed by temporal existence. Self-annihilation terminates the unbearable disruption which is inseparable from the process whereby God becomes himself, but which is destined to be finally remedied.

However, we need not prolong examples unduly. The theme of the contingency of man is found in all pantheistic literature and mystical writing, whether by orthodox Catholics, Protestants, or adherents of no denomination. 'I am not, O my God, that which is,' wrote Fénelon,

Alas! I am almost that which is not. I see myself as an incomprehensible middle point between nothingness and existence; I am one that was and one that will be, I am he who no longer is what he was and is not yet what he will be; and in that betweenness what am I?—a thing, I know not what, which cannot be contained in itself, which has no stability and flows away like water; a thing, I know not what, which I cannot grasp and which slips through my fingers, which is no longer there when I try to grasp it or catch a glimpse of it; a thing, I know not what, which ceases to exist even as it comes to be, so that there is never a moment at which I find myself in a state of stability or am present to myself in such a way that I can say simply 'I am' (*Traité de l'existence et des attributs de Dieu, Œuvres*, 1820, I. 253–4)

The non-sectarian Dutch mystic Jakob Bril (*Alle de Werken ...,* 1715, p. 534) declares that 'All things in nature are what they are except for man, who, considered in himself, is not the thing that he is; for he imagines himself to be something when he is not. All things are what they are, not in themselves but in their creator; man imagines himself to be something in himself, but this is only a false idea of his.' The conception of man as a divided being whose true existence is in God is a common one and is always associated with the hope of return. The view that contingent existence is a negative stage of the evolution of the Absolute clearly entails additional premisses, which can only be found in the writings of those who consciously step outside the confessional orthodoxy of the great Churches or who are branded as apostates.

9. The Enlightenment. *The realization of man in the schema of naturalism*

It might seem that both these schemata could only be products of religious thought, that they involve an interpretation of the physical world as a theophany, and an explanation of man in terms of his relationship—be it positive, negative, or, more usually, twofold—to absolute spirit. But this is not so, for the theory of man's return to himself is also found as a constituent element of the naturalistic philosophy of the Enlightenment. It appears in fact that the theory in question, together with the paradigmatic image of a lost paradise, is an unchanging feature of man's speculation about himself, assuming different forms in different cultures but equally capable of finding expression within a religious or a radically anti-religious framework.

In the literature of the Enlightenment we find the notion of man's lost identity and the summons to recover it, both in utopian writings and in multifarious descriptions of the state of nature. The scepticism and empiricism inculcated by Locke and Bayle provided a negative basis for the notion of an ideal harmony which man has the power and the duty to restore on the natural plane. The acceptance of human finitude proved compatible with the conviction that it was possible to discover what man truly was, or what were the exigencies of his Being. Even though man's existence must be regarded as accidental in the sense that it is not the work of some spirit anterior to nature, nevertheless

Nature herself provides information concerning the perfection of humanity, showing us what man would be if he were fully obedient to his proper calling; so that any particular civilization can be evaluated against the standard observable in nature. Instead of comparing earth to heaven, existing cultures were compared with the natural state of humanity. Whereas the mystics contrasted the general condition of humanity, irrespective of particular cultures, with the true fulfilment of humanity in the Absolute, the naturalists judged every form of civilization, and especially their own, in the light of authentic humanity as prescribed by Nature's imperatives. It is immaterial from this point of view whether they believed that the imperatives were actually fulfilled in some time or place, as in theories of the noble savage, or whether they regarded them as constituting a pattern to be obeyed—not, however, a pattern devised by mere speculation, but one discovered in the laws of nature. The attitude of detachment from one's own culture and criticism of it as 'unnatural' had, it is true, already appeared in the late Renaissance (for example Montaigne) and was transmitted to the Enlightenment by the more or less continuous tradition of Libertinism. It was not till the eighteenth century, however, that it took on such a massive, consistent, and radical form as to constitute a whole new intellectual system. The device of portraying one's own civilization through the eyes of others (Goldsmith's Chinese, Montesquieu's Persians, Swift's Houyhnhnms, and Voltaire's traveller from Sirius) was associated with the belief that there was a true standard for all humanity and that the civilization in question was contrary to nature. An exception should, however, be made for Swift's bitter satire: by placing his ideal state among horses and not men, he showed clearly enough that it was 'utopian' in the full sense of the term's derivation and of its popular use.

The claims of 'natural man' *vis-à-vis* the prevailing civilization involved various assertions of rights and qualitative comparisons. But man's natural equality, his right to happiness and freedom and to the use of reason, were currently accepted themes and were sufficient to constitute an apparatus of criticism. Nevertheless, it turned out before long that the conceptual framework of the Enlightenment ideals was inadequate and that its components did not jibe with one another.

The key concepts of 'nature' and 'reason' proved, on closer analysis, not to combine in a consistent whole: for how was the cult of reason as the gift of nature to be harmonized with the cult of nature as itself reasonable? If, as the materialists held, human reason was a prolongation of animal nature and there was no essential difference between tricks performed by monkeys and the reasoning of mathematicians (de la Mettrie), and if all moral judgements are reducible to reactions of pleasure and pain, then human beings with their abstract reasoning and moral laws are indeed the work of nature but are no more than blind pieces of natural mechanism. If, on the other hand, Nature, as many contend, is a rational, purposeful, protective entity, then she is merely another name for God. Hence either reason is not reason or nature is not nature; we must either regard thought as irrational or credit Nature with God-like attributes. How can it be accepted that human impulses are no less natural than the moral laws that regulate and control them? We are back to the eternal dilemmas with which atheists have taunted worshippers of the gods since Epicurus's time: since the world is full of evil, God must be either evil or powerless or incompetent or all three; and the same can be said of 'good and omnipotent' Nature. If, on the other hand, nature is indifferent to man and his fate, there is no reason to believe that evil can be vanquished: it may be that the only natural law is that of the jungle and that human societies are no better off than plant or animal species. At this point the *idées-forces* of the Enlightenment begin to diverge, and we encounter the pessimistic attitudes of Mandeville, Swift, and the later Voltaire. The notion of a benevolent natural harmony which, once discovered, will remove all conflict and misfortune, begins to falter.

10. Rousseau and Hume. Destruction of the belief in natural harmony

Rousseau, Hume, and Kant are all exponents of this loss of confidence. Rousseau believes in the archetype of man living in self-identity, but he does not believe it possible to erase the effects of civilization and return to natural happiness. Natural man felt no sense of alienation, as his relationship to life was not mediated by reflection; he lived straightforwardly without having to think about life; he accepted, but in an unconscious manner, his own situation and his own limitations. Thus his fellowship with others developed spontaneously and needed no special institutions to preserve it. Civilization introduced man's detachment from himself and ruined the original harmony of society. It made selfishness universal, destroyed solidarity, and degraded personal life to a system of conventions and artificial needs. In this society the self-identity of the individual is unattainable; all he can do is to attempt to escape its pressures and contemplate the world independently of received opinion. Co-operation and solidarity with others does not deprive the individual of a true personal life, whereas the negative bond of selfish interest and ambition destroys both community and true personality. Man's proper duty is both to be himself and to live in willing solidarity with others. Since we cannot undo civilization, we must attempt a compromise: let each man discover the natural state within himself and educate others in the same spirit. There is no historical law to assure us that our efforts will lead to the restoration of true community and that society will be reborn in individuals, but it is not quite impossible that this will be so.

Rousseau does not embark on any historical theodicy or attempt to integrate the evil in the world with the hope of a harmonious order blossoming out of the horrors of the past. The rupture of the original harmony is in his eyes evil pure and simple, without justification and without purpose. There is no dialectic of 'spiral progress' to foster an uncertain hope of improvement.

Rousseau thus has his own model of authentic humanity, but he does not acknowledge any reasons justifying a breach with that model. The fall of man is not, in his view, a self-correcting phase in the advance towards perfection. In this respect he is closer to ordinary Christianity than to theogonists after the manner of Plato: evil is evil, it is the fault of man and has no hidden significance for the history of the cosmos. There is, on the other hand, a summons addressed to man, which is prior to history and not dictated by it; the ultimate reality of that summons is an open question.

Hume's doctrine, in its turn, represents the cleavage between two other basic elements of eighteenth-century thinking: the categories of experience and those of the natural order. When the premisses of empiricism were carried to their logical conclusion, it became clear that the notion of a natural order was untenable. If there can be no knowledge except what is conveyed by the senses, and if our sense-data provide no evidence for any causal connection or law of necessity, then it is clear that our minds are incapable of apprehending reality as anything but a collection of separate phenomena. Nor can we, in that case, perceive any natural order which it would be legitimate to regard as an immanent feature of the universe rather than simply a 'law' of the scientific type, i.e. the subjective fixing in the mind, for reasons of practical convenience, of certain recurrent sequences of events. Nor, again, is there any reason to suppose that we are bound by moral laws possessing a validity independent of our own sensations of pain and pleasure. In short, both the 'physical order' and the 'moral order' are imaginations above and beyond what is or can be conveyed to us by experience. In the same way, it is useless to suppose that there is any human standard, obligation, or purpose independent of the actual course of human history.

Hume does not assert the contingency of either man or the universe: on the contrary, when rebutting the cosmological proofs of the existence of God, he states that we cannot know from experience that the universe is contingent. But this means only that it is not contingent in the schoolmen's sense, i.e. that it has no qualities to indicate that it must be dependent on a necessary creator. To the Scholastics the 'contingency' of the world serves as a demonstration of necessity. Considered in itself the world has nothing necessary about it, but the necessity must be there, since the world exists; its contingency is only apparent, as we see when we relate it to the existence of God. Considered, as it should be, in relation to God, the world is not contingent, since nothing can exist by accident. Thus when Hume tells us that there is nothing in experience to show that the world is contingent, he is really saying that the world *is* contingent, i.e. there is nothing that obliges us to relate it to any necessary or absolute reality. In other words, the expression 'contingent' is only meaningful in opposition to the expression 'necessary'. Hume's position is that the world is what it is, and the antithesis between contingency and necessity has no basis in experience. The universe, for Hume, is contingent in exactly the same sense as it is for Sartre: it is not founded on any 'reason' and does not authorize us to seek for any.

Hume's criticism finally shook the foundations of the eighteenth-century system which had appeared to reconcile empiricism with belief in natural harmony, moral utilitarianism with the belief that man was destined for happiness, reason as a gift of nature with reason as a sovereign power. If the attempt was to be made to restore the legitimacy of belief in the unity and necessity of Being and in an authentic human standard as distinct from empirical and historical humanity, it would have to take account of the devastating implications of Hume's analysis. This was the problem facing Kant, by whom the attempt was actually made.

11. Kant. The duality of man's being, and its remedy

Kant opted for the sovereignty of human reason as against belief in a natural order of which reason is a part or a manifestation. In his philosophy he rejected the hope that reason could discover the natural law, a pre-existing harmony, or a rational God, or could interpret itself within that harmony. This did not mean, as Hume would argue, that all our knowledge is reduced to the contingency of separate perceptions. Not all our judgements are empirical or merely analytic: synthetic *a priori* judgements, i.e. non-empirical ones which tell us something about reality, form the backbone of our knowledge and assure it of regularity and general validity. But—and this is one of the main conclusions of the *Critique of Pure Reason*—synthetic *a priori* judgements relate only to objects of possible experience. This means that they cannot provide a basis for a rational metaphysic, since a metaphysic would have to consist of synthetic *a priori* judgements if it were possible at all. All we can hope for is an immanent metaphysic in the shape of a code of natural laws that are not abstracted from experience but can be ascertained *a priori*. All thought is ultimately related to perception, and the *a priori* constructions that our mind necessarily forms are meaningful only in so far as they can be applied to the empirical world. Thus the order of nature, as far as its constituent determinants are concerned, is not found in nature but is imposed on it by the order of the mind itself. To this order belongs the arrangement of objects in time and space, as the basis of pure apprehension, and also the system of categories, i.e. non-mathematical concepts which give a unity to the empirical world but are not derived from it.

Experience is thus only possible through the unifying force of the intellect. The order of nature bears witness to the mind's sovereignty over it, but the sovereignty is not complete. Every piece of knowledge, apart from analytical judgements which convey no fresh information, has a content derived from two sources. Perception and judgement are radically different activities. In sensual perception objects are simply given to us and we passively undergo their effect, whereas in intellectual activity we exercise the mind upon them. Both these aspects of human presence in the world, the active and the passive, are necessarily involved in every act of cognition. There is no valid thought that is not related to perception, and no perception without the unifying activity of the intellect. The first of these propositions means that there is no legitimate hope of theoretical knowledge extending beyond the empirical world to absolute realities, and also that the variety of experience cannot be wholly subordinated to the power of the intellect. The second proposition brings out the legislative superiority of the mind over nature considered as a system.

The ineradicable duality of human knowledge is not directly perceptible, but once it has been discovered it reveals the basic duality of all human experience, whereby we assimilate the world, at one and the same time, as legislators and as passive subjects. Within the limits of the legitimate use of our intellect, we cannot do away with the inexplicable contingency of the data of experience. That contingency is something given; we are obliged to recognize it and abandon any hope of finally mastering it. Consequently we cannot bestow a final unity on ourselves or on the world. My own ego as I perceive it in introspection is subject to the condition of time and is therefore not identical with my ego in itself, which is inaccessible to theoretical knowledge. Behind the ego of introspection, it is true, we can discern a transcendental unity of apperception, the condition of the uniting activity of the subject, a self-awareness that is capable of accompanying all perceptions; but of this we only know that it exists, and not how it is constituted. In general, the whole of our organized experience presupposes a realm of

unknowable reality which affects the senses but which we perceive not in its real shape but in a form ordered by our
a priori categories. The presence of the world-in-itself is not
deduced from empirical data but is simply known; my awareness
of my own existence is at the same time a direct awareness of
objects. But we can know nothing of independent reality except
the mere fact that it exists; nor can we do away with the contingency of the knowable world or the duality to which the
human intellect is subject.

The human spirit, however, is not content with the knowledge
of its own limitations or with a jejune metaphysic confined to
awareness of the *a priori* conditions of experience. Our minds
are so constituted that they strive unremittingly after the unity
of absolute knowledge; they seek to understand the world not
only as it is but as it must be, and to overcome the distinction,
entailed by the postulates of empirical thought, between what
is possible, what is real and what is necessary. For this
distinction cannot be expelled from the mind: the possible is
everything that is compatible with the formal conditions of
experience, the real is that which is actually given in its material
conditions, the necessary is that part of reality which proceeds
from the general conditions of experience. The realities of the
world thus include contingency, which we could only eliminate
if we had access to unconditional being, to the absolute unity
of the subject and object of knowledge. We strive incessantly
to attain this, although our striving is in vain; the delusions of
metaphysics, even when revealed as such, live on in the human
mind. They find expression in the construction of concepts which
not only are not abstracted from experience (for *a priori* concepts
are legitimate and essential to knowledge) but are not even
applicable to it. These concepts or ideas of pure reason—such
as God, freedom, and immortality—are a perpetual temptation
to the human spirit, although they cannot be significantly used
within the bounds of theoretical reason. In terms of pure reason
they have a certain meaning, but a regulative and not a constitu-
tive one. That is to say, we cannot know any reality that corresponds to these concepts, but can only use them as unattainable
limits or pointers indicating the direction of our cognitive
activity.

The legitimate use of these ideas, therefore, lies in their

constant summons to the mind to exceed its previous efforts; their illegitimate use lies in the supposition that any effort, however great, will enable us to achieve absolute knowledge. For every judgement in a syllogistic chain, the mind seeks to discover a major premiss; the law of the syllogism requires that we seek a premiss for every premiss, a condition for every condition, until we arrive at that which is unconditional. This maxim is a proper one for the purpose of governing the operations of the mind, but it should not be confused with the erroneous supposition that there is in fact a first, unconditioned link in the chain of premisses. For it is one thing to know that every member of an intellectual sequence has a preceding condition, and quite another to maintain that we can comprehend the sequence in its entirety including a first, unconditioned member. (We may elucidate Kant's thought here by pointing out that while it is true to say that, for any particular number, there is a number greater than it, it is not true to say that there exists a number greater than any other number whatsoever.) The failure to distinguish the syllogistic maxim from the fundamental, but false premiss of pure reason is the source of three typical errors corresponding to the three types of syllogism. In the sphere of the categorical syllogism, this premiss states that in seeking successive conditions for predicative judgements we can finally come upon a subject that is not a predicate. In the sphere of the hypothetical syllogism, it tells us that we can come to an assertion that presupposes nothing; and in the sphere of the disjunctive syllogism, that we can discover such an aggregate of the members of the division of a concept as requires nothing further to complete the division. In this way we delude ourselves that we can establish in the domain of knowledge three kinds of absolute unity: in psychology that of the thinking subject, in cosmology that of the sequence of the causes of phenomena, and in theology that of the subjects of thought in general. But within the boundaries of finite experience there is no object corresponding to any of these three ideas. We cannot apprehend theoretically either the substantial unity of the human soul, or the unity of the universe, or that of God.

There can scarcely be any example in history of a philosopher going to so much trouble as Kant to invalidate arguments in favour of propositions to the truth of which he was so deeply attached. Belief in the existence of God, the freedom and immortality of the soul were not, to him, indifferent matters in respect of which he was only concerned to declare his neutrality. On the contrary, he regarded them as of vital importance, but he believed that the mind deludes itself when it is tempted to imagine that it has laid hold of the Absolute. The Absolute is a beacon to the endless progress of knowledge, but cannot itself become a possession of the mind.

To attain to the Absolute by cognition is the same as to become absolute. But the division of man into a passive and an active part, and the corresponding division of the world into what is perceived and what is thought, what is contingent and what is intellectually necessary—this division can only be resolved at the point of infinity; and the same is true of the opposition in our moral life between free will and law, happiness and duty. For our life as beings endowed with will is divided in the same way between two orders in which we participate unavoidably: the phenomenal, natural world subject to causality, and the world of things in themselves, freedom and total independence of the mind. That which is called duty, and expresses itself in the form of an imperative, is not only quite independent of our inclinations but, by its very character as an imperative, must be contrary to them.

For a command that something must be done willingly involves a contradiction, since if we knew of ourselves that we are obliged to do something and also knew that we would do it willingly, the command would be superfluous; while if we performed the action unwillingly and only out of respect for the law, then the command which makes that respect the motive of the maxim would operate in a manner contrary to the disposition commanded. (*Critique of Practical Reason*, I. 3)

But the conformity of will to law must be possible; it is a condition of supreme goodness, which itself must be possible, bringing about a harmonious synthesis of happiness and virtue, which in the empirical world tend, as we all know, to limit each other. The reason perceives the moral law directly, i.e. independently of knowledge of the subjective conditions that make it possible to fulfil the law. This is to say, man knows what he should do before he knows that he possesses freedom of action; from the fact that he *ought*, he learns for the first time that he *can*, i.e. that he is free. But the freedom thus apprehended is an object of practical reason, which has a wider field of action than speculative reason. Everyone will agree that it is in his power to obey the moral imperative, even if he is not certain that he will in fact do so; 'he believes that he can perform an action because he knows that it is his duty, and he recognizes in himself a freedom which would remain unknown to him if it were not for the moral law'. (*Crit. Prac. Reason*, I. 1, 6). The practical reason has its own *a priori* principles which cannot be derived from theoretical knowledge, and their validity makes it necessary to accept certain fundamental truths which are inaccessible to the intellect, whose power to form concepts is limited by their empirical applicability. Since the will, subject to the moral law, has the supreme good as its necessary object, the supreme good must be possible. And, as this good demands absolute perfection, which is only possible as the result of infinite progress, the validity of the moral law necessarily presupposes the infinite duration of the human individual, i.e. personal immortality. Similarly, the postulate of the supreme good requires for its validity that man's happiness should coincide with his duty, but no natural conditions afford evidence that this will certainly be so. Hence the supreme good as a necessary object of the will presupposes the existence of a free, rational cause of nature that is not part of nature, i.e. the existence of God. Thus, thanks to our awareness of the moral law, the ideas of speculative reason acquire an objective reality that theory could never have secured for them. Our immortality, our participation in the world of *intelligibilia*, unconditional freedom, and the Creator's supremacy over nature—all these are shown to be realities whose existence is demanded by the moral law.

To sum up, the division of man between two opposite orders —that of nature and that of freedom, that of desire and that of duty; a passive existence full of contingent things, an active existence in which the contingency of the object disappears—this division is curable, but on the condition of an infinite progress. The prospect before us is that of an unlimited striving towards self-deification, not in the mystic sense of achieving identity with a transcendent God, but in the sense of attaining absolute perfection, which destroys the power of contingency over freedom. The achievement of a God-like condition in which the reason and will are completely dominant over the world is the horizon towards which the infinite progress of each human individual is directed.

Kant's philosophy does not include the history of a lost paradise and the fall of man. It offers a prospect of the realization of essential humanity, not by obedience to nature but by emancipation from it. Kant opens a new chapter in the history of philosophy's attempt to overcome the contingency of human existence, setting up freedom as man's realization and establishing the independence of the autonomous reason and will as the ultimate goal of man's unending pilgrimage towards himself, a self that will then be divine.

12. Fichte and the self-conquest of the spirit

Johann Gottlieb Fichte sought to remove the limitations of the Kantian doctrine of man's summons to freedom and expounded the view that it is within man's power, and is his duty, to achieve a radical awareness of his own unbounded domination over the conditions of Being, the absolute primacy of his own existence, and his complete independence of any pre-existing order. As Fichte said in his address 'On the Dignity of Man' (1794), 'Philosophy teaches us to discover everything in the ego'; 'only through the ego can order and harmony be instilled into the inert, formless mass'; man, 'by virtue of his existence, is utterly independent of everything outside him and exists absolutely in and through himself; ... he is eternal, existing by himself and by his own strength'. However, this awareness by man of his own status as the unconditional author of Being is not something provided ready-made but is a moral precept, a call to incessant self-transcendence and to an ever-repeated effort which regards each form of Being in turn not as a finality but as a fresh obligation.

This philosophical emancipation of spirit from nature, and the conception of the world as a perpetual moral task, were subsequently decried by Marx and others as symptomatic of the weakness of German political radicalism and of a civilization which, lacking the courage for a practical revolutionary effort, transposed action into the realm of thought and envisaged practice in moralistic terms. However, by virtue of its basic position, German philosophy did not regard the world as a source of optimistic anticipation or as the work of a benevolent nature which prescribed values and provided for their vindication, but as a problem and a challenge. Reason was no longer a copy of nature and did not find in it a pre-established harmony. Philosophy was able to discern in man, as a subject of cognition, a part or aspect of the whole man and of his practical being; and in this way cognition came to be interpreted as a form of practical behaviour.

Fichte's opposition to the philosophy of the Enlightenment was based on a Kantian motif. If man is constrained by the pressure of existing nature, to which he himself corporally belongs, then there can be no morality beyond a utilitarian calculation of pleasures and pains, that is to say no morality at all. If the world is to be an object of obligation, man must be free from the determinism of nature. Consequently, metaphysical and epistemological options imply a moral question. We are constantly tempted by what Fichte calls 'dogmatism', i.e. the viewpoint that explains consciousness by means of objects, since this frees us from responsibility and tells us to rely on the supposed laws of causality to be found in nature; anyone who cannot free himself from dependence on objects is, by inclination, a dogmatist. Idealism, on the other hand, treats consciousness as the point of departure and appeals to it for an understanding of the world of things; the idealist is a man who has achieved awareness of his own freedom, accepts his responsibility for the world, and is prepared to grapple with reality. Those who identify self-awareness with man's objective existence among things, i.e. the materialists, are not so much in error as weak and incapable of assuming the role of initiators of Being. Idealism is not only morally superior but is also a natural starting-point for the philosopher, as it avoids unanswerable questions. It does not have to inquire for whom the original fact of experience arises, for from this point of view subject and object coincide; the primary state of Being, self-consciousness, is being-for-itself and requires no explanation.

But this being-for-itself of self-consciousness is not given to our reflective faculty as a thing or a substance: it appears only as an activity. Fichte rejects the observational point of view that substance must precede action, while action presupposes an active substance. On the contrary, action is primary and in relation to it substantial being is only a secondary product or concrement. Consciousness is itself action, the movement of a creative initiative not prescribed from outside; it is *causa sui*. The world of objects has no independent existence; Kant's 'thing in itself' is a relic of dogmatism. In the awareness of his own unlimited freedom man recognizes himself as absolutely responsible for Being, and he recognizes Being as something which, taken as a whole, makes sense because of man. Freedom is also the condition of a true human community, based on voluntary solidarity and not on the negative bond of interest, which is the only bond if we accept the view that man's Being is defined by the needs marked out for him by nature. Fichte's ideal, like Rousseau's, is a society in which the ties between human beings are based on free co-operation and not regulated by a contract imposed from without.

If, however, consciousness is the absolute starting-point, it cannot be the awareness of perceptions, as in Berkeleyan idealism, but must be the awareness of acts of will. Its first and essential postulate is the obligation of thought towards itself, and this requires the ego to create its own counterpart, in which it recognizes itself as its own self-limitation. Consciousness, the ego, brings into existence the non-ego in order to establish itself in creative self-awareness. The mind is not content with its directly given self-identity but demands a reflective self-identity, turned in upon itself and perceived by itself; to achieve this, however, it must first divide into two and objectify itself by creating the world, which then appears to it as something external and enables it to know itself. This dialectic of self-cancelling exteriorization is a direct anticipation of the Hegelian schema, but it is also rooted in the whole history of neo-Platonic theogony and in all doctrines that present God as coming into existence through his own creative activity. In Fichte's version the attributes of the divine Being are transferred to the human mind, which in its boundless autonomy is the standard to which all other reality is related. As far as the ego is concerned, the opposition between activity and passivity is no longer applicable. In the first version of his *Wissenschaftslehre* (*Theory of Science*; 1794; II. 4. E III) Fichte writes: 'Since the essence of the ego consists exclusively in the fact that it posits itself, self-positing and existence are for it one and the same ... The ego can only avoid positing something in itself by positing it in the non-ego ... The activity and passivity of the ego are one and the same.'

The ego is not identical with the empirical, psychological, individual subject: it is a transcendental ego, i.e. humanity considered as a subject, but it cannot be called a collective subject inasmuch as there is no autonomous being (like, for example, the 'universal mind' of the Averrhoists) independent of the individual consciousness. In other words, humanity is present as the nature of every individual man, the consciousness which each must discover in himself. It is thanks to this that human community is possible; the task of every individual is to know himself as Humanity.

The ego must in this way establish the world of things, which is a product of the ego's freedom but which, once established, is a restriction upon it and requires to be lifted. Hence the creation of the world is not a single event but an unceasing effort whereby the objectified products of the mind are reabsorbed into it. In overcoming the resistance of its objectifications—a resistance which is necessary for its own development—the mind thus attains to the state of absolute self-knowledge, through an unending process of setting itself fresh limits to be successfully transcended. The ultimate end of this process is expressed by absolute consciousness, but this end cannot in fact be reached: as in Kant's philosophy, it is a horizon marking the goal of an infinite progress. The positive conquest of freedom in human affairs thus requires a perpetual antagonism of the mind *vis-à-vis* each established form of civilization. The mind is the eternal critic of its own exteriorizations, and the tension between the inertia of established forms and the mind's elemental creative activity cannot cease to be, since it is a condition of the mind's existence or even, we may say, a synonym for it.

The Fichtean philosophy sought in this way to interpret man as a practical being, and introduced into epistemology the supremacy of the practical, i.e. the moral, point of view. Human cognition is determined as to its content by the practical perspective; man's relation to the world is not receptive but creative; the world is given as an object of obligation, not a ready-made source of perceptions. As, however, the true purpose of the ego is to perfect itself, man's true obligations lie in the sphere of education and self-education.

The ego being understood as freedom perpetually overcoming its own limitations, human history can be interpreted as the history of the mind's struggle for freedom. For Fichte, as later for Hegel, history becomes meaningful if it is conceived as progress towards the awareness of freedom. From unreflecting spontaneity via the power of tradition, the domination of individual particularism, and the final discovery of reason as an external governor, history moves towards a state in which individual freedom will coincide completely with universal reason and the sources of human conflict will dry up. History thus considered is a kind of theodicy, or rather anthropodicy: we can either interpret the evil we find in it as a factor of progress in relation to the dynamism of the whole, or we can argue that it is completely irrational and devoid of existential consistency, that in short it is nothing and does not belong to history.

The Fichtean picture of man as freedom, man discovering his proper calling in a perpetual contest with the inertia of his own alienations, provided the basis for a critique of all tradition and appeared to favour aspirations towards freedom in cultural and political life. It turned out, however, that the same philosophy could be made to yield conclusions quite opposite to its apparent intentions; and indeed Fichte himself did so at a later stage of his career. During the Napoleonic Wars his criticism of the utilitarianism of the Enlightenment and his apologia for non-utilitarian bonds between human beings were combined with a cult of the Nation as the embodiment *par excellence* of a non-utilitarian and non-rational community. In this respect Fichte anticipated Romantic thought. The idea that particular nations are the exponents of the main values of each epoch in the march of history led him into German messianism, and the idea of humanity as the essence of man led him to advocate compulsory state education as a means of aiding the discovery by the individual of his own true path in life. The totalitarian Utopia sketched in Fichte's *Der geschlossene Handelsstaat* (*The Closed Commercial State*; 1800) can be basically justified by his philosophy of freedom. The connection may be traced hypothetically as follows. What is required of man is to discover in himself his own absolutely free and creative humanity. This is not an arbitrary ideal but a real, ineluctable call to self-knowledge, progress towards which is identical with human existence itself. Since individuals and peoples do not develop equally towards their destined aim, but differ widely in the degree of self-knowledge they attain, it is quite natural that the education of the less by the more advanced should hasten the development of the former towards full humanity. If it is the task of the state to educate its nationals in the community spirit and in humanity, it is not strange that the rulers, who know the meaning of humanity better than the ruled, should use compulsion to bring out the humanity that lies dormant in every individual. This compulsion will be no more than the social expression of the compulsion that resides in every individual as his own essence, of which he is as yet unaware; it will in fact, therefore, not be compulsion at all, but the realization of humanity. As man is endowed with humanity by nature, compulsion to join in the community is not a violation of the individual's freedom but a release from the prison of his own ignorance and passivity. In this way Fichte's philosophy of humanity as freedom makes it possible to proclaim the police state as the incarnation of liberty.

Fichte was the true author of the immanent dialectic, i.e. the dialectic which does not extend beyond human subjectivity but makes that subjectivity its absolute point of departure (although in the last stage of his work Fichte returned to the idea of an extra-human Absolute, in whose freedom the human mind participates). Subject and object were, in his view, the result of a duality which sought to find a synthesis in infinite progress; however, as the subject was a human one, the synthesis could not realize itself in the contemplation of an extra-human Absolute, but only in the irreplaceable activity of human individuals. Since Fichte regarded humanity as unconditional existence, he could —and indeed, strictly speaking, had to—regard it as practical existence, defined basically by an active attitude towards its own world, which possessed a conditional existence in relation to creative subjectivity. In this way he laid the foundation of the interpretation of human history as the self-creation of a species, the meaningful, unidirectional ascent of freedom to self-knowledge. History is of course the medium through which non-historical consciousness, directly identical with itself, moves towards a reflective self-identity. History, therefore, is not an end in itself; it does not embrace all humanity without exception, but is a bridge between two non-historical realities, viz. consciousness as it was at the beginning and as the final goal of human evolution. The transcendental human subject, rooted in itself as freedom, by its practical exertion divides itself into the world of subject and object, and, through history, returns in an infinite progress to self-conscious freedom—such is the essential content of Fichtean metaphysics.

The possibility of interpreting this doctrine as an apologia for the totalitarian state depends primarily on two of its presuppositions. In the first place, Fichte holds that the purpose of each human individual and that of humanity as a whole are completely identical, that the realization of each and every one of us is exhausted in the realization of the universal humanity which resides in the individual as his own nature although he is not fully aware of it. Further, human beings are more or less advanced according to the degree to which they have realized their own essential humanity. Although, in Fichte's mind, education was to be primarily maieutic and to bring out the human dignity inherent in each individual, nevertheless, given the latitude with which the more enlightened were permitted to specify the kind of humanity to be aimed at, it was easy to interpret his programme as a system whereby everyone was to be coerced into realizing his own freedom. In other words, as freedom is in no way linked with differentiation and as the realization of the individual is no more or less than the realization of undifferentiated humanity, the achievement of freedom does not depend at all on the free self-expression of the individual as an irreducible entity. The transcendental ego is not a product of empirical human experience, but is sovereign *vis-à-vis* human life and can make demands on it by virtue of its own freedom; it can also, like God, hasten the progress of its own freedom by coercing the empirical human being.

13. Hegel. *The progress of consciousness towards the Absolute*

Despite the opposition between Kant's and Fichte's attempts to autonomize human existence, they both maintained an essentially dualistic point of view. In Kant this was a dualism between the contingency of the world of sense and the necessary forms of the intellect, and between duty and nature in man; in Fichte it was the dualism of duty and reality, which is a permanent condition of the development of the mind and is prolonged endlessly in an infinite movement of progress. However, neither Kant nor Fichte overcame the dilemma: either the mind comes to grips with the contingency of existence, and in cognizing it is, so to speak, infected with contingency, or it does away with contingency and thereby does away with the manifoldness of existence.

Hegel's majestic system was intended, among other things, to interpret the nature of Being in such a way as to deprive contingency of its effect while at the same time preserving the richness and variety of the universe. Contrary to Schelling's idealism, Hegel did not wish to reduce Being to the un-differentiated identity of the Absolute, in which the variety and multiplicity of finite reality must be lost or dismissed as an illusion; and again, in opposition to Kant, he refused to regard the thinking subject as abandoned helplessly to the experience of that variety and multiplicity, presented to him endlessly as a datum without reason or meaning. His purpose was to interpret the universe as entirely meaningful without sacrificing its differentiation. This required, as he wrote, 'a self-origination of the wealth of detail, and a self-determining distinction of shapes and forms' (*Phenomenology of Mind*, Preface).

But a Mind free from contingency is the same as an infinite Mind. For in so far as the object is something alien to the subject it is a limitation of it, a negation; a limited consciousness is finite, and the object, as foreign, is, so to speak, its enemy. Only when the Mind perceives itself in the object, thus removing the latter's alienness and objectivity, does it free itself from restrictions and achieve infinity; in this way the variety of Being ceases to be accidental. But, in order for that variety to maintain its richness, the process of removing the alienness and objectivity of the world must not be based on annihilating the created universe or proclaiming it as an illusion that is bound finally to merge into the unity of the all-absorbing Absolute; it must persist even as it passes away, that is to say the negation of it by the Mind must be an assimilating negation. The term *Aufheben* or 'sublation' denotes this particular kind of preserving negation, which safeguards both the independence of the Mind and the manifoldness of Being. But these are to be safeguarded not merely by putting forward an arbitrary definition of the

Mind that meets these conditions, but by means of a historical description comprising the whole development of Being and capable of giving an integral sense to the history of the world and especially of human civilization. This historical system must present the development of Mind, through the travail of history, towards absoluteness. Such is the purpose of Hegel's *Phenomenology*, the most important of those of his works which together contain the germ of Marxism. It presents the successive phases of the necessary development of consciousness, which evolves from pure consciousness to absolute knowledge by way of self-awareness, Reason, Spirit (or Mind), and religion, and in that knowledge fulfils the purpose of the world, which is identical with knowledge of the world.

Apart from its immensely complicated and abstract language, which sets the reader's mind off on conflicting tracks and involves him in monstrous ambiguities, the *Phenomenology* has the further defect that it is not clear in which parts of the work the successive phases of the evolution of Mind are intended to correspond to actual phases of cultural development, and in which they are schemata constructed independently of that development. In some passages Hegel corroborates his account of particular phases by referring to specific events in the history of philosophy, religion, or politics, as when he speaks of Stoicism or scepticism, Greek religion, the Renaissance, the Enlightenment, etc. This might suggest that he is tracing the successive stages of the incarnate Mind in the history of civilization. On the other hand, we find that the phenomenological time-scheme does not correspond with actual history. For example, religion is presented as a phase of development subsequent to the evolution of self-knowledge, Reason, and Spirit—an evolution that comprises many elements of modern times, whereas religion dates back to antiquity. A 'phenomenology', however, is, properly speaking, not a timeless classification but a presentation of the sequence in which phenomena make their appearance and come to maturity. There are many such ambiguities in Helgel's *Phenomenology*, and they affect the question of the proper place of this work in Hegel's system as a whole. None the less we may draw attention to some essential tendencies in the sphere with which our own study is concerned.

Hegel regards it as evident that the spiritual is the starting-point of the whole evolution of Being: in this he is following a tradition that goes back to the beginnings of European philosophy in Parmenides, Plato, and the Platonists. The first principle must be something that depends for its Being on nothing else, that is self-supporting and related only to itself, the manner of this relationship being left for further investigation. It cannot therefore be composed of parts that limit one another or are mutually indifferent; being-in-itself and being self-related is a mode of Being that pertains to the spirit. That which is absolute is by definition free of all restriction or limitation, i.e. it is infinite, and only the Mind can be infinite in this sense. But Hegel goes further: the Mind is not only the first principle but is the only reality. This means that every manifestation of Being, every form of reality is intelligible only as a phase of the development of Mind, as its instrument or as a manifestation of the manner in which it combats its own imperfection.

For the Mind, though self-existent, is not self-sufficient. Hegel avoids the difficulty of the Platonists and Christians who had to account for the finite world while premising the self-sufficiency of the Absolute. He does so on the basis that the Absolute is self-sufficient in the sense that its self-existence does not require support from anything, but not in the sense that it expresses the fullness of its own possibilities. It must also come to exist *for* itself, i.e. as the plenitude of knowledge of itself as Mind. In other words, it must become an object so that it can then do away with its own objectivity and assimilate it completely, when it will be a sublated object, self-directed and existentially identical with self-knowledge. Now—and this is the most distinctive feature of Hegelian thought—our reason, reflecting on the way in which the Absolute comes into being, must regard its own activity as a constituent of that process; for otherwise the evolution of Mind and of our own thinking concerning that evolution would be two separate and disparate realities—our thought would be accidental in relation to the evolution of Mind, or vice versa. This accounts in part for the error of the Kantian criticism, which involves first examining the nature of our cognitive powers and then using them to consider the nature of Being, after Reason has determined the limits of its own validity. This is an impracticable endeavour based on a false assumption. It is impracticable because our finite Reason cannot draw the bounds of its own validity without some prior means of doing so, nor can it exist before it exists. The false assumption is to suppose that man and the Absolute are 'on opposite sides' in the process of cognition, which is represented as a link between them. Reason, thinking of the Absolute, must be able to give a meaning to its own thought by relating itself to the Absolute; otherwise it condemns itself to a contingent role, by the illusory attempt to embrace an Absolute which does not comprise the activity of our intellect concerning it. In thinking about the world we must be aware that our thought is itself part of the evolution of the world, a continuation of the very thing to which it relates. Hegel is not writing *about* the Mind: he is writing the Mind's autobiography.

Thinking in this way, we see that the way to grasp the sense of any process of evolution is to relate the part to the whole. Truth can only be expressed in its entirety; meaning can be understood only in relation to the complete process, 'truth is the whole' (*das Wahre ist das Ganze*). This phrase has a double sense. In the first place, apart from any Hegelian interpretation it means that the knowledge of any part of the universe is significant only in so far as that part relates to the total history of Being. Secondly, the specifically Hegelian meaning is that the truth of every individual being is contained in the concept of that being, and that in realizing itself a being displays the fullness of its nature, which was previously concealed; it conforms progressively to the concept of itself and is finally identical with knowledge of itself. This last point, too, has a different meaning according to whether it is applied to a component of the universe or to the whole. We can say of any particular being that in developing itself it actualizes what was at first only a possibility (but a specific possibility, not a choice of different ones) and in this way attains to its own truth. In this sense, the truth of a seed is the tree that grows out of it, and the truth of an egg is a chicken. By achieving what was only a possibility, an object becomes its own truth. But Hegel goes further: in the development of Being, considered as a single process, truth, or the attainment of conformity with the concept, is not a mere casual conformity, i.e. the coincidence of two realities that the Mind might compare from outside as it compares a picture with its original or the plan of a house with the house itself. Where the whole process of the evolution of Mind is concerned, this conformity consists of the identity of a being with the concept of itself, i.e. the final situation in which the Being of Mind is the same thing as knowledge of that being: the Mind, having cast off its own objectivized form, returns into itself as the concept of itself, but a concept which is not merely abstract but is also awareness of that concept.

The progress of the Mind is thus circular: it ends as it began, which signifies that it is its own truth or has become conscious of what it was in itself. This final state is what is called absolute knowledge.

But this substance in which spirit consists is the development of itself explicitly to what it is inherently and implicitly; and only by this process of reflecting itself into itself is it then essentially and in truth spirit. It is inherently the movement which constitutes the process of knowledge—the transforming of that implicit inherent nature into explicitness and objectivity, of Substance into Subject, of the object of consciousness into the object of self-consciousness, i.e. into an object that is at the same time superseded and transcended—in other words, into the notion [*Begriff*]. This transforming process is a cycle that returns into itself, a cycle that presupposes its beginning, and reaches its beginning only at the end. (*Phen. of Mind*, DD. VIII. 2)

If, however, the operation in which Mind creates the true content of history and finally returns into itself is not an empty one, that is to say if Mind does not simply revert to its original state as though nothing had happened, this is because the final outcome forms an integral whole with the process that has led to it, so that the Mind preserves at the end of the journey all the wealth it has accumulated on the way. The operation is one of continual 'mediation', i.e. the self-differentiation of the Mind, which produces from itself ever new forms which it then reassimilates by de-objectivizing them. At each successive stage the Mind thus proceeds by a continual self-negation; the negation is itself negated, but its values persist although they are absorbed into a higher phase.

But the life of mind is not one that shuns death, and keeps clear of destruction; it endures its death and in death maintains its being. It only wins to its truth when it finds itself in utter desolation [*Zerrissenheit*] ... Mind is this power only by looking the negative in the face, and dwelling with it. This dwelling beside it is the magic power that converts the negative into being. (*Phen. of Mind*, Preface)

The first form of the existence of Mind is awareness that is still not self-awareness. It goes through a phase of sensual certainty, in which consciousness is distinguished from the object, so that for consciousness there is such a thing as being-in-itself. What was an object has become knowledge of an object, so that Being has become being-in-itself-for-consciousness. At the same time consciousness changes in character and gradually frees itself from the illusion that it is burdened by something alien. Then, when consciousness grasps things in their specific character and understands their unity, it becomes a perceiving consciousness, or simply perception. In perception consciousness attains to a new phase, that of apprehending generality in the individual phenomenon. Every actual perception contains a general element: in order to grasp that a present phenomenon is present, we must apprehend the 'now' as something distinct from the perception itself, thus deriving an abstract element from the concrete datum. In the same way, when we apprehend the individuality of things we can do so only by means of an abstract conception of individuality, and we are on the level of generalized knowledge when we become aware of individuality as such. The actual 'thing out there' is inexpressible: language belongs to the realm of generality, and so therefore does every perception as soon as we express it. Perception, by imparting generality to the world of sense, surpasses the concreteness of the given object yet at the same time preserves it. Again, the object is distinguished by its particular qualities from other objects, and this opposition gives it its independence; yet at the same time it deprives it of independence, for the independence that consists in being different from other things is not absolute independence but a negative dependence on something else. The object dissolves into a set of relationships to other objects, so that it is a being-in-itself only in so far as it is a being-for-something-else, and vice versa. The comprehension of this form of generality in the world of sense signifies the entry of consciousness into the domain of intellect. Intellect is capable not only of apprehending the general in the concrete but also of apprehending generality as such, in the full content of its conceptual existence. It comprehends the supersensual world by its opposition to the sensual. In this opposition both worlds are made relative to each other in consciousness: each can only be understood as the negation of the other, each thus contains in itself its own opposite and thereby becomes infinite—for infinity is the lifting of the barriers imposed on Being by anything alien to it; a world becomes infinite by containing within itself what was previously its limitation. Then, when the conception of infinity becomes an object of consciousness, the latter becomes self-awareness or self-reflection.

Self-knowledge is aware that the object's being-in-itself is its manner of existing for another; it endeavours to possess itself of the object and cancel its objectivity. Self-knowledge tends by its nature towards that infinity which it has conceptually made its own. On the other hand, self-knowledge exists in and for itself only by virtue of the fact that it is recognized as such by another self-knowledge. Every self-knowledge is a medium through which every other is linked with itself. In other words, the self-knowledge of a human individual exists only in the process of communication and mutual understanding among human beings; it is a delusion to imagine a self-knowledge that treats itself as an absolute point of departure. But the presence of another self-knowledge as a condition of the first is also a limitation of the latter and a hindrance to its attainment of infinitude. Hence there is a natural tension and antagonism among self-knowledges in one another's presence. It is a fight to the death, in which each self-knowledge voluntarily exposes itself to destruction, and which results in one of them losing its independence and being subdued by the other. There arises a master-and-slave relationship, and this mutual dependence is the beginning of the process of the development of the spirit by human labour. The master has enslaved the independent object, using the slave as an instrument. The slave subjects things to treatment which has first been planned deliberately, i.e. in the Mind; but he performs the role imposed by the master and commander, and it is the latter alone who truly assimilates the object to himself by using it. But in this process, which seems to realize the object as a spiritual extension of the master, there occurs the reverse of what one would expect from the master–slave relationship. Labour signifies abstention from enjoyment, the repression of desire; in the slave's case it is a perpetual abstention for fear of the master, but in that fear the slave's self-knowledge achieves being-in-itself, and the repression gives form to objects; the slave regards the

Being of things as an exteriorization of his own consciousness, and in this way being-in-itself is restored to consciousness as its own property. In labour, which is as it were the spiritualization of things, the slave's self-knowledge discovers its own meaning, although it appeared only to be actualizing the meaning of another. In servile work man perfects himself in humanity by the active spiritual assimilation of the object and by aptitude for ascesis. This phase, however, is not one of freedom or of the unity of subject and object: self-knowledge as an independent object is distinct from the independent object as self-knowledge.

The next form of self-knowledge is the thinking consciousness which apprehends itself as infinite and is therefore free. When I think, I am within myself and am free; the object becomes my being-for-myself. This form of free self-knowledge is that of Stoic philosophy, which refuses to recognize slavery and holds that spiritual freedom is independent of external conditions. The essence of this freedom is thought in general; thought withdraws into itself, relinquishes the attempt to assimilate the object, and declares itself indifferent to the question of natural existence. This moral negation of things is carried to an extreme by scepticism, which denies them intellectually as well, declares the non-existence of everything 'other', and annihilates the multiplicity of the universe. The sceptical consciousness would destroy both the object and its own relation to it. It suffers from a contradiction, however, since it purports to achieve self-identity by denying the fact of differences in the world, yet in that very act it becomes aware of its own contingency, which is the opposite of self-identity. When this contradiction is perceived we have an unhappy consciousness, torn between awareness of itself as an autonomous being and as a contingent one. This divided state is exemplified by Judaism and early Christianity. Consciousness is confronted by the otherworldly Being of God, in which it beholds itself indeed, but in opposition to God's immutability; it humbly acknowledges its own individual contingency in the presence of divinity, but does not know its own individuality in its truth and universality. Powerless individuality perceives renunciations on God's part even in the results of its own activity, but in the consequent acts of thanksgiving it rediscovers its own reality and attains to the next stage of spiritual evolution, that of Reason.

Reason is the affirmation of the individual consciousness as a consciousness that is antonomous and sure of itself; it expresses this certainty in idealistic doctrines which aim at regarding the whole of reality as something comprehended by the individual consciousness. However, this rationalistic idealism is unable to find room within its boundaries for all the variety of experience, and declares this variety to be of no concern to it. In so doing it falls into a contradiction, for, while seeking to affirm the independence of Reason, it acknowledges, if only by its indifference, the existence of something that, as in Kant's doctrine, is outside the unity of apperception. In addition, it is obliged to take cognizance of another's ego as different from its own and thus as a limitation of its Being. Reason, however, is confident of discovering itself in the world and removing the 'otherness' of natural being; it sets about doing this, firstly in scientific observation (Reason as observer), with the purpose of turning the evidence of the senses into concepts, and then seeks to establish laws that will eliminate sensual being, so as to recognize as real only that which purely and simply fulfils the conditions of the law. However, unassimilated reality cannot be invalidated in this fashion. Reason is constantly faced with a contradiction between its demands and the world as it is encountered. Consciousness is thus again subjected to an inner conflict, due to the chronic opposition between that which is given and the purposes devised by reason. The issue is between individuality and universality, between law and the individual, between virtue and the actual course of history.

This last point is of especial importance, as it raises the whole general question of the relation between moral imperatives and existing reality. In the contest between virtue and history, the former is bound to succumb.

Virtue will thus be overpowered by the world's process, because the abstract unreal essence is in fact virtue's own purpose . . . Virtue wanted to consist in the fact of bringing about the realization of goodness through sacrificing individuality; but the aspect of reality is itself nothing else than the aspect of individuality. The good was meant to be what is implicit and inherent, and opposed to what *is*; but the implicit and inherent, taken in its real truth, is simply *being* itself. The implicitly inherent element is primarily the abstraction of essence as against actual reality: but the abstraction is just what is not true, but a distinction merely for consciousness; this means, however, it [the implicitly inherent element] is itself what is called actual, for the actual is what essentially is for another—or it is being. But the consciousness of virtue rests on this distinction of implicitness and explicit being, a distinction without any true validity ... The way of the world is, then, victorious over what, in opposition to it, constitutes virtue; it is victorious over that whose nature is an unreal abstraction. (*Phen. of Mind*, C. v. B, c, 3)

This is an expression, in more complicated terms, of the classic aphorism in the Preface to the *Philosophy of Right*: 'What is rational is actual, and what is actual is rational.' Hegel regards it as a delusion of Reason to set up a basic opposition between the actual course of history and the 'essential' demands of the world—an opposition in the form of a conflict between the normative ideal derived from Reason itself and the realities of the spirit as it evolves into Being. This criticism of Hegel's is levelled both against Fichte and against the Romantics: the error of postulating an eternal conflict between the rational imperative and the existing world lies in the fact that Reason is not yet able to comprehend reality as the gestation of Reason, so that reality constantly appears to it as something contingent and to be overcome. It is over this question that the most important conflicts of interpretation have arisen among Hegel's successors. Did Hegel mean to declare that it is in accordance with Reason to assent *ex animo* to reality as we encounter it in all its detail, the world at any given moment being simply a necessary stage in the evolution of Mind, so that his doctrine is a logodicy devoted to proving that 'whatever is, is right'? Or, on the contrary, is it the duty of Reason to investigate what parts of existing reality are truly in accordance with the principles of its evolution, and thus reserve to itself the right to judge any particular situation? It is difficult to resolve the ambiguity of Hegelianism on this fundamental point. Hegel does not in fact seek to apply moral judgements to past history, but rather to understand it, with all its horrors, as the travail of Mind struggling to be free. On the other hand, he restricts philosophy to awareness of the past historical process and denies it the right to peer into the future, while claiming that his own philosophy represents the final emancipation of Mind from the trammels of objectivity. We may thus say that as regards the past his philosophy is a reasoned justification of history in relation to its final goal, while as regards the future he, as it were, chooses to suspend judgement.

This point of view is confirmed when it comes to transferring the accord between developing essence and actual existence to the case of a human individual. The individual knows himself only through his own action: his nature is revealed in his attitude to the world and the way he expresses that attitude in practice. What he does *is* himself: activity is merely the bringing of possibility into existence, the awakening of latent potential. But if so, we cannot find rules in the Hegelian system that enable us, in a practical situation, to distinguish what is 'essential', either in the individual's case or in the context of history, from what is a distortion and corruption of that essence. It appears natural to suppose, indeed, that factual reality is generally the fulfilment of the developing possibilities of Mind, which exists just in so far as it manifests itself ('The essence must reveal itself', as Hegel says in his *Logic*), and which is not presented with a choice between different roads of development, but brings to fruition the single possibility that it contains within itself.

To proceed: when Reason has attained the certainty that it is itself its own world and that the world is it, and when it thus knows that it is objective reality and that this reality is at the same time its being-for-itself, Reason then becomes Mind in the narrower sense of the term, i.e. as limited to the developmental phase of consciousness. Reason in the form of Mind recognizes itself in the world, i.e. it sees the world as rational and frees it of contingency, but at the same time it does not regard the world as a delusion but as a reality in which it actualizes itself. It is not the kind of Reason which separates itself from the world and places itself above or beside it, which is not prepared either to entrust its own contingency to the world of Being or, on the other hand, to obtain for itself an illusory autonomy by declaring the world to be a mere appearance. It stands in opposition to Kant's solution as to those of the Romanticists and idealists. The spirit actualizes itself in the world of ethics, culture, and the moral conscience. 'But only spirit which is object to itself in the shape of Absolute Spirit is as much aware of being a free and independent reality as it remains therein conscious of itself'

(*Phen. of Mind*, CC. VII, Introduction). The Mind conscious of itself as Mind is the mind that functions in religion, that is to say in the action of the Absolute Being in the guise of Mind's self-knowledge. The first actuality of Mind is natural religion; the elimination of that naturalness leads to the religion of art, and when the one-sidedness of both these stages is done away with there appears a revealed synthetic religion in which the 'I' of the spirit is directly present and reality is identified with it. Religion, however, is not the final fulfilment of the activity of Mind, for in it the Mind's self-knowledge is not an object of its consciousness, its own consciousness has not yet been overcome. The final form of Mind is absolute knowledge, i.e. the pure being-for-itself of self-knowledge. Being, truth, and certainty of truth have all become one; the full content of Mind, accumulated in the course of history, takes on the form of the ego. Objectivity has been eliminated as such, and Mind discourses with itself, imbued with the fullness of variety created through history and at the same time freed from all 'otherness' by which it was limited and from all the differences that arose at particular stages between Being, concept, and conceptual consciousness.

Despite all the ambiguities of the *Phenomenology of Mind*, the uncertainty as to the relationship between the developmental necessities of consciousness and the actual history of civilization, and the immense difficulty of following the transitions between the successive phases of Mind's self-denial and the reassimilation of its own exteriorizations—despite all this, Hegel's metaphysical epic affords sufficient clues as to its general intentions. Hegel insists that our acts of cognition include not only the object of knowledge but also the fact that it is known; in the cognitive act whereby the Mind assimilates things, it must understand its present relation to them. He thus aims at an observational standpoint from which reality and the thought of reality are alike explicable, a standpoint that comprehends both Being and the understanding of Being. Only from such a standpoint, if it can be achieved, will the world and the intellect lose their contingency; alternatively, one or other of them must be inexplicable or arbitrarily written off as mere appearance. At the same time we perceive that even the phrase 'observational standpoint' is incorrect; if the intellect were able to contemplate its own relation to the world, this contemplation would be a new kind of relationship, not comprised by self-understanding, and there would be no end to the process of ascending to higher and higher vantage-points, while consciousness would always remain in some inexplicable place outside the world and outside itself. Therefore the final elimination of the estrangement between Mind and object must at the same time be the effective elimination of the object's own objectivity, and not a mere theoretical understanding of the object as of an alienated consciousness; the object and knowledge of it must coincide in unity.

If the elimination of the opposition between subject and object were merely a regulative ideal for the purposes of thought, and not a state of affairs that can actually be attained in the course of a finite development, then the operation of Mind would be in vain. Progress might go on for ever, but it would be no real progress, as the goal would still be infinitely far away. From this point of view Hegel, especially in his *Logic*, denounces the idea of 'spurious infinity' which he finds in the doctrines of Kant and Fichte. In their view of progress the antagonism between the order of nature and that of freedom, between duty and Being, is eternalized, so that finitude becomes something absolute and unvanquishable.

The understanding persists in this sadness of finitude by making non-being the determination of things and at the same time making it imperishable and absolute. Their transitoriness could only pass away or perish in their other, in the affirmative; their finitude would then be parted from them; but it is their unalterable quality, that is, their quality which does not pass over into its other, that is, into its affirmative; it is thus eternal ... But certainly no philosophy or opinion, or understanding, will let itself be tied to the standpoint that the finite is absolute; the finite is only finite, not imperishable; this is directly implied in its determination and expression. (*Science of Logic*, Bk. 1, 2, B, c, α)

If we regard infinity as merely the negation of finitude, then the very concept of the former is dependent on finitude considered as the basic reality; infinity is merely the extremity of finitude, from which it cannot free itself, and is therefore a finite or 'spurious' infinity. As against this, an affirmative, true infinity is the negation of finitude conceived as a negation: it is thus the negation of a negation, a real victory over finitude, an issuing of finitude beyond itself. Only when finitude by reasons of its contradictions shows itself to be infinite, and when the finite by becoming truly itself puts on infinity—only then will infinity assume a positive sense. Hence 'progress to infinity', or the idea of unlimited self-perfection, the eternal approximation of reality to an ideal, is an internal contradiction but an inert one, repeating itself over and over again without change and leading nowhere. It is the tedium of monotonous non-fulfilment; whereas authentic infinity, 'as the consummated return into self, the relation of itself to itself, is being—but not indeterminate, abstract being, for it is posited as negating the negation; ... the image of true infinity, bent back into itself, becomes the circle, the line which has reached itself, which is closed and wholly present, without beginning and end' (ibid., Ch. 2, C, c).

As we have seen, Hegel regards the notion of infinite progress as encumbered by an internal, non-dialectical contradiction. If the idea of an ascending development is to make any sense in general, it must be a development with an effective terminus in view. The elimination of the contingency of Mind and the conquest of freedom must be actually possible; to say that they can be reached at infinity is the same as to say that they cannot be reached at all. If the history of Being is intelligible, if any sense can be given to the dialectic in which Mind wrestles with its own objectifications, this can only be in relation to a *real* Absolute—not an Absolute which is merely a signpost to a place that the Mind knows it will never reach, that is to say a place that does not exist.

To sum up, the Hegelian dialectic is not a method that can be separated from the subject-matter to which it is applied and transferred to any other sphere. It is an account of the historical process whereby consciousness overcomes its own contingency and finitude by constant self-differentiation.

14. Hegel. Freedom as the goal of history

This overcoming of contingency is the same as freedom of the spirit. From this point of view the evolution of Mind is dealt with especially in Hegel's *Lectures on the Philosophy of History*. Published after his death, the *Lectures*, along with the *Philosophy of Right*, are the most popular and most often read of his works.

Unlike the *Phenomenology of Mind*, they are written in fairly clear and uncomplicated language, and have thus had a major effect in forming the stereotype of Hegel's doctrine. His philosophy of history is an account of the spirit's search for freedom through the variety of past events.

According to Hegel the meaning of history can be discovered, but it is a meaning that is not indicated by history: rather it uses history as an instrument. Freedom is proper to the Mind, as gravity is to matter; but Mind must first realize its own nature by elevating its freedom to the dignity of freedom-for-itself, self-knowing freedom. This freedom is equivalent to being-within-itself, i.e. the state of being unlimited by any alien objectivity. In the course of human history the Mind becomes that which it was in itself; it does not, however, throw away the riches it has accumulated on the journey, like a ladder that is no longer needed after the ascent, but preserves them all. 'The life of the ever-present Spirit is a circle of progressive embodiments [*ein Kreislauf von Stufen*], which looked at in one aspect still exist beside each other, and only as looked at from another point of view appear as past. The grades [*Momente*] which Spirit seems to have left behind it, it still possesses in the depths of its present' (*Lectures on the Philosophy of History*, Introduction).

Nature does not contain in itself the element of freedom, and consequently there is no progress in it, only changes and endless repetition of the same thing. Nature is an indispensable condition for the operation of the human spirit, and as such has its place in the divine economy. But the actual progress of Mind takes place in human history and particularly in the evolution of civilization, in which the human spirit attains to an increasing self-knowledge of freedom. History becomes intelligible as a whole if we regard it as the development of the consciousness of freedom, a development which in its main lines is necessarily determined. In the ancient Orient only one man, the despotic ruler, enjoyed freedom, and all that world knew of freedom was expressed in the tyrant's whim. Ancient Greece and Rome had an elementary notion of freedom in general and knew that some of their citizens were free, but they did not ascend to the concept that man as such is free. This concept was recognized only in Christian-Germanic civilization, and is an essential, inalienable conquest of the human spirit.

World history is also the history of Reason: i.e. its course follows a rational design which the philosopher's eye is capable of perceiving. At first sight, it is true, history appears to be a chaos of surging passions and confused struggles, in which the collision of individual or group interests produces irrational, accidental results; the mass of human suffering and misfortune serves no useful purpose and is engulfed by the indifference of time. But in fact the situation is quite otherwise. Individual passions, which are the mainspring of human activity, play their part, independently of anyone's intention, in the progress of evolution and are instruments of the wisdom of history, which cunningly uses for its own purpose actions motivated by private designs. History is thus not intelligible if we present it in a psychological light by examining the motives of particular agents; its meaning consists in the process that is not contained in any of these motives, but uses them to fulfil the purpose for which Mind exists. The subjective motives of human acts are not accidental, inasmuch as they relate to an overriding purpose that precedes both history and the individual subject. Hegel says, it is true, that 'reason is immanent in historical existence and fulfils itself in and through it' (*Lectures*, Introduction), but this does not mean that the rules of the operation of universal Reason were created for the first time by empirical history. Reason is immanent in history in the same way as the Christian God, incarnate in human form; his purpose is fulfilled only through history, which is, as it were, the body of the Deity, but it is not by history that the purpose is determined.

It is by no means the purpose of the operation of the spirit in history to satisfy human desires. 'The history of the world is not a scene [*Boden*] of happiness. Periods of happiness are blank pages in it, for they are periods of harmony—periods when the antithesis is in abeyance' (ibid.). Humanity undergoes struggles and antagonism, suffering and oppression in order to fulfil its own calling, which is also that of the universal spirit. For 'man is an object of existence in himself only in virtue of the Divine that is in him—that which was designated at the outset as Reason and which, in view of its activity and power of self-determination, was called Freedom' (ibid.).

Once we understand this we can evaluate for ourselves the utopias or ideals which men, following their own whims, have opposed to the poverty of reality. Reason justifies history when discerned in it, and condemns to vanity and ineffectiveness all arbitrary models of a perfect society. Even if these are in accordance with the just demands and rights of the individual, 'the claim of the world-spirit rises above all particular claims'. And this right of the spirit actualizes itself with inexorable necessity, according to the self-determination to which the spirit is subject.

All forms and aspects of civilization—law and the state, art, religion, philosophy—have their defined place in the progress of Mind towards freedom. Thanks to them, the rational consciousness of the individual is not condemned, like that of the Stoics, to the kind of freedom that consists of withdrawing helplessly into oneself and accepting the inevitability of external, alien, accidental, and uncontrollable events. Hegelian freedom is the understanding of necessity, but it is quite different from what the Stoics meant by this. The human spirit desires to reconcile itself with reality, but not through humble resignation which eternalizes the opposition between a closed-off, autarkic self-awareness and the indifferent course of events. The subjective human will has a means of reconciling itself to the world by understanding it and realizing itself in it, rather than turning away from it in a spurious dignity which is merely a cloak for despair. This means consists of civilization and especially the state. The state is the 'ethic whole' in which the individual can realize his own freedom as a part of the community, at the price of giving up the whims of self-will and the making of arbitrary demands on the world as his fancy may dictate. The state is not merely an institution invented for the settling of conflicts or the organization of collective enterprises in accordance with a social contract. As the locus of the reconciliation of the subjective will with universal Reason it is the realization of freedom, an end in itself, 'the divine idea as it exists on earth' and the reality which alone gives value to the individual life. 'Every value and every reality that man possesses, he owes to the state alone' (ibid.). As the highest form of objectivization of Mind, the state represents the general will, and the freedom of the individual is a reality when it is based on obedience to the law, for then the will is obeying itself. In this subordination the opposition between freedom and necessity ceases to exist, since the necessity prescribed by the Reason of history comes about not through compulsion but through free will. Hegel did not assert that the private sphere must be completely absorbed into the collective will embodied in the organs of the state: he believed, on the contrary, that the state is a mediator between the spheres of private and collective life and that its institutions are the embodiment of that mediation, for the private interest of the state's servants is identical with the collective interest. In the case of other members of society, the restrictions imposed on their personal wishes and impulses, far from being a limitation of freedom, are a condition of it. The state, it is true, has no other reality than its citizens, but this does not mean that the will of the state can be determined by the collectivity of their private, individual opinions. The general will is not the will of the majority but the will of historical Reason.

Hegel's historiosophy was criticized from the beginning, as it still is today, on two main grounds. In the first place, there was the complaint that it denied the independent value of personal human life, allowing to the individual only the role of complying with the demands of universal Reason, and, in the name of those demands, authorizing the state to coerce individuals as much as it liked for the sake of a higher freedom. Secondly, critics pointed out that the doctrine justified every actual reality as praiseworthy by the very fact of its existence, which proved it to have been planned by the divine Mind. The first of these objections relates chiefly to the introduction to *Lectures on the Philosophy of History*, the second to the introduction to the *Philosophy of Law*.

The objection which represents Hegel as an apologist of the totalitarian state is weakened to a certain extent by the fact that he regarded the development of society not only as the development of absolute spirit through historical events, but also as the gradual reconciliation of the subjective will with the general will. This means that no state acting by means of violence could fulfil the supreme commands of Reason. It is true that in the earlier stages law appears, from the individual's point of view, as an external system of restriction and constraint, but the whole tendency of the development of the spirit is to overcome this opposition and so interiorize the general will. The course of history does not begin with a Golden Age; the mythology of a happy state of nature or a lost paradise is completely alien to Hegel's thought. On the contrary, the state of nature is one of barbarity and lawlessness, and it is by the gradual perfecting of politico-legal institutions that this gives place to rational thought and the subduing of private impulses. But, in Hegel's view, Reason tolerates only the compulsion of reasoning itself. That is to say, the systematic coercion of individuals is the mark of an immature society, and progress leads to a situation in which the subjective will and the general will coincide spontaneously as a result of acts of understanding of the world by those who form the collectivity of the state. It is impossible for Reason to rule in a situation where it has to assert its demands by violence so as to triumph finally in opposition to the consciences of individuals; its triumph can only be assured by the intellectual maturity and reformed consciousness of the state's citizens.

If it is true, however, that Hegel was by no means a champion of the tyrannic power which forces its subjects to obey the dictates of historical Reason, the practical application of his doctrine means that in any case where the state apparatus and the individual are in conflict, it is the former which must prevail. For as long as the individual consciousness is not fully transformed and is still subject to egoistical impulses, so that there is not yet a complete and voluntary accord between the subjective will and universal Reason, the question must arise: who is to decide, in a particular situation of conflict, what the universal will requires? Since there is no other institution but the state which could assume this role, and since the state by definition is the incarnation of Reason, in cases of conflict it must play the role of the medieval Church, i.e. the sole authorized interpreter of the divine message. Hence, although Hegel's ideal was certainly the complete interiorization of historical Reason in the soul of every individual, and the perfection of the state as an institution was to manifest itself by the disappearance of the need for compulsion, nevertheless in actual cases of conflict where there is no practical question of appealing to the 'majority will' or the voice of the people, the state apparatus, independent of the changing opinions of citizens, must be the final court from which there is no appeal. Hegel supposes, of course, that it is an apparatus functioning according to law and not by the whim of a tyrant or a civil servant; but in cases where the law is ambiguous or it is a question of changing the laws, the state apparatus for the time being has the last word. In this sense, despite Hegel's emphasis on legal and constitutional forms of communal life, the state apparatus is privileged in his eyes and is entitled to assert itself not only against any individual but against all together, since the force of Reason resides in it and not in the will of the majority. Historians have pointed out, it is true, that Hegel's apologia for the Prussian monarchy as the ideal state is qualified inasmuch as he describes institutions that Prussia at that time did not possess. Nevertheless, and although he recognized legality as the essential feature of the state, in which all men were to be equal before the law (though not in making the laws), Reason as embodied in particular individuals or even a majority of them was bound to be nonsuited when it came into conflict with the authorities. Thus, while Hegel demanded that reality should be answerable to the tribunal of Reason, there was no possibility of finding Reason in this sense elsewhere than in the state apparatus.

Hegel does not answer clearly the question whether, and to what extent, the value of the human individual is preserved in the triumphal progress of Mind through history. On the one hand, Mind in the course of becoming itself loses none of the wealth of its exteriorization, and the instruments it uses for its purposes are not simply cast aside, but endure as part of its infinite richness. It might seem therefore that individual life is permanently valuable in itself. But, on the other hand, the value of the individual consists only in the 'element of divinity' within him, and it thus actualizes itself *as a value of the Absolute*; moreover, it seems to disappear completely in the final consummation of the destinies of being. Mind, in completing its progress, attains to infinitude, i.e. the removal of all limitations by anything that is not itself; it appears therefore that in Hegel's view the final destiny of every separate individual is to be absorbed in universal Being, since otherwise the Absolute would be limited by the self-knowledge of individuals and would not accomplish its own purpose. At this central point Hegel appears once again to follow the tradition of neo-Platonic pantheism: the abolition of the contingency of man and the fulfilment of man in his essence, his self-reconciliation, must signify his total absorption into universal Being. It is not clear how individuality as such could be preserved in all its richness when all difference between subject and object disappears; in other words, how is it possible for an infinite being, which has attained to full self-knowledge and reabsorbed all its own objectifications, to be other than *one* being? We must finally conclude that in the Hegelian system humanity becomes what it is, or achieves unity with itself, only by ceasing to be humanity.

We can, of course, consider Hegel's philosophy of history in the light of its partial conclusions, concentrating on the rationalistic determinism of the historical process, its indifference to individual human desires, and its development through successive negations, while abstracting our minds from the final result. But to ignore the eschatological perspective is to deprive the doctrine of its specifically Hegelian character: neither Hegel's dialectic nor its application to history make sense without the eschatology, the vision of the final salvation of Being in a return into itself.

The question of the rationality of the world as such, in all its details, also requires some differentiation. Hegel in fact believes that only the actual historical process is creative of values—that is to say, it is vain and foolish to imagine ideals independently of the actual state of history, or to postulate a radical opposition between the world as it ought to be and as it is. In this respect his anti-utopianism is emphatic and unambiguous. Those who defend him against the charge of conservative bias point out with truth that he believes in a tribunal of Reason to distinguish between what is truly real and what appears real but is no longer 'essential', maintaining a purely empirical existence and destined soon to be swept away. 'Reality', to Hegel, does not mean any fact that happens to present itself: for instance, he excludes from his definition of the civilizing process various forms of behaviour, such as personal caprices, which are not rooted in the will of history. What appears to be an overwhelmingly evident and inescapably real feature of the present situation may be, from Hegel's point of view, no more than the empty shell of a bygone reality, while something that is barely emerging from a dormant or virtual state, and is hardly accessible to empirical inquiry, may actually contain more reality. In the same way an egg which is about to hatch into a chicken looks as though it would remain an egg

for ever, but in fact it is on the point of giving birth to a new
form which, though invisible, is already mature and is the one
important thing about it. In this sense Hegel taught that we
must look for what is 'truly' real as opposed to the superficial
reality that is passing away. This distinction is, in his view, a
matter of scientific reflection and does not call for any value-
judgements opposed to the actuality of the facts. Such evaluation, abstracted from historical necessity, was in Hegel's opinion
a symptom of the sterile recalcitrance of Fichte and the
Romantics. This does not mean that his prescription was to
find out what was necessary and at once infer that it was desirable,
but rather that he rejected the dichotomy of facts and values.
It is not necessary first to discover what is real and then to evaluate
it. Acts of comprehension of the world are undivided: in the
very act in which we perceive something as a portion of
evolving Reason, we accept that something. The positivistic distinction between judgements of fact and of value does not arise
in Hegel's system any more than in dogmatic religion: once we
know what the will of God is, we do not have to express our
approval of it by a separate intellectual act. The perception of
the world which relates every detail to the will of the Absolute
does not involve any such dichotomy either: it consists of acts
of comprehension merged with practical acts of affirmation. The
submission of the intellect to the authority of the Absolute is
an indivisible whole composed of simultaneous understanding
and trust in its wisdom.

While, however, it would be wrong to think of Hegel as
approving any and every part of existing reality simply because
it is there, the question immediately arises: by what criteria are
we to judge whether a particular feature is 'real' or not? Who
is to decide, and on what basis, whether a particular state of
affairs is a sham from which the energy has gone out, or is still
full of vitality? Certainly purely empirical criteria will not suffice.
How then are we, in practice, to appeal to universal Reason
to tell us, for example, whether an institution or form of state
has outlived its usefulness or is still rational? The Hegelian
system does not provide an answer to this question. As Mind
endeavours to interiorize its freedom in individual intellects, it
might seem that a form against which empirical individuals are
in rebellion must for that reason be irrational, and we should

therefore condemn political systems that are unquestionably opposed by the generality of people. But, on the other hand, we are told that the *consensus omnium* is not a valid criterion and that all men, or nearly all, may be in opposition to right Reason, since 'the affairs of the state are matters of knowledge and education and not of its people' (*Lectures*, Introduction). We are thus led back to a conservative apologia for existing institutions as such, from which there is no appeal to any other empirical reality for the purpose of interpreting the decrees of Reason.

Although Hegel's thought on this fundamental point is beset by ambiguity, it is clear that fewer difficulties and glosses are required if we choose to interpret it in a conservative sense: for this provides an indication of the principles on which things are to be condemned, whereas if we seek to adopt the 'critical principle' we are left uncertain which criteria to apply.

The question can be seemingly avoided by reference to the famous passage in the preface to the *Philosophy of Right* where Hegel says that philosophy always comes too late and can only interpret a completed process: 'By philosophy's grey in grey, [a shape of life] cannot be rejuvenated but only understood'. From this point of view our thoughts about the world are of no significance for the purpose of practical evaluation, since we cannot judge the future but can only try to understand the past. There is no point in debating whether we should accept the present as simple reality or judge its empirical qualities by the transcendental demands of Reason, since we are concerned as philosophers only with what is irrevocably past and not with the present world or its prospects. But for practical purposes this attitude itself amounts to a conservative acceptance of the *status quo*, since it prohibits us from speculating as to what might be better. The final message of Hegelianism, therefore, is not the opposition between Reason and an unreasonable world, but contemplation of the world as *a priori* reasonable. We do not know what parts of the existing world are or are not true instruments of Mind: we cannot be sure, for example, that it has ceased to use criminals for its purposes. The individual has no rules of morality which he can oppose to the supremacy of the historical process. In Hegel's system, rebellion against the existing world *may* be justified in a particular case, but we have no means of telling whether it is or not until its destiny is accomplished. If it proves successful, this shows that it was historically right; if crushed, it will evidently have been only a sterile reaction of 'what ought to have been' (*Sollen*). The vanquished are always wrong.

Up to this stage of our exposition we have been concerned with doctrines which presuppose that man is not the same in his empirical being as he is in reality or in essence, and that the basic imperative is that the two should once more become identical. This leads to two alternatives: either the essence of man is not only outside empirical human life but outside humanity altogether, so that man's 'return to himself' is not a return to himself but a realization of the Absolute, in which the particular character of humanity disappears without trace; or else, as in Kant and Fichte, the realization of man's essence is an infinite process. In both cases the progress of humanity towards fulfilment was either dictated by the Absolute, as preceding humanity, or by humanity as preceding actual human nature: human existence was not rooted in itself as a natural form of Being. A new philosophical possibility and a new eschatology came into view with the conception of humanity self-present as an Absolute in its own finitude, and the rejection of all solutions that involve man realizing himself by the actualization, or at the command, of an antecedent absolute Being. This new philosophical prospect is that displayed in the work of Marx.

CHAPTER II

The Hegelian Left

1. The disintegration of Hegelianism

LIKE all other such philosophies, Hegel's attempt at a universal synthesis of Being soon led to discordant results. Immediately after his death in 1831 it became clear that both the general theory of consciousness and its application to the meaning of history and to problems of law and politics were capable of different and contradictory interpretations. In particular, it was not at all clear how far Hegel's political conservatism was the natural consequence of his philosophy of history, or whether it could be distinguished from it as a private and personal opinion. To Hegel's radically minded interpreters it seemed evident that a philosophy which proclaimed the principle of universal negativism, treating each successive phase of history as the basis of its own destruction, a philosophy which presented the critical and self-annihilating process as the eternal law of spiritual development, could not consistently tolerate the endorsement of a particular historical situation, or recognize any kind of state, religion, or philosophy as irrefutable and final.

Hegel's doctrine, apart from the explicit political views it contained, embodied two essential themes which seemed hard to reconcile and likely to prove contradictory in some at least of their consequences. On the one hand, Hegelianism was inexorably anti-utopian: it expressly condemned the viewpoint which, in the face of a particular historical reality, puts forward exigences based on arbitrary normative ideals, moralistic presumptions, and notions of how the world ought to be. Hegel's dialectic was a method of understanding past history but did not pretend to gaze into the future; in fact it condemned any such extrapolation and did not aspire to shape the course of human affairs. From this point of view it might seem that

Hegelianism amounted to recognizing history and the *status quo* as realities no less unshakeable than the rules of logic, so that any protest against the present world in the name of an imaginary one must be rejected as the caprice, understandable certainly but sterile, of an immature consciousness. On the other hand, the Hegelian apologia for Reason could equally be taken as postulating a reasonable world, as demanding that reality should be made rational and empirical history coincide with the requirements of the spirit struggling to be free. On the first interpretation the Hegelian system tended towards contemplative acceptance of the historical process as something natural and inevitable, any revolt against which was condemned to futility. On the opposite interpretation it encouraged a spirit of mistrust and criticism, requiring the confrontation of any existing world with the imperatives of Reason, and contained within itself standards which entitled mankind to judge and criticize reality and to demand that it be reformed.

For a few years after Hegel's death his system functioned as, in effect, the official doctrine of the Prussian state; apologists for that state drew on its wealth of theory, and the authorities began to fill university chairs with Hegelians. This situation rapidly altered in the middle thirties when it became clear that the most active of Hegel's disciples had ideas that were unpalatable to the Prusso-Christian monarchy, and that the intricacies of his thought involved elements of radicalism as regards, in particular, the critique of established religion. The celebrated and often-interpreted aphorism 'what is actual is rational' could be taken as sanctifying any factual situation simply because it existed, or contrariwise as meaning that an empirical fact only deserved to be called 'actual' if it conformed to the demands of historical Reason: on this view, elements contrary to Reason were not truly actual, although they might empirically be more obvious than the rational ones. This interpretation was the one that finally prevailed, chiefly owing to the works of the Hegelian Left; but it does not answer the question, by what signs are we to distinguish actual and rational features of the universe from illusory, irrational ones? Are these criteria to be established independently of the facts of history, according to the arbitrary dictates of pre-historical Reason, or are they to be inferred from history? And, in the latter case, how do we define the role of historical knowledge in the opinion-forming or normative operation of the spirit? In other words, how far and in what sense can rules be derived from a knowledge of history so as to enable us to judge the rationality of the world as it now is? If rules cannot be so derived, they are as emptily formal as the Kantian imperative.

The Young Hegelian movement, as it is called, singled out as the dominant theme of Hegel's philosophy the principle of permanent negation as the ineluctable law of spiritual development. This led by degrees to an attitude of radical criticism in politics, certain forms of which supplied the philosophical basis of communism. Engels observes in one of his early writings that the Hegelian Left was the natural approach to communism, and that Hegelian communists such as Hess, Ruge, and Herwegh were a proof that the Germans must adopt communism if they were to remain faithful to their philosophical tradition from Kant to Hegel. This remark, it is true, belongs to a time when Engels was himself connected with the Young Hegelians, and is contrary to opinions he expressed after those ties were broken; nevertheless, it is typical of the hopes that were cherished in the early stages for a radicalization of the master's system.

Young Hegelianism was the philosophical expression of the republican, bourgeois-democratic opposition which criticized the feudal order of the Prussian state and turned its eyes hopefully towards France. Prussia's western provinces, the Rhineland and Westphalia, had been under French rule for the best part of two decades and had benefited from the Napoleonic reforms— abolition of feudal estates and privileges, equality before the law. After their annexation to Prussia in 1815 they were a natural centre of lively conflict with the monarchical system. In the domain of literature the opposition was led in the early thirties by the group known as Junges Deutschland (Heine, Gutzkow, Börne), and later by the Hegelian radicals who, at that time, were mostly concentrated in Berlin. They included a club of young philosophers and theologians (Köppen, Rutenberg, Bruno Bauer) who reinterpreted Christianity in a Hegelian spirit, and with whom Marx came into contact at the time when he was beginning to formulate his own ideas.

2. *David Strauss and the critique of religion*

One of the chief literary manifestations of the Hegelian Left was David Strauss's *Life of Jesus* (*Das Leben Jesu*, 1835), which attempted to apply Hegelianism to a philosophical reconstruction of the origins of Christianity. For the generation brought up on Kant, Fichte, and Hegel the fact that the universe is ruled by Spirit was so obvious as scarcely to need proof, but it had to be explained exactly how that rule was exercised. The Young Hegelians, especially in their later phase (1840–3), were to 'Fichteanize' Hegel, if we may so put it, by reintroducing the aspect of obligation (*Sollen*) in their approach to history. That is to say, they regarded Hegelian Reason as having above all a normative sense: all social realities should be subjected to the irrefragable criteria of rationality. Christianity was the first victim of this attack. Strauss used Hegelian premisses to overthrow the Hegelian belief in the absolute character of the Christian religion; he thus applied the Hegelian method against its inventor in a particular question of exceptional importance. His argument was that no one religion, the Christian or any other, could claim to be the bearer of absolute truth. Christianity, like other faiths, was only a transitional phase, though a necessary one, in the evolution of the spirit. The Gospels were not a system of philosophical symbols, but a collection of Jewish myths. In his mythical interpretation of the Gospels, Strauss went so far as to question the existence of a historical Jesus. At the same time he was convinced of the complete immanent presence of God in history, and rejected whatever remained in Hegelianism of the notion of a personal God. In particular, the myth of a single incarnation of the Absolute in a historical person was absurd: infinite Reason could not express itself fully in any finite human being.

Strauss's critique and the polemics it aroused led to the crystallization of the Hegelian Left and made it conscious of its separate identity. This was expressed first and foremost in the conviction that Hegel's dialectical method could not, without contradicting itself, permit of belief in the finality of history or of any one civilization. (The rejection of the Christian belief in an incarnate God was an essential instance of this view, though only a particular one.) Accordingly, the dialectic of negation could not stop at the interpretation of past history but must address itself to the future, being not merely a clue to understanding the world but an instrument of active criticism; it must project itself into unfulfilled historical possibilities, and be transformed from thought into action.

3. *Cieszkowski and the philosophy of action*

In the transformation of Hegel's dialectic of negation into a 'philosophy of action', or rather a call to abolish the difference between action and philosophy, an essential role was played by a Polish writer, Count August Cieszkowski, especially in his early work *Prolegomena zur Historiosophie* (1838). Cieszkowski (1814–94) studied in Berlin from 1832 onwards and became interested in Hegelianism through Karl Ludwig Michelet, whose lectures he attended and whose life-long friend he became.

The *Prolegomena* were intended as a revision of the Hegelian philosophy of history, breaking with its contemplative and backward-looking tendencies. Philosophy was to become an act of will instead of merely reflection and interpretation, and was to turn itself towards the future instead of the past. According to Cieszkowski, Hegel's rationalism had forbidden philosophy to consider what would be, and commanded it to content itself with what had been. But Hegel's universal synthesis was itself only a particular historical phase of intellectual development, which it was now necessary to surmount. Cieszkowski divided human history into three phases after the fashion of medieval millenarians such as Joachim of Fiore, to whom he refers in his later works. The period of antiquity had been dominated by feeling: the spirit then lived in a state of pre-reflective, elemental immediacy and unity with nature, and expressed itself pre-eminently in art. The spirit was 'in itself' (*an sich*) and had not yet experienced the division of mind and body. The second era, lasting to the present time, was that of Christianity—a period of reflection in which the spirit turns toward itself, moving from natural, sensual immediacy to abstraction and universality. In spite of all changes and reversals, since the advent of Christ humanity has essentially remained at the level of the spirit 'for itself' (*für sich*). The supreme and final work of the spirit in this phase is Hegel's own philosophy, the absolutization of thought and universality at the expense of individual existence, will, and matter. Throughout the Christian centuries humanity has been in a state of intolerable duality, in which God and the temporal world, spirit and matter, action and thought have been opposed to each other as antagonistic values. But that era has now come to an end. It is time for a final synthesis surmounting both Christianity and Hegelianism—but surmounting them in a Hegelian sense, preserving all the wealth of past ages. This will put an end to the dualism of matter and spirit, cognition and will. Philosophy, properly speaking, came to an end with Hegel: that is to say, in future the spirit will not express itself in philosophical speculation, but what was hitherto manifested as philosophy will coincide with the creative activity of man. It is not so much a question of the 'philosophy of action', i.e. philosophy glorifying action, as of the real merging of philosophical activity in the synthesizing practice of life. The spirit, developing its potentialities out of itself (*aus sich*), will assimilate to itself both nature, which was despised in the Christian era, and thought, which that era one-sidedly worshipped. The new era of the final synthesis will also mean a rehabilitation of the body: it will reconcile subjectivity with nature, God with the world, freedom with necessity, elemental desires with external precepts. Heaven and earth will join in eternal friendship, and the spirit, fully aware of itself and completely free, will no longer distinguish its active life in the world from its thought concerning it.

If the Christian centuries plunged humanity into a painful state of disruption, this does not mean that that suffering could have been avoided. History unfolds itself according to the innate necessity of the spirit, and original sin—*felix culpa*—had to precede the great resurrection that is to come. In the light of the final synthesis all past events will be seen as tending to salvation, and all the conflicting manifestations of the spirit will appear as contributions to the future rebirth.

Cieszkowski's main part in the evolution of Hegelianism consisted in the idea of identifying philosophy with action, and thus superseding the former as it had hitherto been understood. It is debatable how far, if at all, he should be considered as belonging to the Hegelian Left. Inasmuch as the identification of philosophy with action appeared subsequently in the work of Hess and became, through him, a cornerstone of Marxism, it seems natural to regard Cieszkowski as a Left Hegelian, and some writers, such as A. Cornu, have done so. Others, such as J. Garewicz, have objected on the ground that in his later works (*Gott und Palingenesie*, 1842, and especially *Ojcze nasz* (*Our Father*), vol. i, 1848) Cieszkowski formulates his triad in terms of sacred history (the periods of God the Father, of the Son, and of the Holy Ghost) and thus comes down in favour of a personal God (who, however, achieves perfection in human history) and of personal immortality, or rather reincarnation. In Germany the Hegelian Left and Right were distinguished above all by their respective attitudes to religion and Christianity, and from this point of view Cieszkowski could clearly not be ranked with the Left. Nor did the latter regard him as one of their own, even though the unity of philosophy and action soon became a radical war-cry. Michelet, on the other hand, defended Cieszkowski, while taking the view that his ideas did not go beyond orthodox Hegelianism. Cieszkowski himself, when attacking Feuerbach, treated the latter's naturalism and atheism as natural consequences of Hegelianism, and by so doing placed himself, according to German criteria, to the 'right' of Hegel. As for Hess, while following Cieszkowski on the crucial issue, he did not accept his historiosophy altogether. He held, in particular, that syntheses of thought and action have taken place ever since the beginning of history and that the new era is not simply a matter of the future, but was inaugurated by the German Reformation.

Some scholars, such as A. Walicki, observe that while in Germany the Left and Right were distinguished by their attitude towards religion, this was not the case in France, whence Cieszkowski derived much of his inspiration. The religious interpretation of socialism and the conception of a new epoch as the fulfilment of the true content of Christianity were, in fact, current coin in the French socialism of the 1830s and 1840s. Cieszkowski was much influenced by Fourier and the Saint-Simonists, and incorporated into his soteriology an elaborate system of social reforms.

The question of Cieszkowski's place on the map of post-Hegelian disputation is not especially significant to the history of Marxism; nor, from this point of view, does much importance attach to his later philosophical fortunes and his contribution to Polish culture. It is true that his division of history into three phases and his belief in a future, final synthesis of spirit and matter were not new, but were quite common in French philosophical literature. Nevertheless, he played an essential part in the prehistory of Marxism by expressing in Hegelian language and in the context of the Hegelian debates the idea of the future identification (not merely reconciliation) of intellectual activity and social practice. It was out of this seed that Marx's eschatology grew. Marx's most frequently quoted saying—'The philosophers have only interpreted the world in various ways; the point, however, is to change it'—is no more than a repetition of Cieszkowski's idea.

4. *Bruno Bauer and the negativity of self-consciousness*

The idea of the spirit which, as simply spirit, is always opposed to the existing world, always creative, critical, and in a state of disquiet, served the Left Hegelians as an instrument of political and religious criticism. The Hegelians hoped and expected that the irresistible force of their ideas would eventually do away with anachronistic institutions and bring the state into conformity with the demands of Reason. On the political side their criticism was of a general and abstract kind, largely inspired by the ideals of the Enlightenment. But the hopes of an early transformation, to be brought about by philosophical criticism alone, were soon disappointed. The authorities gradually withdrew their support for Hegelianism as the Young Hegelian movement came to show its destructive attitude towards the system, and its philosophers were subjected to increasing harassment.

Bruno Bauer (1809–82), who began as an orthodox Protestant theologian, forsook the orthodox line in 1838 (*Die Religion des Alten Testaments*) and was soon writing pamphlets of a more anti-Christian character than anything else in Germany at the time, including the works of Feuerbach. He moved from Berlin to Bonn, where he was a *Privatdozent* at the university and where his attacks on Christianity grew even sharper. Bauer interpreted history in general, in the Hegelian style, as an expression of the developing self-consciousness of Mind. At the same time the whole of empirical reality presented itself to him on Fichtean lines as a collection of negatives, a kind of necessary resistance for the spirit to overcome in the course of its infinite progress. The meaning of everything that empirically *is* consists in the fact that it can and must be overcome, that it constitutes a centre of resistance against which the critical activity of the spirit is directed. The principle of this activity is a never-resting negation, a perpetual criticism of what exists simply because it does exist. History is determined by the permanent antagonism between what is and what ought to be, the latter being expressed by the spirit in its quest for self-consciousness. This principle, which is eminently Fichtean and non-Hegelian, formed the nucleus of Bauer's critique of religion. The Gospel narrative, in his view, contained no historical truth whatever but was merely the expression of a transient stage of self-consciousness, a fanciful projection of the latter's own vicissitudes into historical events. Christianity was of service to the development of spirit in that it awakened a consciousness of values that belongs to every human individual; but at the same time it created a new form of servitude by requiring individuals to accept subjection to God.

The growth of state power in imperial Rome obliged men to recognize their impotence *vis-à-vis* the outside world. Self-consciousness withdrew into itself and declared that world to be contemptible, as the only way of escaping its pressure. (It should be noted that in Bauer's view the idea of Christianity was itself a product of Roman culture; he minimized the part played in its origins by the Jewish tradition, ascribing a much more important role to popular Stoic philosophy.) In Christianity, religious alienation reaches an extreme form: man divests himself of his own essence and entrusts it to mythical forces to which he regards himself henceforth as subservient. The main task of the present phase of history is to restore to man his alienated essence, by liberating the spirit from the bonds of Christian mythology and freeing the state from religion. A practical consequence of Bauer's historiosophy was the call for the laicization of public life. He was never an adherent of communism, however; on the contrary, he held that if it were possible to create a system based on communist principles, it would tend to subject all human activity and thought to itself, so that freedom of thought and human individuality would be destroyed and the creative activity of mind replaced by a code of official dogma.

In 1841, during his lectureship at Bonn, Bauer published

anonymously a satirical pamphlet entitled *Die Posaune des jüngsten Gerichts über Hegel den Atheisten und Antichristen. Ein Ultimatum* (*The Trump of the Last Judgement upon Hegel, the Atheist and Anti-Christian. An Ultimatum*). Marx had a share in this work, but it is not known how much of it is his: probably not a large amount, as the work is full of biblical quotations and references to theological literature which are evidently due to Bauer's erudition. The book was ostensibly a critique of Hegel from the standpoint of orthodox Protestant theology, denouncing the atheistic implications of his doctrine. With pretended indignation the author showed that Hegel's pantheism was bound to develop in the direction of radical atheism and that its true purport had been revealed by the Young Hegelians, the only faithful expounders of his philosophy. Hegel was an enemy of the Church, of Christianity, and of religion altogether. Even his pantheism was a mere show: religion played no part in his system except as the relationship of self-consciousness to itself, and anything that differed from self-consciousness must be interpreted as an element (*Moment*) in it. Hegel's criticism of the 'sentimental religion' of Jacobi or Schleiermacher was misleading: he accused it of subjectivism, as though he himself were a champion of the reality of God's existence, but this was quite untrue. By representing the finite spirit as a manifestation of universal spirit, Hegel made the latter a projection of historical self-consciousness, while infinity appeared as merely the self-negation of finitude—i.e. God, in the last analysis, is merely a creation of the human ego, which with diabolic pride lays claim to almighty power. Hegel's 'world spirit', too, acquires reality only thanks to the operation of human historical self-consciousness. Human history is thus self-sufficient and has no significance beyond its own self-development. So, according to Hegel, God is dead and the only reality is self-consciousness. All this, Bauer continues, fits perfectly with the other ingredients of Hegel's system: his glorification of Reason and philosophy, his violent critique of all that exists simply because it exists, his worship of the French Revolution, his love of the Greeks and French, his hatred and contempt for the Germans (as a nation of cowards, incapable of doing without religion even in their most radical and rationalistic moods), even his dislike of Latin. Religion, the Church, and belief in God are presented as obstacles that the spirit must overcome in order to achieve absolute mastery; humanity must realize in the end that when it thinks it is contemplating God it is only looking at its own face in a mirror, and behind that mirror there is nothing.

Although Bauer's work purported to be the lament of a believing Christian at the wickedness of a blasphemer, its basic argument was perfectly sincere: Hegel was interpreted as an *alter ego* of Bruno Bauer, a mocker and an atheist, a worshipper of nothing but Self-Knowledge. Hegel's Absolute Idea was merely the self-consciousness that the spirit strives to attain through successive manifestations of itself. The *Weltgeist* is actualized only in the human spirit; each stage of its operation ends in the assumption of a form that begins to encumber it, and requires to be surmounted, as soon as it comes to fulfilment. Every form of the life of the spirit soon becomes anachronistic and irrational, by its very existence challenging the spirit to a fresh effort of criticism and opposition. Philosophy is the criticism which knows how things ought to be, and on the strength of that knowledge it is philosophy's business to condemn and destroy the world as it finds it, attacking especially the established forms of religious mythology. These were Bauer's own views, and it is not surprising that, holding the destruction of Christianity to be the most urgent task of mankind, he was looked at askance in the faculty of Protestant theology and was eventually deprived of his lectureship.

As can be seen, Bauer's philosophy treats the operation of the intellect as a purely negative one. Whereas Hegel's philosophy of history sought to maintain a positive link between the Idea and empirical reality, Bauer and other Hegelians of his school reintroduced a radical dualism between the critical mind and the existing universe. In this interpretation, the spirit is no more than the agent of an eternal dissolution to which every feature of the empirical world is bound to be subjected. The spirit has no positive support in reality itself: its only such support consists in the imperatives of reason which are at all times in advance of reality. The Idea is a tribunal judging the world in accordance with its own suprahistorical laws: every empirical reality is an object of condemnation in the eyes of the spirit. The spirit is defined by its destructive function, and the world is essentially the inertia that opposes criticism; thus the spirit and the world are defined negatively by their relation to each other. History cannot of itself unfold the principles whereby each of its stages is to be judged, but, in order to be changed, it must be judged on the basis of suprahistorical demands. The grounds of historical change lie outside history. The spirit must break through the shell imposed on it by the empirical world, but it cannot derive from that world the strength it needs for its destructive task.

Bauer's critique of religious alienation was strongly reflected in Marx's early thought, including the famous comparison of religion to opium. At the same time the philosophy of self-knowledge was one of the main points in opposition to which Marx came to adopt his own distinctive philosophy.

5. *Arnold Ruge. The radicalization of the Hegelian Left*

Other Left Hegelian writers reinterpreted the master's philosophy on similar lines. Arnold Ruge, as the editor of a journal, did most to consolidate Young Hegelianism as a political movement. Along with others he went through the evolution which gradually radicalized the anti-religious critique and transferred its impact to the sphere of politics. In 1838–41 he edited the Young Hegelian philosophical journal, *Hallische Jahrbücher*, which at first shared Hegel's delusion that Prussia was the embodiment of historical Reason. The Young Hegelians originally believed that historical self-consciousness was the prerogative of the Prussian system, and that the development of freedom which historical Reason demanded could take place there gradually by means of peaceful reforms. The ideal towards which Prussia should evolve was, in the journalists' opinion, a Protestant constitutional monarchy; its Protestantism, however, should not signify the domination of any organized Church, but the conformity of all public institutions to the demands of Reason, and the voluntary subjection of religion to scientific principles. The Young Hegelians' philosophy was reflected in anti-feudal postulates: the abolition of privileged estates, public office open to all, freedom of speech and property—in short, a bourgeois egalitarian state. They envisaged a rational state in accordance with the ideas of the Enlightenment, as expressed in the career of their hero Frederick the Great; this did not appear to them as a mere speculative Utopia, but as part of the natural course of history in which Prussia for the time being enjoyed a special mission. From this point of view they attacked Catholicism as a religion of bygone times, exalting dogma above reason; they also attacked orthodox Protestantism and pietistic sentimentalism, as well as Romantic philosophy, which set reason below emotion and subjected the spirit to a cult of unreasoning Nature.

The change in the Young Hegelians' political orientation brought with it a modification of the belief in historical Reason. The Prussian government showed no enthusiasm for their vision of itself as the embodiment of Reason which was to sweep away feudal inequality and political slavery. The Young Hegelians' appeals were met with repressive measures especially after 1840, when the new king, Frederick William IV, on whom the radicals had pinned their hopes, proved to be a staunch defender of the old, class-ridden order and the Prussian hereditary monarchy, and curtailed political freedom and religious tolerance more drastically than ever. Arnold Ruge and other contributors to the *Hallische Jahrbücher* (later the *Deutsche Jahrbücher*, edited by Ruge from 1841 to 1843) ceased to believe that Prussia was evolving of its own accord towards the kingdom of Reason, and they saw what a gulf lay between their ideals and the stagnant social situation. It was then that they adopted the theory of an inevitable disharmony between the demands of Reason and the empirical world. Reason was no longer an instrument of reconciliation with reality, the latter being rational by definition; it was a source of obligation, a standard with which to confront the world. Practical action and conscious criticism were categories expressing the opposition between the world as it should be and as it was. Ruge proclaimed that Hegel had betrayed his own idealism when he absolutized particular forms of social and spiritual life (the Prussian state, Protestant Christianity) as the ultimate fulfilment of the demands of reason; he had abandoned the principle of eternal criticism and turned his system into an apologia for a merely contemplative, conformist attitude towards the universe.

The radicalization of Young Hegelianism took three main forms. In philosophy it appeared as a breach with Hegel's doctrine of the self-fulfilment of history, and an acceptance of the opposition between the facts of history and normative Reason. In the religious sphere the Young Hegelians rejected the Christian tradition even in diluted, pantheistic forms and adopted a position of out-and-out atheism, first formulated by Bauer and Feuerbach. In politics they abandoned reformist hopes and accepted the prospect of revolution as the only way to regenerate humanity and Germany in particular. However, if we leave aside Hess and the uninfluential Edgar Bauer, this radicalism did not have any socialist content: the expectation of revolution was confined to political change and was unconnected with any hope of a transformation in the system of property and production. Unlike Hegel, who saw an inevitable division between the state with its political institutions and 'civil society' as the totality of private and particular interests, the Young Hegelians in their radical phase believed that in the perfect society of the future the division and even the difference between these two aspects would disappear. Hegel himself did not think it possible to do away with all tension between the general interest and the conflicting private interests of individuals, but only that this tension could be lessened by the mediation of the official machine identifying its own interest with that of the state. In Hegel's view the state as a mode of collective being did not have to justify itself by the interest of the individuals who composed it; on the contrary, their highest and absolute good consisted in membership of the state. It followed that the state's function of overruling the discordant interests of civil society could be justified by the value of the state in and for itself. Hegel's political doctrine expressed the ideology of the Prussian bureaucracy, and in his view the general good, i.e. the good of the state, was independent of private interests and did not derive from them; on the contrary, the individual's interest and his essential value lay in being a citizen of the state. The Young Hegelians, however, completely rejected this view. In proclaiming their own republican ideal and demanding the general participation of the people in political life, with universal suffrage on a basis of equality, freedom of the Press and public criticism, and a freely elected government which truly represented the whole community—in advocating all this, they believed that when it came to pass there would be no difference between the general good and private interests. When political institutions were a free emanation of the people, they could not appear to individuals as alien forces; a state in which education aroused the universal consciousness of every individual citizen and made him aware of the dictates of Reason would signify the unity of private and public interests. In this way the Young Hegelians revived the republican idealism of the eighteenth century, believing that education and political liberties would solve all social problems without any need to alter the system of property on which material production and economic exchanges were based.

The Young Hegelians played an important part in awakening Germany intellectually and spreading democratic ideas. In spite of the attention they aroused, however, they did not succeed in making their philosophy the nucleus of a political movement in which the country's significant social forces were involved. The break-up of the Hegelian Left, which began after the *Deutsche Jahrbücher* were suppressed in 1843, took the form of ideas which postulated a general opposition between abstract thought and politics. The beginning of this dissolution of the movement coincided in time with the early thinking of Marx, who grew up amid the Hegelian Left but, even when he still accepted its philosophical categories and its identification of the problems to be solved, showed clearly that he took an essentially different view of history.

CHAPTER III

Marx's Thought in its Earliest Phase

1. Early years and studies

WHEN Marx came into contact with the Hegelian Left, it was already aware of itself as an independent movement. At the university he was able to witness the conflict between Hegelian rationalism and the conservative doctrine of what was called the Historical School of Law (Historische Rechtsschule). His upbringing and his own critical temperament were conducive to the early development of a radical outlook.

Karl Marx was born at Trier on 5 May 1818, the child of Jewish parents with a long rabbinical tradition on both sides. His grandfathers were rabbis; his father, a well-to-do lawyer, changed his first name from Herschel to Heinrich and adopted Protestantism, which in Prussia was a necessary condition of professional and cultural emancipation. The young Marx was brought up in a liberal democratic spirit. In the autumn of 1835, after leaving the Trier Gymnasium, he enrolled as a law student at Bonn University. The influence of Romantic philosophy, popularized at the university by August von Schlegel, can be seen in Marx's early poetic efforts. However, he received his first real intellectual stimulus at Berlin University, to which he moved in the following year. Although still a law student, he was more absorbed by lectures on philosophy and history. The former subject was taught by, among others, Eduard Gans, who was regarded as belonging to the liberal centre of the Hegelian movement. Hegelianism, in his view, was the interpretation of history as a progressive rationalization of the world in accordance with the ineluctable laws of the spirit; the chief purpose of philosophical thought was to observe this evolution, in which empirical reality was seen gradually to conform to universal reason. Gans was also one of the few Hegelians of his time to profess socialist views, which he had absorbed in Saint-Simon's version. Thus Marx was introduced from the outset to a form of Hegelianism that by no means required obedient acceptance of the *status quo*, but demanded that it be judged by the dictates of Reason.

A directly opposite viewpoint was represented at Berlin University by Friedrich Karl von Savigny (1779–1861), the main theoretician of the Historische Rechtsschule and the author of works on the history of Roman law; he also wrote the pamphlet *Vom Beruf unserer Zeit für Gesetzgebung und Rechtswissenschaft* (*On the Mission of Our Times for Legislation and Jurisprudence*, 1814). Savigny's philosophy expressed the view that obligation should be derived from actual being, and in particular that all law should be based on positive enactments, customs, and rules sanctified by tradition. His conservatism was sharply opposed to the Enlightenment doctrine that laws and institutions must be justified by abstract norms before the tribunal of sovereign Reason, regardless of which laws and institutions are actually valid by force of historical tradition. Political radicalism was expressed in the cult of reason and the systematic refusal to acknowledge the authority of history, and the republican ideals exhibited the vision of a world as it ought to be. For Savigny, on the other hand, the factual, positive institutions and customs that were 'given' by history and rooted in it were authoritative for that very reason. The source of law, from this point of view, could not be an arbitrary legislative act based on the assumed needs of a rational social order; the rightful source of all legislation was customary law and history. This conservative doctrine provided a justification and sanctification of the existing political order simply because it did exist in a positive form, while any attempt to improve it in the name of a better, imagined order was condemned *a priori*. All the feudal elements of backward Germany deserved to be revered, their very antiquity proving them to be legitimate. Savigny combined this irrational cult of 'positiveness' with a belief in the 'organic', suprarational nature of a social community and particularly a national one. Human societies were not instruments of rational co-operation but were welded together by a non-rational bond that is its own justification irrespective of any utilitarian purpose.

The legislating subject is the nation, which spontaneously

evolves and modifies its laws. The nation is an indivisible whole, and the laws, like its customs and language, are only one expression of its collective individuality. There cannot be, as the utopians would have it, a single 'rational' form of legislation for all peoples, regardless of their different traditions. Lawgiving is not an arbitrary matter: the legislator finds a particular legal system in being, and can only formulate changes in legal awareness that have occurred as a result of the community's organic growth. In sharp opposition to utilitarian and rationalistic theories, and in close alliance with Romantic philosophy, Savigny was the true promulgator of the notion that 'whatever is, is right'—the doctrine, in fact, that was ascribed to Hegel by some of his disciples and opponents as a possible interpretation of his dictum that what is actual is also rational. In point of fact Savigny did not derive his inspiration from Hegel, who criticized his conservative views. Hegel, although he refused to oppose the arbitrary dictates of Reason to the real process of history, was not willing to accept the existing order as rational and worthy of respect simply because it existed.

The Young Hegelian radicals, who claimed to judge empirical reality by the abstract requirements of reason, and Savigny, who demanded that reality should be accepted as given, represent opposite solutions of the problem on which Marx's early thought was focused. Hegel, with his ambiguities and incomplete utterances, stood between the two extremes, and Marx's position on this key question was closer to Hegel than to the Young Hegelians. The conservative outlook of the Historical School of Law was quite foreign to Marx, and in the summer of 1842 he satirized it directly in an article in the *Rheinische Zeitung* on Gustav Hugo's philosophy of history. ('Everything that exists is an authority in [Hugo's] eyes, and every authority is an argument ... In short, an eruption on the skin is no less valuable [*positiv*] than the skin itself'.) But equally Marx never adopted in an extreme form the Young Hegelian, or rather Fichtean, opposition between what ought to be and what historically is, between the dictates of Reason and the actual social order, though this point of view was certainly more congenial to him than the other. At a very early stage he endeavoured to interpret the revolutionary principle of the permanent negativity of the spirit in such a way as not to imply the latter's absolute sovereignty.

He did not accept the Absolute in the form of a rational standard imposing itself on the world from outside and taking no account of historical fact; he sought to preserve Hegel's anti-utopian standpoint and to safeguard the respect for the undeniable factual characteristics of the world as we know it.

2. *Hellenistic philosophy as understood by the Hegelians*

Marx's efforts to find a position for himself between the rationalist utopia and the conservative cult of 'positiveness' can be traced in his early studies of post-Aristotelian Greek philosophy. There was a sound reason for his interest in this subject. The Young Hegelians were much interested in Hellenistic philosophy, as they perceived an analogy between the period after Alexander the Great, marked by the twilight of pan-Hellenistic ideas and the decay of the great Aristotelian synthesis, and their own time, which had witnessed the failure of the Napoleonic union of Europe and the collapse of Hegel's attempt at a universal philosophy. The Young Hegelians, as it were, rehabilitated the post-Aristotelian schools—the Epicureans, Sceptics, and later Stoics—and brought to light their values which Hegel had neglected. Hegel, in fact, had accused all these schools (which he mostly considered in their Roman forms) of eclecticism and irrelevance, declaring that their purpose was merely to teach the soul indifference in the face of a cruel and hopeless social reality. They provided an imaginary reconciliation with the world by means of thought which turned in upon itself and lost contact with the object, and a will whose only purpose was to have no purpose. The Hellenistic philosophies were a purely negative defence against the despair aroused by the dissolution of political and social ties in imperial Rome. In Hegel's view, a mode of being in which the intellect withdrew into sterile self-contemplation was condemned to an abstract individuality, whereas concrete individuality was bound to refresh itself by constant contact with universality and with the outside world.

In Bruno Bauer's opinion, on the other hand, these 'philosophies of self-consciousness' were far from being mere negative expressions of impotence. If they made it possible for the individual, engulfed by the collapse of his former world, to achieve a certain spiritual emancipation by returning into himself, and if they could in some degree protect his consciousness against the onslaughts of the world, then, by providing a basis for spiritual autonomy, they opened up a new and necessary phase in the development of spirit; they endowed the individual mind with autonomy, enabled it to assert itself against the world, universalized and liberated it, and made it conscious of its own freedom in and by the critical faculty which it opposed to the corruption of reality. In short, Hegel and the Young Hegelians gave a similar interpretation to the Hellenistic philosphies, but took a different view of their historical and philosophical importance. According to Hegel the absolutization of individual self-consciousness merely showed the impotence of the philosophical spirit, whereas to Bauer it represented the victory of the critical intellect over the pressure of the external world.

3. *Marx's studies of Epicurus. Freedom and self-consciousness*

At an early stage in his Berlin studies Marx underwent a conversion to Hegelianism and frequented a club of young graduates who interpreted the master's doctrine in a radical spirit. For his doctoral thesis he orginally intended to analyse all three schools of Hellenistic thought. However, the subject grew unmanageable and he finally restricted himself to a single aspect of Epicureanism, viz. a comparison between the natural philosophy of Epicurus and the atomism of Democritus. He worked on his dissertation from the beginning of 1839, and in April 1841 was awarded a doctor's degree by Jena University. He intended to print the dissertation, but was soon absorbed by other occupations. The work remained in a manuscript state, with several gaps; it was published in part by Mehring in 1902, and appeared in 1927, with introductory notes, in the *Marx–Engels Gesamtausgabe*.

The work was entitled *On the Differences between the Natural Philosophy of Democritus and of Epicurus*; it is written in a romantic style and in accordance with the categories of Hegelian logic. It is clear that as regards the relationship between the spirit and the world, Marx was a long way from articulating the viewpoint that he was to express three or four years later. Nevertheless, if we compare the dissertation with his subsequent writings we can trace the beginnings of a departure both from the Young Hegelian faith in the supremacy of the critical spirit and from Hegelian conservatism. Using Marx's writings of 1843–5 as a key to the dissertation, we can see it as an attempt to declare his connection with a particular philosophical tradition: that which requires that the spirit should not remain submissive to existing facts, nor yet believe in the absolute authority of normative criteria which it discovers freely in itself without regard to those facts, but should make of its own freedom a means wherewith to influence the world. Marx criticizes Epicurus but is a great deal more severe upon the latter's critics, especially Cicero and Plutarch who, he thinks, quite misunderstood the Epicurean philosophy. In some passages he appears carried away by the rhetorical sweep of Lucretius, his revolt against religion and his Promethean faith in human dignity rooted in freedom.

In opposition to the tradition which, following Plutarch and Cicero, regarded Epicurus' atomism as a corruption of Democritean physics by the arbitrary and fantastic theory of deviations in the movement of atoms (*parenclisis, clinamen*), Marx argues that the apparent similarity of the two philosophers conceals a deep and fundamental difference. Epicurus' theory of accidental deviation is not a mere caprice but an essential premiss of a system of thought centred upon the idea of the freedom of self-consciousness. On the basis of laboriously collected material (editions of the scattered writings of Greek philosophers, such as those of Diels and Usener, did not exist at the time) Marx sought to prove that the philosophical intentions of Democritus and Epicurus were quite different from each other. Democritus opposes the world of atoms, which is inaccessible to the senses, to perception, which is inevitably illusory. He turns to empirical observation, although aware that it does not contain the truth; truth, to him, is empty, however, because the senses cannot grasp it. He prefers to rest content with an illusory knowledge of nature, which he treats as an end in itself. Epicurus' view is different: he regards the world as an 'objective phenomenon' and accepts the evidence of perception uncritically (exposing himself to undeserved mockery by the champions of 'common sense'). His concern, however, is not to know the world but to achieve the ataraxia of self-knowledge through the consciousness of individual freedom. *Parenclisis* is the actualization of freedom, which is essential to the atom. The actualization is fraught with a contradiction, since in Epicurus' view the atom entails the negation of all qualities; but its actual existence is necessarily subject to all qualitative determinations, such as size, shape, and weight. The atom—as the principle of being, not as a physical unit—is, to Epicurus, a projection of the absolute freedom of self-knowledge, but at the same time he is concerned to point out the unreality and fragility of nature conceived as a world of atoms. His theory of meteors was intended to show, according to Marx, that, contrary to traditional belief, the heavenly bodies were not eternal, unchangeable, and immortal; if they were, they would overwhelm self-knowledge by their majesty and permanence and would deprive it of freedom. Their movements can be explained by many causes, and any non-mythical explanation is as good as any other. Epicurus thus robs nature of its unity and makes it feeble and transient, because the serenity of self-consciousness would otherwise be disturbed. The down-grading of nature (in which Epicurus is not interested from the viewpoint of physical science) signifies the removal of a source of disquiet; it gives the mind a sense of its own supremacy and of total freedom from the world. The atom, which is a metaphysical principle, is degraded in the most perfect form of its existence, namely the heavens. The most important source of terror is removed by the destruction of the myths which oppose the frailty of self-consciousness to the immortality of superterrestrial nature. The enemy, to Epicurus, is any form of definite being that is relative to, or determined by, something other than itself. The atom is being-for-itself, and its very nature therefore entails the necessity of deviating from a straight course. Its law is the absence of law, i.e. chance and spontaneity. *Parenclisis* is not a sensual quality (as Lucretius says, it does not occur in any particular place or time), but is the soul of the atom, the resistance that is inseparable from it and therefore from ourselves.

Marx sees Epicurus as a destroyer of the Greek myths and as a philosopher bringing to light the break-up of a tribal community. His system destroyed the visible heaven of the ancients as a keystone of political and religious life. Marx allies himself, so to speak, with Epicurean atheism, which he regards at this stage as a challenge by the intellectual élite to the cohorts of common sense. 'As long as a single drop of blood pulses in her world-conquering and totally free heart, philosophy will continually shout at her opponents the cry of Epicurus: "Impiety does not consist in destroying the gods of the crowd but rather in ascribing to the gods the ideas of the crowd." '

The dissertation, moreover, introduces the theme of religious alienation in its analogy with the alienation of economic life. Referring in an incidental (and critical) passage to Kant's refutation of the ontological argument for the existence of God, Marx writes:

The ontological proof merely amounts to this: 'What I really imagine is for me a real imagination' that reacts upon me, and in this sense all gods, heathen as well as Christian, had a real existence. Did not Moloch reign in antiquity? Was not the Delphic Apollo a real power in the life of the Greeks? Kant's *Critique* does not make any sense here. If someone imagines that he has one hundred pounds, and if for him this is no arbitrary fancy but he really believes in it, then the one hundred imagined pounds have the same worth for him as one hundred real ones. He will, for example, contract debts on the strength of his imaginings, they will have an effect, just as the whole of humanity has contracted debts on the strength of its gods ... Real pounds have the same existence as imagined gods. Surely there is no place where a real pound can exist apart from the general, or rather collective, imagination of men. Take paper money into a land where the use of such money is not known, and everyone will laugh at your subjective imagination. Go with your gods into another land where other gods hold sway, and it will be proved to you that you are suffering from fanciful dreams. Rightly so ... What a particular land is for particular foreign gods, the land of reason is for God in general, an area in which his existence ceases.

As this passage shows, the image *à la* Feuerbach of a man who is ruled by his own imaginations and is not aware that he is their creator, so that their domination over him is real and not merely supposed, is linked in Marx's mind at this period with the necessary role of 'imagination' involved in the power of money. This is a kind of early, obscure prefiguration of Marx's later theory of 'commodity fetishism'.

While, however, Marx paid tribute to Epicurus and Lucretius for liberating the ancient world from the terrors of alien deities and an alien Nature, and restoring the mind's awareness of its own freedom, he regarded Epicurean freedom as a flight from the world, an attempt by the mind to withdraw to a place of refuge. In Epicurus' philosophy the ideal of the sage and the hope of happiness are rooted in a desire to sever links with the world. They are an expression of the mind of an unhappy era when

their gods are dead and the new goddess has as yet only the obscure form of fate, of pure light or of pure darkness . . . The root of the unhappiness, however, is that the soul of the period, the spiritual monad, being sated with itself, shapes itself ideally on all sides in isolation and cannot recognize any reality that has come to fruition without it. Thus, the happy aspect of this unhappy time lies in the subjective manner, the modality in which philosophy as subjective consciousness conceives its relation to reality. Thus, for example, the Stoic and Epicurean philosophies were the happiness of their time; thus the moth, when the universal sun has sunk, seeks the lamplight of a private person.

Marx regards the monadic freedom of Epicurus as escapism: he objects not to the belief in the freedom of the spirit, but to the idea that this freedom can be attained by turning one's back on the world, that it is a matter of independence and not of creativity. 'The man who would not prefer to build a world by his own strength, to create the world and not simply remain in his own skin—such a man is accursed by the spirit, and with the curse goes an interdict, but in the reverse sense: he is cast out from the sanctuary of the spirit, deprived of the delight of intercourse with it and condemned to sing lullabies about his own private happiness, and to dream of himself at night.'

Marx's first work is almost wholly within the limits of Young Hegelian thought. The sweeping attack on religion, and the conviction of the creative role of the spirit in history, do not go beyond the Young Hegelian horizon, nor does the critique of Epicureanism as a philosophy in which the mind seeks to shake off the yoke of nature and envelop itself in a purely subjective autonomy. For the Young Hegelians, too, the supremacy of the spirit was not related to a desire for isolation, but was the precondition of a critical attack on the irrationality of the empirical world. However, we can perceive in Marx's dissertation the germ of what later emerged as the 'philosophy of praxis' in contrast to the critical philosophy of the Young Hegelians. The crucial difference between the latter and Marx's philosophy in its full development may be described as follows. In critical philosophy the free spirit enters the world as a permanent negation of it, a normative act of judgement upon factual life, a statement of what reality ought to be, irrespective of what it actually is. Thus understood, critical philosophy is unalterably supreme over the world. It does not seek to be separate from it, but to affect it and break up its stability; but at the same time it preserves the autonomy of a judge, and the standards by which it measures reality are not derived from that reality but from itself. The philosophy of praxis, on the other hand, declares that in so far as philosophy is purely critical it is self-destructive, but that its critical task is consummated when it ceases to be mere thinking about the world and becomes part of human life. Its function is thus to remove the distinction between history and the intellectual or moral critique of history, between the praxis of the social subject and his awareness of that praxis. As long as theory is the sovereign judge of practice there is a rift between the individual mind and its surroundings, between thought and the world of men. To remove this division is to do away with philosophy and with 'false consciousness'; for as long as consciousness signifies the understanding of an irrational world from outside, it cannot be the self-understanding of that world or the self-awareness of its natural development. If the identification of self-awareness with the historical process is to be a real prospect, that self-awareness must emerge from the immanent pressure of history itself and not from extra-historical principles of rationality. We must therefore find in history itself conditions that can make it rational—i.e. conditions thanks to which its empirical development can coincide with the consciousness of its participants and do away with false consciousness, with that consciousness which contemplates the world but is not as yet the world's self-consciousness.

Some passages of Marx's doctoral thesis contain an embryonic expression of the 'philosophy of praxis' in this sense. He observes, for instance, that when philosophy *as will* turns itself against empirical reality, then it is an enemy of itself *as a system*: in its active form, it is opposed to its own ossified self. This contradiction is resolved by a process in which the world 'philosophizes itself' while philosophy turns into world history. In this conflict philosophical self-consciousness takes a double form: on the one hand positive philosophy, which seeks to cure philosophy of its deficiencies and turns in upon itself, and on the other hand a liberal attitude, which addresses itself critically to the world and, while affirming itself as an instrument of criticism, tends unconsciously to eliminate itself as a philosophy. It is only this latter method which is capable of bringing about real progress. An antique sage who sought to oppose his own free judgement to 'substantial' reality was bound to suffer defeat because he could not escape from that 'substantiality' and, in condemning it, was all the time unconsciously condemning himself. Epicurus tried to free mankind from dependence on nature by, in effect, transforming the immediate aspect of consciousness, its being-for-itself, into a form of nature. But in fact we can only become independent of nature by making it the property of reason, and this in turn requires us to recognize the rationality of nature in itself.

When we consider these remarks in the dissertation we perceive the rudiments of a new outlook: the prospect of philosophy being incorporated into history and thus abolished, and the conviction that mind must look to the 'rationality' of the world as a support for its own emancipation, that is to say its absorption by the reality upon which it is directed. We can see in this outline the future ideal in which there ceases to be a difference between life and thinking about life, so that man is set free by the reconciliation of self-consciousness with the empirical world. We see also the germ of what was to become the theory of false consciousness. Marx was aware that philosophers have, in addition to the overt structure of their ideas, a substructure which is unknown to them; their thought as they themselves present it is different from the crystallization of systems in which the mole-like activity of true philosophical knowledge, finds expression. To discover this unconscious, underlying structure is the proper task of the historian of philosophy, and is the task that Marx set himself in regard to Epicurus.

However, this early work contains no reference, even in general terms, to the social causes that lead philosophers to deceive themselves, or the social conditions that may eliminate false consciousness and restore the unity of experience and self-knowledge. Marx is still thinking in terms of an abstract opposition between the spirit and the world, self-consciousness and nature, man and God. His philosophy did not crystallize further until he had made closer contact with political realities and engaged in the political journalism of his day.

CHAPTER IV

Hess and Feuerbach

IN the year 1841, in which Marx completed his dissertation on Epicurus, there were published at Leipzig two important books by different authors which were to influence his early activity and enable him gradually to free himself from the current schemas of Young Hegelian thought. Moses Hess, author of *The European Triarchy*, made the first attempt to integrate the Hegelian philosophical inheritance with communist ideals; Ludwig Feuerbach, author of *The Essence of Christianity*, rescued the Hegelian Left from its bondage to the philosophy of self-consciousness and not only led the critique of religious belief to its ultimate conclusion, but extended it to all forms of philosophical idealism and unequivocally espoused the point of view which treats all spiritual life as a product of nature.

1. Hess. The philosophy of action

Moses Hess (1812–75), the self-educated son of a Rhineland merchant, was brought up in strict Jewish orthodoxy. In youth he was attracted by the writings of Spinoza and Rousseau; the former taught him to believe in the unity of the world and the identity of reason and will, while from the latter he imbibed a conviction of the natural equality of man. He came into contact with socialist ideas in France, but was soon drawn into the Young Hegelian movement, and from all these sources composed his own, communist philosophy. His writings, including those of the period when he was active in the German socialist movement and was under Marx's influence, are always stamped with a visionary quality. The gaps in his education, and his enthusiastic temperament, prevented him from giving his thoughts a coherent and methodical shape; but many of his ideas helped considerably to form Marx's conception of scientific socialism.

In his first book, *The Sacred History of Mankind* (1837), Hess predicted a new era of man's covenant with God, when, by the operation of unfailing historical laws embodied in the conscious acts of men, there was to be a final reconciliation of the human race, a free and equal society based on mutual love and the community of goods. He suggested for the first time that the social revolution would come about as the result of an inevitable deepening of the contrast between growing wealth on the one hand and misery on the other. In *The European Triarchy* (1841) he based his communism on a Hegelian schema, while endeavouring to strip Hegelianism of its contemplative, backward-looking tendencies and transform it into a philosophy of action. Like other Young Hegelians (including, as we shall see, the young Marx) he desired to see an alliance between the German speculative genius and French political sense, so that German philosophy could take on a substantial form instead of remaining in the field of theoretical meditation. The 'philosophy of action' as Hess conceived it was a development of Cieszkowski's ideas. The history of humanity was divided into three stages. In antiquity, spirit and nature were allied to each other, but unconsciously; the spirit operated in history without any intermediary. Christianity introduced a division whereby the spirit withdrew into itself. In our own day the unity of spirit and nature is being restored; it will, however, no longer be elemental and unreflecting, but conscious and creative. The inaugurator of the modern era is Spinoza, whose Absolute realized, though still only in theory, the unity of being-in-itself and being-for-itself, the identity of subject and object. In Hegelianism this understanding of the identity between subject and object reached its height, but only as an act of understanding: Hegel confined himself to interpreting past history and had not the strength to make philosophy an instrument for the conscious moulding of the future. The transition from the philosophy of the past to that of the future, from interpretation to action, is the work of the Hegelian Left. The essence of the last stage is that what is planned by the spirit to take place in history should be the result of free action. In this stage, human freedom and historical necessity coincide in a single act: that which *must* take place by virtue of historical law, *can* only take place through absolutely free activity. Sacred history, or the work of the spirit in human history, then becomes the same thing as history *tout court*. The surmounting of Hegelianism consists primarily in this, that philosophy henceforth lays claim to the future—aware of historical necessity, but also of the fact that only through freedom can this necessity embody itself in actual history. In this way past history will likewise be sanctified through its relationship to the future, which will be the accomplishment of mankind's historical mission. Hegel ruled out this relationship by decreeing that the dialectic could not apply to the future, and consequently he was unable to sanctify the past even though he desired to do so. The freedom of the spirit, initiated by the German Reformation and brought to theoretical perfection by German philosophy, must ally itself with the freedom of action inaugurated by the French Revolution. When it does, Europe will undergo a swift regeneration, the fulfilment of Christianity and the authentic religion of love. The religion of the new world will not need churches or priests, dogmas or a transcendent Deity, belief in immortality or education through fear. God will not help men from outside, punishing or instructing, but will manifest himself in them in the spontaneity of love and courage. The separation of Church and State will have no purpose, since, unlike the medieval situation in which their unity was only contingent, they will henceforth be identified with each other in a fundamental social unity: secular and religious life will be the same, and particular creeds will be revealed as anachronisms. In a society united voluntarily from within, without coercion, there will no longer be any antagonism between public order and freedom, which will support instead of limiting each other. A prior necessity is that the principle of love should triumph in human life, and Hess regards the transformation of minds as a pre-condition of communism. 'Moral and social slavery proceeds only from spiritual slavery; and, contrariwise, legal and moral emancipation is bound to result from spritual liberation.' The society of the future will not need to protect itself by any repressive laws or institutions, as it will be based on a voluntary harmony and on the identity of the individual and collective interest as a result of the development of self-awareness.

2. *Hess. Revolution and freedom*

In later articles and books Hess gave a clearer description of the communist society he envisaged, and attempted a deeper analysis of the economic causes of contemporary ills; he also expressed his atheistic outlook more emphatically. He remained convinced that the perfect society was simply the realization of the essence of humanity, i.e. that it would consist of bringing empirical existence into harmony with the normative pattern contained in the conception of man; this, he believed, would remove all possibility of social conflict, as the essence of humanity is equally binding upon all. He sought to show that the principle of social unity combines absolute freedom of the individual with the perfect equality aimed at by Fourier, and that the ideal of authentic freedom excludes private property as contradictory to the universal essence of humanity. For this reason he believed that communism, as a scheme for the abolition of property, was justified by faith in the community of man as a species. This community, when realized in practice, would do away with the need for religion and politics (i.e. political institutions) at a stroke, since both are instruments and manifestations of the servitude endured by men and women who are set at variance by conflicting egoisms. When man is conscious of himself as essential man there ceases to be a distinction between thought and action, which are absorbed in the undifferentiated process of living; in Hess's (quite arbitrary) interpretation of Spinoza, the identity of reason and will is the philosophical basis of this identity between action and thought. The free spirit will recognize itself in all objects of its own thought and action and will thus make the whole world its own; the alienness between nature and man, or one man and another, will cease to exist; man will truly be 'at home' in the universe. In the world as it has been till now, 'generalities' have dissipated real human contact into religious and political abstractions; communism will remove the contradiction between the individual and the community, by enabling the individual to regard the general patrimony as the work of his own hands. There will no longer be alienation, i.e. the domination of human products over men and women who do not realize that this is what they are dominated by. Negative freedom, which is no more than a margin secured by struggling against coercion, will be replaced by the voluntary self-limitation that constitutes true freedom; for 'freedom consists in overstepping external boundaries by means of self-limitation, the self-awareness of active spirit, the replacement of natural determination by self-determination ... In humanity every self-determination of the spirit is merely a degree of development that oversteps itself.' But freedom is indivisible: social servitude and spiritual servitude, i.e. religion, go hand in hand; misery and oppression lead to the illusory panacea of religious 'opium'. Consequently servitude cannot be abolished in only one of its forms, such as religion: the evil must be exterminated at the roots, and it is primarily a social evil. Criticizing Feuerbach, who regarded the illusions of religion as the root of social servitude, Hess argued that money was a no less primary form of alienation than God. The influence of Proudhon can be seen here. The alienated essence of man, dominating its own creator, is not only and not primarily God, but money—the blood and sinew of the working man, turned into an abstract form and acting as a standard of human value. Proletarians and capitalists are alike obliged to sell their own vital activity and to feed, like cannibals, on the product of their own blood and sweat in the abstract shape of a medium of exchange. The money-alienation is the most complete inversion of the natural order of life. Instead of the individual being a means, as nature requires, and the species being an end in itself, the individual subordinates the species to himself and makes of his generic essence an unreal abstraction, which takes the form of God in religion and money in social life.

Hess's work bears the mark of hasty and ill-digested reading and of transient influences which affected his thought but were not brought into a synthetic harmony. It is not clear how we are to reconcile the Young Hegelian belief in the species-essence of man, realizing itself in time in every individual and, as Rousseau hoped, removing the possibility of a conflict between the individual and society, with Hegel's own principle of the primacy of the species over the individual. Nor is it clear whether, in Hess's view, spiritual liberation is in the last resort a prior condition of social liberation, or the other way about. His ideal of communism as a perfect harmony ensured by the abolition of private property and the right of inheritance appears clear enough; but his Utopia does not extend beyond themes that were current in France in his day, if not in Prussia. Socialism as a social movement was, in his view, mainly the result of poverty, although the opposition between rich and poor already ceases to dominate his picture of society, yielding to the opposition between capitalist and proletarian.

Hess was the first writer to express certain ideas which proved especially important in the history of Marxism, even though in his works they appear only in a generalized and aphoristic form. Above all, he expressed the conviction that the social revolution would be the result of polarized wealth and poverty, with a gradual disappearance of the middle classes. He suggested the analogy between religious and economic alienation, which was the germ of later Marxist analyses of commodity fetishism. He attempted to resolve the philosophical opposition between necessity and freedom, especially in the philosophy of action which proclaims that in the new phase of history what is necessary will come about through free creative activity, and which identifies self-consciousness with the historical process. This thought was expressed in the context of the philosophical self-consciousness of humanity as such, but it recurred with Marx in the form of a conviction of the identity of class-consciousness with the historical process in the privileged case of the proletariat. The prospect of philosophy being absorbed in its own realization is also found in Marx and is already present in Hess's work: 'When German philosophy becomes practical it will cease to be philosophy.' The importance of Hess is that he made the first attempt to synthesize Young Hegelian philosophy with communist doctrine and, in the name of social revolution, took issue with Young Hegelian expectations of a purely political change. Hess's work is linked with the German movement of 'true socialism' (Karl Grün, Hermann Pütmann, Hermann Kriege), which was more than once branded by Marx (for example in *The German Ideology* and *The Communist Manifesto*) as a reactionary Utopia: the movement regarded actual economic conditions as mere manifestations of spiritual servitude, and looked for the arrival of socialism as men became aware of their own species-essence. Hess, who met Marx in the autumn of 1841 and became his friend and collaborator for some years, later adopted to some extent the class orientation of Marxian socialism. They thus influenced each other, but Hess did not keep up with the theoretical development of socialism over which Marx presided, and did not adopt either the materialistic interpretation of history in the Marxist sense, or the Marxist theory of proletarian revolution.

3. Feuerbach and religious alienation

Ludwig Feuerbach (1804–72) was already a well-known writer in 1841. He had studied at Berlin under Hegel and Schleiermacher, but abandoned Hegelian idealism and Christianity at an early stage. In *Thoughts on Death and Immortality* (1830) he criticized theories of eternal life, and in *A History of Modern Philosophy from Bacon to Spinoza* (1833) and studies of Bayle and Leibniz he manifested his sympathy with the free-thinking tradition. He opposed independent reason to all forms of dogmatism, called for the philosophical rehabilitation of nature, and criticized Hegelianism on the ground that, as it began with the spirit, it was bound to confine itself to the spirit and to define nature as a secondary manifestation (*Anderssein*) of the latter. But he first became celebrated with *The Essence of Christianity* (1841), a naturalistic critique of religion expressed in Hegelian language. Feuerbach did not care for the term 'materialism' because of its unfavourable moral associations, but on the basic issue he adopted a materialistic standpoint. He argued that 'the secret of theology is anthropology', i.e. that everything men have said about God is an expression in 'mystified' terms of their knowledge about themselves. If the real truth of religion is uttered it will prove to be atheism, or simply the positive affirmation of humanity. In general everything that man can apprehend in thought is the objectification of his own essence. 'Man becomes self-conscious in the object: consciousness of the object is man's self-knowledge ... the object is the manifest essence of man, his true, objective self. And this applies not only to spiritual objects but also to those of sense. Even the objects that are furthest from man, in so far as they are objects to him, are a manifestation of his essence.' This of course does not mean that things owe their being to human consciousness, but only that the definitions whereby we apprehend them are definitions of ourselves projected on to the object, so that things are always perceived in human terms and are a projection and image of our self-consciousness. On the other hand, 'man is nothing without an object': it is only in objectivity that he recognizes himself. This idea of the mutual dependence of subject and object (the subject constituting itself in self-knowledge through the object, the object constituting itself in the projection of self-knowledge) is not further analysed by Feuerbach; in general he expresses it in the formula that man belongs to the essence of nature (despite vulgar materialism) and nature to the essence of man.

But Feuerbach was interested above all in the particular kind of objectification that occurs in religious alienation. When men relate to the object in an essential and necessary manner, when they affirm in it the fullness and perfection of their species-essence, that object is God. God is thus the imaginative projection of man's species-essence, the totality of his powers and attributes raised to the level of infinity. Every species-essence is 'infinite', i.e. as an essence it is full of perfection and is a model or standard for individual beings. Man's knowledge of God is an attempt to perceive himself in the mirror of exteriority; man exteriorizes his own essence before he recognizes it in himself, and the opposition between God and man is a 'mystified' version of the opposition between the species and the individual. God cannot in principle have any other predicates than those which men have abstracted from themselves; he is real in so far as these predicates are real. Religion, however, inverts the relationship between subject and predicates, giving human predicates—in the shape of Deity—the primacy over what is real, human, and concrete. Religion is a self-dichotomy of man, his reason, and feelings, the transference of his intellectual and affective qualities on to an imaginary divine being which asserts its own independence and begins to tyrannize over its creator. Religious alienation, the 'dream of the spirit', is not only an error but an impoverishment of man, since it takes away all his best qualities and faculties and bestows them on the Deity. The more religion enriches the essence of God, the more it devitalizes man; the nature of religion is most clearly symbolized in the ritual of blood-sacrifice. Humanity must be humiliated, degraded and stripped of its dignity in order that the Deity may be revealed in majesty. 'Man asserts in God what he denies in himself.' Moreover, religion paralyses men's ability to live together in concord, for it diverts the energy of love on to the divinity and rejects the real fellowship of man into an imaginary heaven. It destroys the feeling of solidarity and mutual love, encourages egoism, depreciates all the values of earthly life, and makes social equality and harmony impossible. To overthrow religion is to realize the true values of religion, which are those of humanity. When people come to themselves and realize that the personifications of religion are the fruit of their own childish imagination, they will be able to form genuinely humanistic societies, in the light of Spinoza's principle *homo homini Deus*. The cult of fictitious other-worldly beings will give place to the cult of life and love. 'If the essence of man is the supreme essence for man, then the first and supreme law of action must be man's love for man.'

The Essence of Christianity was an attempt to apply the Hegelian category of alienation to the formulation of a purely naturalistic and anthropocentric viewpoint. Unlike Hegel, Feuerbach regarded alienation as an altogether negative phenomenon. In Hegel's view, Being realizes its essence by first excluding and then reabsorbing it in a process of self-enrichment; what it potentially contains must be made external before it can be actualized. The Absolute Idea attains to self-consciousness through its own alienated manifestations; it is not pure act, like the God of the Scholastics, but comes to fulfilment only through history and the successive phases of alienation. To Feuerbach, on the other hand, alienation is purely evil and erroneous and possesses no positive value. Religious mystification divides man from his species and opposes the individual to himself; it wastes human energy in the cult of unreal beings and distracts it from the one true value, that of man in and for himself.

4. *Feuerbach's second phase. Sources of the religious fallacy*

Feuerbach's later writings display an ever-widening rift with Hegelianism and a more and more explicit materialism of the eighteenth-century kind. In the preface to the second edition of *The Essence of Christianity* (1843) he rejects the theory that subject and object condition each other, declaring that our apprehension of things is primarily sensual and passive, and only secondarily active and conceptual. In *Lectures on the Essence of Religion* (1848–9, published in 1851) he repeats this view and also emphasizes that religious imaginations result from man's feeling of dependence on nature, whereas he had previously treated them as an objectification of the 'essence of humanity'. Originally he envisaged that the overthrow of religion would put an end to human egoism; now he asserted that egoism is a natural and inevitable human trait, present in even the most altruistic of actions, i.e. he reverted to the Enlightenment stereotype of 'natural egoism'. In his earlier work he had described the process of God-creating projection but had not explained its causes. He now tried to fill the gap, but did not go further than stating that the source of religious imagination is generally ignorance and man's inability to interpret aright his own situation in nature. Realizing his dependence on nature, which is eternal and inescapable, man fails to comprehend this dependence in rational categories; instead he devises anthropomorphic fancies to express his fear of Nature's incalculable caprices and the positive feelings of gratitude and hope that she arouses in him. Religion is an ersatz satisfaction of human needs that cannot be met in any other way: men seek to compel nature to obey them by using magic or appealing to divine goodness, i.e. they try to achieve by imagination what they cannot have in reality. As knowledge increases, religion, which is an infantile state of mind, gradually yields to a rational world-view, and men are able, by the arts of civilization and technology, to control forces that were previously untameable. At the same time Feuerbach draws attention to the sources of religious imagination that lie in the very nature of the cognitive processes, especially that of abstraction. Since we can only think or express ourselves in terms of abstracts, we are apt to credit them with an independent existence *vis-à-vis* individuals, which are in fact the only reality. In the same way God and other religious figments, personifying human ideas, feelings, and abilities, are an illegitimate autonomization of the legitimate instruments of cognition. 'The idea or generic concept of God in the metaphysical sense is based on the same necessity and the same foundations as is the concept of things or of fruit ... The gods of polytheists are nothing but names and collective or generic concepts imagined as actual beings'; but 'in order to understand the meaning of general concepts it is not necessary to deify them and turn them into independent beings that differ from individual essences. We can condemn wickedness without at once personifying it as the devil.'

In *Lectures on the Essence of Religion* there is no longer any trace of Feuerbach's Hegelian upbringing. Religion is explained in simple Enlightenment terms as the result of fear and ignorance, and Feuerbach also takes over the Enlightenment theory of perception as merely sensual and empirical. The *Lectures* were a novelty in German philosophy, dominated as it was by the categories of Kant and Hegel, but to the rest of Europe they were merely a repetition of well-known theories. The essential feature was that Feuerbach consistently saw religion as the root of all social evil. He believed that when religious mystification was done away with there would be an end of the sources of social inequality, exploitation, egoism, and slavery. Religion was the source and epitome of all the evil in history; and he expected that public enlightenment, sweeping away religious prejudice, would at the same time eradicate social servitude. This was one of the main points, though not the only one, on which Marx soon took up a sharply critical attitude towards Feuerbach's philosophy.

By the end of the 1840s Feuerbach had completely rejected Hegelianism, regarding it, like all other forms of idealism, as nothing but a continuation of religious fiction. All the creations of classic German philosophy, such as Hegel's Idea, Fichte's Ego, or Schelling's Absolute, appeared to him simply as substitutes for the Deity, reduced to a more abstract form by philosophical imagination. Interpreting humanity in purely zoological categories, he saw the social community as a form of natural co-operation within the species, distorted or depraved by religious prejudice, while in moral speculation he did not advance beyond the eudaemonistic schema of the Enlightenment. The sweep of his rhetoric, with its humanistic and free-thinking ideas, won him many adherents. *The Essence of Christianity* had immense influence in Germany and played a large part in transforming the Young Hegelian camp by radicalizing its anti-religious orientation. To Marx in particular, Feuerbach's philosophy was not only a point of repulsion but also one of the main stimuli that enabled him to reject Hegelian categories in his own thinking. He also owed much to Feuerbach as regards knowledge of the history of philosophy, especially in the sixteenth and seventeenth centuries. He adopted the critique of Hegelianism as a philosophy that 'put the predicate where the subject should be' and gave human creations priority over man himself, and made use of it in his analysis of Hegel's philosophy of law.

It might be thought that after Marx's criticisms the philosophy of Feuerbach was completely outdated, especially in view of his rather dull and repetitive style. Yet it continues to excite interest among seekers after a universal humanistic formula, and even among theologians. The radical anthropocentrism which is its chief feature may be summed up as follows. Firstly, man is the only value; all others are instrumental and subordinate. Secondly, man is always a live, finite, concrete entity. Thirdly, there are permanent features of human nature which make it possible for men to live in a harmonious community based on mutual love and respect for life. Fourthly, the abolition of religion in the dogmatic and mystical forms in which it has hitherto been known will open the way to a new, authentic religion of humanity, enabling men to attain what has been their true object in all religions, namely the satisfaction of their need for happiness, solidarity, equality, and freedom.

CHAPTER V

Marx's Early Political and Philosophical Writings

AFTER completing his studies Marx returned to Trier in the spring of 1841 and afterwards moved to Bonn, where he began to write for Young Hegelian journals. His first article, on the Prussian government's new decree concerning Press censorship, was written for the *Deutsche Jahrbücher*, the issue of which was confiscated as a result; it appeared in 1843 in a collective work published in Switzerland. However, Marx was able to publish a series of articles on this subject in the *Rheinische Zeitung*, a liberal bourgeois journal founded at Cologne at the beginning of 1842 and dominated by Young Hegelian writers: among its contributors were Adolf Rutenberg, Friedrich Engels, Moses Hess, Bruno Bauer, Karl Köppen, and Max Stirner. Marx himself edited the paper from October 1842 to March 1843. During this time, apart from articles on the freedom of the Press, he wrote analyses of the debates in the Landtag (provincial assembly), in which for the first time he devoted his attention to economic questions and the standard of living of the deprived classes. Adopting a radical democratic standpoint, he denounced the pseudo-liberalism of the Prussian government and stood up for the oppressed peasantry.

1. *The state and intellectual freedom*

From the point of view of the development of Marx's theories, his early journalistic writings are important for two main reasons. In his sharp attacks on the censorship law he spoke out unequivocally for the freedom of the Press, against the levelling effect of government restriction ('You don't expect a rose to smell like a violet; why then should the human spirit, the richest thing we have, exist only in a single form?'), and also expressed views concerning the whole nature of the state and the essence of freedom. Pointing out that the vagueness and ambiguity of the Press law placed arbitrary power in the hands of officials, Marx went on to argue that censorship was contrary not only to the purposes of the Press, but to the nature of the state as such.

The freedom of the Press has quite a different basis from censorship, for it is the form of an idea, namely freedom, and is an actual good; censorship is a form of servitude, the weapon of a world view based on appearances against one based on the nature of things. Censorship is something purely negative . . . Freedom lies so deeply in human nature that even the opponents of freedom help to bring it about by combating its reality ... The essence of a free press is the rational essence of freedom in its fullest character. A censored press is a thing without a backbone, a vampire of slavery, a civilized monstrosity, a scented freak of nature. Is there any further need to prove that freedom is in accordance with the essence of the Press, and that censorship is contrary to it?

Thus, Marx continues, 'Censorship, like slavery, can never be rightful, even though it existed a thousand times in the form of laws', and a Press law is a true law only when it protects the freedom of the Press. Censorship is contrary to the very nature of law and of the state, for a free Press is an indispensable condition of a state fulfilling its own nature: it is embodied civilization, the individual's link with the state, a mirror of the people. A censored Press depraves public life and means that the government hears only its own voice. Freedom does not require arguments to justify it, for it is part and parcel of man's spiritual life. 'In a free system every separate world revolves round the central sun of freedom as, and only as, it revolves upon itself ... For is it not a denial of one person's freedom to demand that he should be free after the manner of another?' The written word is not a means to an end, but an end in itself, and must not be confined by laws that have in view any interest other than spiritual development.

In this argument, as will be seen, Marx distinguishes the 'real' law and state, those which correspond to their own proper nature, from laws and institutions which are maintained by police methods but are only binding in an external sense. This distinction belongs to the Hegelian tradition: a state and a law which are not the realization of freedom are contrary to the very concept or essence of state and law and are thus not truly such, even though upheld by force. Marx, however, unlike Hegel, denies that freedom of speech and writing can be limited by the overriding interest of the 'true' state, since he claims that this freedom is an essential part of the concept of a state. Thus, while he uses the normative concept of the state as a model with which existing states are to be compared so as to determine whether they are 'real' or merely empirical, in applying this method he parts company with Hegel by asserting that the freedom of diversity is an essential human value which carries with it its own justification.

Another main theme appears at this time in Marx's comments on the Landtag debate on the law concerning the theft of timber (this was a revocation of the custom allowing peasants to gather brushwood in the forests without payment). Defending the peasantry and the customary law, Marx adopted a philanthropic viewpoint but also argued that the Landtag was degrading the laws and the authority of the state to the role of an instrument of the private interests of landowners, and was thus contravening the very idea of the state. In this way he opposes the state, representing the whole community, to institutions which turn it into the agent of one sectional group or another. At this stage, however, it is not clear that he has any answer to the question as to how state institutions can be brought into conformity with the general interest, or how, if at all, the state is capable of solving social questions, especially poverty and the inequality of incomes.

2. *Criticism of Hegel. The state, society, individuality*

Marx's concern with politics led him to make a deeper study of Hegel's philosophy of law. His lengthy *Critique of Hegel's Philosophy of Right*, written in 1843 (and first published in 1927), remained unfinished, but some of its main ideas can be found in two articles entitled *On the Jewish Question* and *Introduction to a Critique of Hegel's Philosophy of Right*. These were written towards the end of 1843 and appeared in the *Deutsch-Französische Jahrbücher*, which Marx was then editing in collaboration with Arnold Ruge and Hess. He had moved to Paris in the autumn of that year, accompanied by his newly wedded wife Jenny, the daughter of Baron Ludwig von Westphalen, a City Councillor of Trier, and was in contact with local socialist organizations of French and German workers. Before this he probably knew something of communist propaganda in France from Lorenz von Stein's *Socialism and Communism in Present-Day France* (1842). Stein, a conservative Hegelian, had been investigating socialist movements on the instructions of the Prussian government, which was interested in subversive activity among German workers in Paris. He was anti-socialist and regarded the class hierarchy as a precondition of organized society, but his book, which contained a large amount of information, was widely known in radical circles in Germany.

In his long critique of Hegel, Marx attacked especially the idea that the state was, in its origin and value, quite independent of the empirical individuals who composed it. Hegel had argued that the functions of the state were connected with the individual in an accidental manner, whereas in fact there was between them an essential link, a *vinculum substantiale*. Hegel conceived the functions of the state in an abstract form and in themselves, treating empirical individuals as an antithesis to them. But in fact

the essence of a human person is not that person's beard or blood or abstract physical nature, but his or her social character, and the state's functions are nothing more or less than the forms of existence and operation of man's social characteristics. It is reasonable, therefore, to consider individuals as representatives of the functions and authority of the state, from the point of view of their social and not their private character.

In the second place Marx, following Feuerbach, criticizes the 'inversion of the relationship of predicate and subject' in Hegel's philosophy, where human individuals, who are real subjects, are turned into predicates of a universal substance. In reality everything that is general is merely an attribute of individual being, and the true subject is always finite. To Hegel, the individual man is a subjective and secondary form of the existence of the state, whereas 'democracy starts from man and makes the state into objectified man. Just as religion does not make man, but man makes religion, so the constitution does not make the people, but the people makes the constitution.' Marx thus endeavours to reduce all political institutions, as a matter of theory, to their actual human origins. At the same time he seeks to subordinate the real state to human needs, and to strip it of the appearance of independent value apart from its function as an instrument to serve the needs of empirical individuals. The aim of democracy as Marx understands it is to make the state once more an instrument of man, i.e. to de-alienate political institutions. Only a state which is a form of the existence of its people, and not a foreign body over against them, is a true state, one which conforms to the essence of statehood. An undemocratic state is not a state. Hegel perpetuates the gulf between man and state by regarding society not as the realization of personality but as a goal approached by the state; the empirical human being is thus the supreme reality of the state, but not its creator. 'In Hegel it is not subjects who objectify themselves in the common cause, but the common cause itself becomes a subject. It is not subjects who need the common cause as their own true cause, but the common cause is in need of subjects as a condition of its own formal existence. It is the business of the common cause to exist as a subject also.' The purport of this criticism is clear: if human individuals are only 'moments' or stages in the development of universal substance, which through them attains the supreme form of being, then they are mere instruments of that universal substance and not independent values. The Hegelian philosophy thus sanctions the delusion that the state as such is the embodiment of the general interest, which is only the case if the general interest is completely alienated from the interests and needs of actual individuals. This question is closely connected with that of state bureaucracy. Hegel believed that the spirit of the state and its superiority to the particular interests of its citizens were embodied in the consciousness of officialdom; for officials identified their particular interest with that of the state as a whole and thus, as an organ of the state, effected a synthesis between the common good and that of particular sections or corporations. To Marx this was an illusion, a reflection within Hegelianism of the ideology of the Prussian bureaucracy which persuaded itself that it was the supreme embodiment of the general good. The fact was, on the contrary, that

wherever the bureaucracy is a principle of its own, where the general interest of the state becomes a separate, independent and actual interest, there the bureaucracy will be opposed to the corporations in the same way as every consequence is opposed to its own premisses ... The purpose of a true state is not that each citizen should devote himself to the general cause as though to a particular one, but that the general cause should be truly general, i.e. the cause of every citizen.

Hegel distinguished two separate spheres of contemporary life, viz. civil society and the political state. In this division, which Marx accepted, civil society was the totality of divergent particular interests, individual and collective—empirical daily life with all its conflicts and disputes, the forum in which every individual carried on his day-to-day existence. At the same time, as a citizen he was a participant in the organization of the state. Hegel believed that the conflicts within civil society were held in check and rationally synthesized in the supreme will of the state, independent of particular interests. On this point Marx strongly opposes the Hegelian illusion. The division into two spheres is real, but a synthesis between them is not possible. The state in its present form is not a mediator between particular interests, but a tool of particular interests of a special kind. Man as a citizen is completely different from man as a private person, but only the private person who belongs to civil society possesses real, concrete existence; as a citizen he is part of an abstract creation whose apparent reality is based on a mystification. This mystification did not exist in the Middle Ages, for in those days the division into estates was also an immediate political division: the articulation of the civil community coincided with the political division. Modern societies, which have altered or abolished the political significance of the division into estates, have brought about a dualism which affects every human existence and creates within every human being a contradiction between his private capacity and his capacity as a citizen. Marx does not set out, however, merely to describe this contradiction but to explain its origin.

3. *The idea of social emancipation*

In his essay *On the Jewish Question* Marx repeats this theme more clearly, in the form of a description and a plan of action. Commenting on Bauer's critique of the subject, he expresses his own idea of human as distinct from political emancipation.

Bauer, in Marx's opinion, turned social questions into theological ones; he called for religious emancipation as the chief precondition of political emancipation and was content with a programme for liberating the state from religion, i.e. disestablishing the latter. But, Marx objected, religious restrictions were not a cause of secular ones, but a manifestation of them. By freeing the state from religious limitations we do not free mankind from them: the state may free itself from religion while leaving the majority of its citizens in religious bondage. In the same way the state may cancel the political effect of private property, i.e. abolish the property qualification for voting, etc., and may declare that differences of birth and station have no political significance, but this does not mean that private property and differences of birth and station will cease to have any consequences. In short, a purely political and therefore partial emancipation is valuable and important, but it does not amount to human emancipation, for there is still a division between the civil community and the state. In the former, people live a life which is real but selfish, isolated and full of conflicting interests; the state provides them with a sphere of life which is collective, but illusory. The purpose of human emancipation is to bring it about that the collective, generic character of human life is real life, so that society itself takes on a collective character and coincides with the life of the state. Bauer does not penetrate to the real source of antagonism between individual and collective life; he combats only the religious expression of that conflict. The freedom he proclaims is that of a monad, the right to live in isolation; as in the Declaration of the Rights of Man, it is based on mutual self-limitation (my freedom is bounded by the freedom of someone else). Given the separation of the two spheres, the state does not help to abolish the egoistic character of private life but merely provides it with a legal framework. Political revolution does not liberate people from religion or the rule of property, it merely gives them the right to hold property and to profess their own religion. Political emancipation thus confirms the dichotomy of man. 'The actual individual man must take the abstract citizen back into himself and, as an individual man in his empirical life, in his individual work and individual relationships, becomes a species-being; man must recognize his own forces as social forces, organize them and thus no longer separate social forces from himself in the form of political forces. Only when this has been achieved will human emancipation be completed.'

In this way Marx came upon the idea which, in the political context, enabled him to go beyond the purely political, republican, anti-feudal programme of the Young Hegelians and to proclaim the objective of a social transformation which would remove the conflict between private and political life. From the philosophical point of view this was based on the idea of an integrated human being overcoming his own division between private interest and the community. Marx's conception of humanity goes far beyond Feuerbach, since the mystification of religion appears to him merely as a manifestation, not a root, of social servitude. He does not, like Feuerbach, regard man from a naturalistic point of view; he does not imagine a return to innate rules of co-operation which would, of their own accord, prevail in human society once the religious alienation was overcome. On the contrary, he regards the emancipation of man as a specifically human emancipation made possible by the identification of private with public life, the political with the social sphere. The conscious absorption of society by the individual, the free recognition by each individual of himself as a bearer of the community is, in Marx's view, the way in which man rediscovers and returns into himself.

However, as these postulates are expressed in the *Critique of Hegel's Philosophy of Right* and the essay *On the Jewish Question*, they remain utopian (in the sense in which Marx later used this word) inasmuch as they simply oppose the actual state of man's dichotomy to an imaginary unity, described in very abstract terms. The question of how and by means of what forces that unity is to be attained remains open.

4. *The discovery of the proletariat*

The *Introduction to a Critique of Hegel's Philosophy of Right* is regarded as a crucial text in Marx's intellectual development, as it is here that he expresses for the first time the idea of a specific historical mission of the proletariat, and the interpretation of revolution not as a violation of history but as a fulfilment of its innate tendency.

The latter idea appears in a letter from Marx to Ruge written in September 1843.

Let us develop new principles for the world out of its own principles. Let us not say to it 'Cease your nonsensical struggles, we will give you something real to fight for.' Let us simply show the world what it is really fighting for, and this is something the world *must* come to know, whether it wishes to or not. The reform of consciousness consists only in the world becoming aware of its own consciousness, awakening it from vague dreams of itself and showing it what its true activity is ... Then it will be seen that the world has long been dreaming of things that it only needs to become aware of in order to possess them in reality.

It may be seen that the tremendous role that Marx ascribes to the awakening of consciousness does not signify—as it did with most of the Young Hegelians, with Feuerbach, and the majority of socialist writers of the thirties and forties—that people could be offered an arbitrary ideal of social perfection, so sublime and irresistible that they would at once seek to put it into practice. In Marx's view, a reformed consciousness was a basic condition of social transformation because it was, or could be, the revealing and explication of what had been merely implicit; because it gave recognizable form to what had all along been the aims of the struggle for liberation, and thus converted an unconscious historical tendency into a conscious one, an objective trend into an act of will. This is the basis of what Marx later called scientific socialism, as opposed to the utopian variety which confined itself to propounding an arbitrarily constructed ideal. In calling for a revolution as a result of men coming to understand the meaning of their own behaviour, Marx turned his back on the utopianism of contemporary socialists and on Fichte's opposition, which the Young Hegelians had taken over, between obligation and reality.

In the *Introduction* Marx pursued this theme and at the same time emphatically opposed Feuerbach's critique of religion. He accepts that man is the creator of religion, but adds that

Man is the world of man, the state, society. This state, this society, produce religion's inverted attitude to the world, because they are an inverted world themselves. Religion is ... the imaginary realization of human being, because human being possesses no true reality. Thus the struggle against religion is indirectly the struggle against that world whose spiritual aroma is religion.... Religion is the opium of the people. The real happiness of the people requires the abolition of religion, which is their illusory happiness. In demanding that they give up illusions about their condition, we demand that they give up a condition that requires illusion.... Once the holy form of human self-alienation has been unmasked, the first task of philosophy, in the service of history, is to unmask self-alienation in its unholy forms. The criticism of heaven is thus transformed into criticism of earth, the criticism of religion into the criticism of law, and the criticism of theology into the criticism of politics.

Having thus exposed the delusions of the anti-religious criticism which claims to possess in itself the power to abolish human servitude, Marx repeats his critique of conditions in Germany, where the only revolutions have been philosophical ones—a state of political anachronism, with all the drawbacks and none of the advantages of the new order. The liberation of Germany can only be brought about by ruthless awareness of its true position. 'We must make the actual oppression even more oppressive by making people conscious of it, and the insult even more insulting by publicizing it. ... We must force these petrified relationships to dance by playing their own tune to them. To give people courage, we must teach them to be alarmed by themselves.' A German revolution would mean the realization of German philosophy by its own abolition. But philosophy can only be realized in the sphere of material action.

The weapon of criticism is no substitute for criticism by weapons: material force must be opposed by material force. But theory, too, will become material force as soon as it takes hold of the masses. This it can do when its proofs are *ad hominem*, that is to say when it becomes radical. To be radical is to grasp the matter by the root; and for man, the root is man himself.

The social revolution can only be carried out by a class whose particular interest coincides with that of all society, and whose claims represent universal needs. That class is the proletariat,

which has a universal character by reason of the universality of its sufferings, and which does not lay claim to any specific rights because the injustice to which it is subjected is not particular but general. ... It cannot liberate itself without breaking free from all the other classes of society and thereby liberating them also ... It stands for the total ruin of man, and can recover itself only by his total redemption.

Thus the liberation of the proletariat signifies its abolition as a separate class and the destruction of class distinctions in general by the abolition of private property. Marx believes that Germany is the destined birthplace of the proletarian revolution because it is a concentration of all the contradictions of the modern world together with those of feudalism. To abolish a particular form of oppression in Germany will mean the abolition of all oppression and the general emancipation of mankind. 'The head of this emancipation is philosophy, its heart is the proletariat. Philosophy cannot realize itself without transcending the proletariat, the proletariat cannot transcend itself without realizing philosophy.'

It is noteworthy that the idea of the proletariat's special mission as a class which cannot liberate itself without thereby liberating society as a whole makes its first appearance in Marx's thought as a philosophical deduction rather than a product of observation. When Marx wrote his *Introduction* he had seen very little of the actual workers' movement; yet the principle he formulated at this time remained the foundation of his social philosophy. He also formulated at this early stage the idea of socialism, not as the replacement of one type of political life by another but as the abolition of politics altogether. In articles published in the Paris journal *Vorwärts* in the summer of 1844 he declared that there could not be a social revolution with a political soul, but there could be a political revolution with a social soul. Revolution as such was a political act, and there could be no socialism until the old order was overthrown; but 'When the organization of socialism begins and when its true purpose and soul are brought to the forefront, then socialism will cast off its political integument.'

It should be observed that from start to finish Marx's socialist programme did not, as his opponents have often claimed, involve the extinction of individuality or a general levelling for the sake of the 'universal good'. This conception of socialism was indeed characteristic of many primitive communist doctrines; it can be found in the utopias of the Renaissance and the Enlightenment, influenced as they are by traditions of monastic communism, and in socialist works of the 1840s. To Marx, on the other hand, socialism represented the full emancipation of the individual by the destruction of the web of mystification which turned community life into a world of estrangement presided over by an alienated bureaucracy. Marx's ideal was that every man should be fully aware of his own character as a social being, but should also, for this very reason, be capable of developing his personal aptitudes in all their fullness and variety. There was no question of the individual being reduced to a universal species-being; what Marx desired to see was a community in which the sources of antagonism among individuals were done away with. This antagonism sprang, in his view, from the mutual isolation that is bound to arise when political life is divorced from civil society, while the institution of private property means that people can only assert their own individuality in opposition to others.

From the outset, then, Marx's criticism of existing society makes sense only in the context of his vision of a new world in which the social significance of each individual's life is directly evident to him, but individuality is not thereby diluted into colourless uniformity. This presupposes that there can be a perfect identity between collective and individual interests, and that private, 'egoistic' motives can be eliminated in favour of a sense of absolute community with the 'whole'. Marx held that a society from which all sources of conflict, aggression, and evil have been thus extirpated was not only thinkable, but was historically imminent.

CHAPTER VI

The Paris Manuscripts. The Theory of Alienated Labour. The Young Engels

In Paris in 1844 Marx was engaged in composing a critique of political economy in which he attempted to provide a general philosophical analysis of basic concepts: capital, rent, labour, property, money, commodities, needs, and wages. This work, which was never finished, was published for the first time in 1932 and is known as the *Economic and Philosophical Manuscripts of 1844*. Although merely an outline, it has come to be regarded as one of the most important sources for the evolution of Marx's thought. In it he attempted to expound socialism as a general world-view and not merely a programme of social reform, and to relate economic categories to a philosophical inter-pretation of man's position in nature, which is also taken as the starting-point for the investigation of metaphysical and epistemo-logical problems. In addition to German philosophers and socialist writers, Marx addressed himself in this work to the writings of the fathers of political economy, whom he had begun to study: Quesnay, Adam Smith, Ricardo, Say, and James Mill.

It would, of course, be quite wrong to imagine that the Paris Manuscripts contain the entire gist of *Capital*; yet they are in effect the first draft of the book that Marx went on writing all his life, and of which *Capital* is the final version. There are, moreover, sound reasons for maintaining that the final version is a development of its predecessor and not a departure from it. The Manuscripts, it is true, do not mention the theory of value and surplus value, which is regarded as the corner-stone of 'mature' Marxism. But the specifically Marxist theory of value, with the distinction between abstract and concrete labour and the recognition of the labour force as a commodity, is nothing but the definitive version of the theory of alienated labour.

1. Critique of Hegel. Labour as the foundation of humanity

Marx's negative point of reference is contained in Hegel's *Phenomenology*, in particular the theory of alienation and of labour as an alienating process. The greatness of Hegel's dialectic of negation consisted, in Marx's view, in the idea that humanity creates itself by a process of alienation alternating with the transcendence of that alienation. Man, according to Hegel, manifests his generic essence by relating to his own powers in an objectified state and then, as it were, assimilating them from the outside. Labour, as the realization of the essence of man, thus has a wholly positive significance, being the process by which humanity develops through externalization of itself. Hegel, however, identifies human essence with self-consciousness, and labour with spiritual activity. Alienation in its original form is the alienation of self-consciousness, and all objectivity is alienated self-consciousness; so that the transcendence of alienation, in which man reassimilates his own essence, is the transcendence of the object and its reabsorption into the spiritual nature of man. Man's integration with nature takes place on the spiritual level, which makes it, in Marx's view, an abstraction and an illusion.

Marx, following Feuerbach, bases his own view of humanity on labour, understood as physical commerce with nature. Labour is the condition of all spiritual human activity, and in it man creates himself as well as nature, the object of his creativity. The objects of human need, those in which he manifests and realizes his own essence, are independent of him; that is to say, man is also a passive being. But he is a being-for-himself, not merely a natural being, so that things do not exist for him simply as they are, irrespective of their being human objects. 'Human objects are not natural objects, therefore, in the form in which they are immediately given; and human sense, as it is immediately given in its objective form, is not human sensibility and human objectivity.' Consequently, the transcendence of the object as alienated cannot be, as Hegel maintained, the transcendence of objectivity itself. In order to show how man can reabsorb nature and the object into himself, it is necessary first to explain how the phenomenon of alienation actually arises, through the mechanism of alienated labour.

2. *The social and practical character of knowledge*

Since, in Marx's view, the basic characteristic of humanity is labour, i.e. contact with nature in which man is both active and passive, it follows that the traditional problems of epistemology must be looked at from a fresh standpoint. Marx denied the legitimacy of the questions posed by Descartes and Kant: it is wrong, he argued, to inquire how the transition from the act of self-consciousness to the object is possible, since the assumption of pure self-awareness as a starting-point rests on the fiction of a subject capable of apprehending itself altogether independently of its being in nature and society. On the other hand, it is equally wrong to regard nature as the reality already known and to consider man and human subjectivity as its product, as though it were possible to contemplate nature in itself regardless of man's practical relation to it. The true starting-point is man's active contact with nature, and it is only by abstraction that we divide this into self-conscious humanity on the one hand and nature on the other. Man's relationship to the world is not originally contemplation or passive perception, in which things transmit their likeness to the subject or transform their inherent being into fragments of the subject's perceptual field. Perception is, from the beginning, the result of the combined operation of nature and the practical orientation of human beings, who are subjects in a social sense and who regard things as their proper objects, as designed to serve some purpose.

Man assimilates his many-sided essence in a many-sided way, and thus as a whole man. His whole human attitude to the world—sight, hearing, smell, taste and touch, thought, contemplation and sensation, desire, action and love—in short, all the organs of his personality, and those that in their form are directly social organs, constitute in their objective relationship, or their relation to the object, an assimilation of that object, an assimilation of human reality; their relation to the object is a manifestation of human reality ... The eye became a human eye, and its object became a social and human object, created by man and destined for him. Thus the senses, in practice, became directly theoreticians. They relate themselves to things with regard to the thing itself, but the thing itself is an objective, human relationship to itself and also to man, and vice versa ... An object is not the same to the eye as it is to the ear, and the object of the eye is different from that of the ear. The peculiarity of every essence is its own peculiar essence, and hence also its peculiar way of objectifying it, its own objectively real, live being ... To the unmusical ear the finest music means nothing—it is not an object for it. To be an object for me, a thing must be a confirmation of one of the forces of my being; it can exist for me only as a force of my being exists for itself, as a subjective faculty, since the significance of the object for me can extend no further than my own senses extend. The senses of a social man, then, are different from the senses of an unsocial one.

Marx, it may be seen, takes up the basic question posed by Kant and Hegel, viz. how can the human mind be 'at home' in the world? Is it possible, and if so how, to bridge the gulf between the rational consciousness and the world which is simply 'given' in a direct, irrational form? If the question is put in such general terms as this, we may say that Marx inherited it from classical German philosophy; but the specific questions he asks are different, especially from those of Kant. In the latter's doctrine, the alienness of nature *vis-à-vis* the free and rational subject is insuperable: the duality of the subject-matter of cognition, i.e. the fundamental difference between what is given and *a priori* forms, cannot be overcome in real terms, the manifoldness of the data of experience cannot be rationalized. The subject, which is self-determining and therefore free, encounters nature, which is constrained by necessity, as something other than itself, an irrationality which it has to tolerate. In the same way ideals and moral imperatives cannot be derived from the world of irrational data, so that the real and the ideal are bound to be in conflict. The unity of the world comprising subject and object, sense and thought, the freedom of man and the necessity of nature—such unity is a limiting postulate which reason can never actually bring about, but towards which it must strive without ceasing. Thus reality is an unceasing limitation for the subject, its mental faculties, and its moral ideals. In Hegel's view the Kantian dualism was an abdication of rationalism, and the postulate of unity as the unattainable limit of endless striving represented an anti-dialectic view of the world. If the gulf between the two worlds to which man belongs remains equally wide in every single cognitive or moral act, then the striving to overcome it is a sterile infinitude in which man's inability to heal his internal rift is endlessly reproduced. Hegel therefore seeks to represent the process whereby the subject gradually assimilates reality as a progressive discovery of its latent rationality, i.e. its spiritual essence. Reason is impotent if it cannot discover rationality in the very facticity of Being, if it wraps itself up in its own perfection and is at the same time encumbered with an irrational world. But when it discovers emergent rationality in that world, when it perceives reality as a product of self-consciousness and of the self-limiting activity of the Absolute, then it is able to recover the world for subjectivity; and this is the task of philosophy.

It was perhaps Feuerbach who first made Marx aware of the arbitrary and speculative character of the solution proposed by Hegelian idealism to the dualism of Kant. Hegel presupposes that actual existence is alienated self-consciousness, merely in order to recover the world for the thinking subject. But self-consciousness cannot, by alienating itself, create more than an abstract semblance of reality; and if, in human life, products of this self-alienation come to acquire power over men, it is our task to put them back in their proper place and recognize abstractions for what they are. Man is himself part of nature, and if he recognizes himself in nature it is not in the sense of discovering in it the work of a self-consciousness that is absolutely prior to nature, but only in the sense that, in the process of the self-creation of man by labour, nature is an object *for* man, perceived in a human fashion, cognitively organized in accordance with human needs, and 'given' only in the context of the practical behaviour of the species. 'Nature itself, considered in the abstract, fixed in separation from man, is nothing as far as he is concerned.' If the active dialogue between the human species and nature is our starting-point, and if nature and self-consciousness as we know them are given only in that dialogue and not in a pure inherent sense, then it is reasonable for nature as we perceive it to be called humanized nature, and for mind to be called the self-knowledge of nature. Man, a part and product of nature, makes nature a part of himself; it is at once the subject-matter of his activity and a prolongation of his body. From this point of view there is no sense in putting the question as to a creator of the world, since it presupposes the unreal situation of the non-existence of nature and man, a situation which cannot be posited even as a fictive starting-point.

By enquiring as to the creation of nature and man, you abstract from both man and nature. You assume their non-existence and yet wish to have it proved that they exist. Let me tell you, then, that if you give up your abstraction you will likewise abandon your question ... Since, for socialist man, the whole of so-called universal history is nothing but the formation of man by human labour, the shaping of nature for man's sake, man thus possesses a clear, irrefutable proof that he is born of his own self, a proof of the process whereby he has come to be. Since the essentiality of man and nature has become something practical, sensual and evident; since man has become for man practically, sensually and evidently the being of nature, and nature has become for man the being of man, the question of a foreign being over and above both man and nature ... has become a practical impossibility. Atheism, as a denial of that non-essentiality, is also meaningless, for atheism consists in denying the existence of God and establishing man's being on that denial. But socialism as such no longer needs such assistance; it takes as its starting-point the theoretical and practical sensual awareness of man and nature as an essence. It is the positive self-consciousness of man which no longer stands in need of the abolition of religion, just as real life is the positive reality of man which no longer stands in need of the abolition of private property—that is to say, communism.

Thus, in Marx's view, epistemological questions in their traditional form were no less illegitimate than metaphysical ones. A man cannot consider the world as though he were outside it, or isolate a purely cognitive act from the totality of human behaviour, since the cognizing subject is an aspect of the integral subject which is an active participant in nature. The human coefficient is present in nature as the latter exists for man; and, on the other hand, man cannot eliminate from his intercourse with the world the factor of his own passivity. Marx's thought on this point is equally opposed to the Hegelian theory of self-consciousness constituting the object as an exteriorization of itself, and to the versions of materialism which he encountered, in which cognition was, at its source, a passive reception of the object, transforming it into a subjective content. Marx describes his own view as consistent naturalism or humanism, which, he says, 'differs equally from idealism and materialism, being the truth which unites them both'. It is an anthropocentric viewpoint, seeing in humanized nature a counterpart of practical human intentions; as human practice has a social character, its cognitive effect—the image of nature—is the work of social man. Human consciousness is merely the expression in thought of a social relationship to nature, and must be considered as a product of the collective effort of the species. Accordingly, deformations of consciousness are not to be explained as due to the aberrations or imperfections of consciousness itself: their sources are to be looked for in more original processes, and particularly in the alienation of labour.

3. The alienation of labour. Dehumanized man

Marx considers the alienation of labour on the basis of capitalistic conditions in their developed form, in which land-ownership is subject to all the laws of a market economy. Private property is, in his view, a consequence and not a cause of the alienation of labour; however, the Paris Manuscripts as they have survived do not examine the origins of this alienation. In the developed conditions of capitalist appropriation the alienation of labour is expressed by the fact that the worker's own labour, as well as its products, have become alien to him. Labour has become a commodity like any other, which means that the worker himself has become a commodity and is obliged to sell himself at the market price determined by the minimum cost of maintenance; wages thus tend inevitably to fall to the lowest level that will keep the workmen alive and able to rear children. The situation which thus arises in the productive process is analogous to that which Feuerbach described in connection with the invention of gods by the human mind. The more wealth the worker produces, the poorer he gets; the more the world of things increases in value, the more human beings depreciate. The object of labour is opposed to the labour process as something alien, objectified, and independent of its producer. The more the worker assimilates nature to himself, the more he deprives himself of the means of life. But it is not only the product of labour that is alienated from the subject: labour itself is so alienated, for instead of being an act of self-affirmation it becomes a destructive process and a source of unhappiness. The worker does not toil to satisfy his own need to work, but to keep himself alive. He does not feel truly himself in the labour process, i.e. in that form of activity which is specifically human, but only in the animal functions of eating, sleeping, and begetting children. Since, unlike animals, 'man produces even when he is free from physical need, and indeed it is only then that he produces in the true sense', the alienation of labour dehumanizes the worker by making it impossible for him to produce in a specifically human manner. Work presents itself to him as an alien occupation, and he forfeits his essence as a human being, which is reduced to purely biological activities. Labour, which is the life of the species, becomes only a means to individual animalized life, and the social essence of man becomes a mere instrument of individual existence. Alienated labour deprives man of his species-life; other human beings become alien to him, communal existence is impossible, and life is merely a system of conflicting egoisms. Private property, which arises from alienated labour, becomes in its turn a source of alienation, which it fosters unceasingly.

The reification (as it would be called later) of the worker—the fact that his personal qualities of muscle and brain, his abilities and aspirations, are turned into a 'thing', an object to be bought and sold on the market—does not mean that the possessor of that 'thing' is himself able to enjoy a free and human existence. On the contrary, the process has its effect on the capitalist, too, depriving him of personality in a different way. As the worker is reduced to an animal condition, the capitalist is reduced to an abstract money-power: he becomes a personification of this power, and his human qualities are transformed into aspects of it.

My power is as great as the power of money. The attributes and essential strength of money are those of myself, its owner. It is not my own personality which decides what I am and what I can afford. I may be ugly, but I can buy the prettiest woman alive; consequently I am not ugly, since money destroys the repellent power of ugliness. I may be lame, but with money I can have a coach and six, therefore I am not lame. I may be bad, dishonest, ruthless and narrow-minded, but money ensures respect for itself and its possessor. Money is the supreme good, and a man who has it must be good also.

The effect of the alienation of labour is to paralyse man's species-life and the community of human beings, and therefore it paralyses personal life also. In a developed capitalist society the entire social servitude and all forms of alienation are comprised in the worker's relationship to production; the emancipation of workers is therefore not simply their emancipation as a class with particular interests, but is also the emancipation of society and humanity as a whole.

However, the emancipation of the worker is not simply a question of abolishing private property. Communism, which consists in the negation of private property, exists in different forms. Marx discusses, for instance, the primitive totalitarian egalitarianism of early communist utopias. This is a form of communism which seeks to abolish everything that cannot be made the private property of all, and therefore everything that may distinguish individuals; it seeks to abolish talent and individuality, which is tantamount to abolishing civilization. Communism in this form is not an assimilation of the alienated world but, on the contrary, an extreme form of alienation that consists in imposing the present condition of workers upon everybody. If communism is to represent the *positive* abolition of private property and of self-alienation it must mean the adoption by man of his own species-essence, the recovery of himself as a social being. Such communism resolves the conflict between man and man, between essence and existence, the individual and the species, freedom and necessity. In what, however, does the 'positive' abolition of private property consist? Marx suggests an analogy with the abolition of religion: just as atheism ceases to be significant when the affirmation of man is no longer dependent on the negation of God, in the same way socialism in the full sense is the direct affirmation of humanity independent of the negation of private property: it is a state in which the problem of property has been solved and forgotten. Socialism can only be the result of a long and violent historical process, but its consummation is the complete liberation of man with all his attributes and possibilities. Under the socialist mode of appropriation man's activity will not be opposed to him as something alien, but will be, in all its forms and products, the direct affirmation of humanity. There will be 'wealthy man and wealthy human need'; 'a rich man is at the same time a man needing the manifestations of human life in all their fullness'. Whereas in conditions of alienated labour the increase of demand multiplies the effect of alienation—the producer strives artificially to arouse demand and to make people dependent on more and more products, which in such circumstances only increase the volume of servitude—in socialist conditions the wealth of requirements is indeed the wealth of mankind.

While the Paris Manuscripts thus attempt to establish socialism as the realization of the essence of humanity, they do not present it as an ideal pure and simple but as a postulate of the natural course of history. Marx does not regard private property, the division of labour, or human alienation as 'mistakes' that could be rectified at any time if men came to a correct understanding of their own situation; he regards them as indispensable conditions of future liberation. The vision of socialism outlined in the Manuscripts involves the full and perfect reconciliation of man with himself and nature, the complete identification of human essence and existence, the harmonization of man's ultimate destiny and his empirical being. It may be supposed that a socialist society in this sense would be a state of complete satisfaction, an ultimate society with no incentive or need for further development. While Marx does not express his vision in these terms, he does not rule out such an interpretation either, and it is encouraged by his view of socialism as the removal of all sources of human conflict and a state in which the essence of humanity is empirically realized. Communism, he says, 'is the solution to the riddle of history and is aware of that fact'; the question then arises whether it is not also the termination of history.

4. *Critique of Feuerbach*

The philosophy of the Manuscripts is confirmed and completed by Marx's *Theses on Feuerbach*, written in the spring of 1845. Published by Engels in 1888, after Marx's death, they were regarded as an epitome of the new world-view and are among the most frequently quoted of their author's works. They contain the most trenchant formulation of Marx's objections to Feuerbachian materialism, especially his opposition of a purely contemplative theory of knowledge to a practical one and the different meaning he ascribed to religious alienation. The reproach that Marx levelled against Feuerbach and all previous materialists was that they envisaged objects merely in a contemplative way and not as 'sensuous, practical, human activity,

not subjectively', with the result that it was left to idealism to
develop the active side—'but only abstractly, since of course
idealism does not know real, actual, sensuous activity as such'.
This objection repeats the thought expounded in more detail
in the Manuscripts: perception is itself a component of man's
practical relationship to the world, so that its object is not
simply 'given' by indifferent nature but is a humanized object,
conditioned by human needs and efforts. The same practical
standpoint appears in Marx's refusal to enter into a speculative
dispute on the conformity of thought with its object. 'In practice
man must prove the truth, that is, the reality and power, the
this-sidedness [*Diesseitigkeit*] of his thinking. The dispute over
the reality or non-reality of thinking which is isolated from
practice is a purely scholastic question.' As might be expected,
and as *The German Ideology* later confirms, the cognitive function
of practice does not merely signify that the success of an activity
confirms the accuracy of our knowledge, nor that practical life
expresses the range and purpose of human interests; it means
also that veracity is itself the 'reality and power' of thought,
i.e. that those ideas are true in which man confirms himself as
a 'species-being'. On this ground Marx dismisses as 'scholastic'
the Cartesian question as to the conformity between a pure act
of thought and reality. The epistemological question is not a
real question, because the pure act of perception or thought
which it premises is a mere speculative fiction. Since the mind,
having achieved self-understanding, apprehends itself as a
coefficient of practical behaviour, it follows that the questions
that may legitimately be put to it as to the meaning of its acts
are also questions as to its effectiveness from the point of view
of human society.

Marx also repeats in the *Theses* his criticism of Feuerbach's
theory of religion, viz. that it reduces the world of religion to
its secular basis but does not explain the duality in terms of the
internal disharmony of man's situation in the world, and is therefore unable to offer an effective cure: the mind can only be
freed from mystification if the negativities of social life from which
it arises are removed by practical action.

Further, Marx criticizes Feuerbach's conception of the essence
of man as 'an abstraction inherent in a particular individual',
whereas it is in fact 'the totality of social relationships'. The effect

of Feuerbach's conception is that he takes as his point of departure the individual in his species-characteristics, and reduces the tie between human beings to a natural tie. The same thought appears in the tenth *Thesis*, having previously been expressed in Marx's essay *On the Jewish Question*: 'The standpoint of the old materialism is "civil" society; the standpoint of the new is human society, or socialized humanity.' This corresponds to Marx's previous contention that civil society must coincide with political society, so that both of them cease to exist in the old form: no longer will the first be a mass of conflicting egoisms, the second an abstract, unreal community; man, himself a true community, will absorb his own species-nature and realize his personality as a social one.

In the important third *Thesis* Marx expounds his opposition to the doctrines of utopian socialism based on eighteenth-century materialism. It is not sufficient to say that human beings are the product of conditions and upbringing, since conditions and upbringing are also the work of human beings. To assert only the former proposition amounts to 'dividing society into two parts, of which one is superior to society (in Robert Owen, for example). The coincidence of the changing circumstances and of human activity can be conceived and rationally understood only as revolutionary praxis.' This statement means that society cannot be changed by reformers who understand its needs, but only by the basic mass whose particular interest is identical with that of society as a whole. In the revolutionary praxis of the proletariat the 'educator' and 'educated' are the same: the development of mind is at the same time the historical process by which the world is transformed, and there is no longer any question of priority between the mind and external conditions or vice versa. In this situation of revolutionary praxis the working class is the agent of a historical initiative and is not merely resisting or reacting to the pressure of the possessing classes.

The same 'practical' viewpoint is dominant in Marx's conception of the cognitive functions of the mind and its role in the historical process; 'practical' is always regarded as implying 'social', and 'social life is practical by its very essence'. So is the task of philosophy as defined in the eleventh *Thesis*, in what are perhaps Marx's most-quoted words: 'The philosophers have only interpreted the world in various ways; the point, however, is to change it.' It would be a caricature of Marx's thought to read this as meaning that it was not important to observe or analyse society and that only direct revolutionary action mattered. The whole context shows that it is a formula expressing in a nutshell the viewpoint of 'practical philosophy' as opposed to the 'contemplative' attitude of Hegel or Feuerbach—the viewpoint which Hess, and through him Cieszkowski, suggested to Marx and which became the philosophical nucleus of Marxism. To understand the world does not mean considering it from outside, judging it morally or explaining it scientifically; it means society understanding itself, an act in which the subject changes the object by the very fact of understanding it. This can only come about when the subject and object coincide, when the difference between educator and educated disappears, and when thought itself becomes a revolutionary act, the self-recognition of human existence.

5. *Engels's early writings*

The year 1844 saw the beginning of Marx's friendship and collaboration with Friedrich Engels, whom he had already met briefly in Cologne. Engels had been through a similar spiritual evolution to Marx, though their early education was different. Born on 28 November 1820, Engels was the son of a manufacturer at Barmen (Wuppertal, near Düsseldorf). He grew up in a stifling atmosphere of narrow-minded pietism, but soon escaped from its influence, leaving school before his final year to work in his father's factory; in 1838 he was sent to Bremen to gain business experience. As a result of practical contact with trade and industry he soon became interested in social questions. In the course of private study he imbibed liberal-democratic ideas and was attracted to Young Hegelian radicalism. His first Press articles were written in 1839 for the *Telegraph für Deutschland*, published by Gutzkow at Hamburg, and the Stuttgart *Morgenblatt*. He attacked German bigotry and the hypocrisy of petty-bourgeois pietism, but also described industrial conditions and the oppression and poverty of the workers. He was attracted by the sentimental pantheism of Schleiermacher and did not at first wholly abandon Christianity, but became an atheist under the influence of Strauss's *Life of*

Jesus. During his military service at Berlin in 1841 he joined the philosophical radicals and wrote three pamphlets criticizing Schelling from a Young Hegelian point of view. Later, when he regarded himself as a communist, he declared that communism was the natural fruit of German philosophical culture. Towards the end of 1842 he went to his father's works in Manchester for further commercial training, and spent much time observing the conditions of the British working class and studying political economy and socialism. The number of the *Deutsch-Französische Jahrbücher* to which Marx contributed his articles on Hegel's *Philosophy of Right* and *The Jewish Question* also contained an essay by Engels entitled *Outline of a Critique of Political Economy.* This argued that the contradictions of capitalist economy could not be resolved on the basis of that economy; that periodical crises of overproduction were the inevitable consequence of free competition; that competition led to monopoly, but monopoly in turn created new forms of competition, etc. Private property led necessarily to antagonism between classes and between individuals in each class, and to an incurable conflict between private and public interests; it was also bound up with anarchy in production and the resultant crises. Economists who defended private property could not understand this chain of causes and were driven to invent groundless theories, such as that of Malthus which blamed social evil on the fact that population grew faster than production. The abolition of private property was the only way to save humanity from crises, want, and exploitation. Planned production would do away with social inequality and the absurd situation in which poverty was caused by an excess of goods. 'We shall liquidate the contradiction', wrote Engels, 'simply by removing it. When the interests that are now in conflict are merged into one, there will be no contradiction between the excess of population at one end of the scale and the excess of wealth at the other; we shall no longer experience the amazing fact, more extraordinary than all the miracles of all religions put together, that wealth and an excess of prosperity cause peoples to die of hunger; we shall no longer hear the foolish assertion that the earth cannot maintain the human race.'

Engels remained rather less than two years at Manchester, and published his observations at Leipzig in *The Situation of the Working Class in England* (1845). In this book, which was a revelation for its time, he painted a broad picture of the results of the industrial revolution in Britain and described graphically the cruel poverty of the urban proletariat and the starvation, brutality, and hopelessness of working-class life. He did not write as a moralizer or philanthropist, but inferred from the conditions of the working-class that the latter was bound to bring about a socialist revolution by its own efforts within a few years. His prediction of socialism was thus based not on general ideas about human nature or the need to bring human existence into conformity with the essence of humanity, but on actual acquaintance with working-class conditions and trends of development. He was convinced that the middle classes would disappear, that capital in Britain would concentrate more and more and that there would soon be an inevitable and blood-thirsty war between the needy and the rich. Engels set his pre-diction within the framework of a clear-cut division of classes, the proletariat being not only the most oppressed and afflicted but also destined to put an end to all oppression. At the same time, while describing with a wealth of detail the villainy of the English bourgeoisie, Engels did not treat their behaviour as being due simply to moral depravity, but as an inevitable effect of the situation of a class of men obliged by cut-throat competition to exploit their fellows to the maximum degree.

CHAPTER VII

The Holy Family

MARX's meeting with Engels in Paris in August 1844 was the beginning of forty years' collaboration in political and literary activity. While Marx's powers of abstract thought were superior to those of his friend, Engels surpassed him in relating theory to empirical data, whether social or scientific. Their first joint work, entitled *The Holy Family, or a Critique of Critical Criticism: against Bruno Bauer and Co.*, was published at Frankfurt-on-Main in February 1845; only a small part of it was the work of Engels, who had returned to Barmen after a short stay in Paris.

The Holy Family is a radical and, one may say, ruthless challenge to Young Hegelianism. It is a virulent, sarcastic, and unscrupulous attack on Marx's former allies, especially Bruno and Edgar Bauer. The work is diffuse and full of trivial mockery, puns on his adversaries' names, etc. It sets out to display the naïvety and intellectual nullity of the Hegelian 'holy family' and the speculative character of its criticism; unlike *The German Ideology*, it contains little in the way of independent analysis. Nevertheless it is an important document, bearing witness to Marx's final break with Young Hegelian radicalism: for its proclamation of communism as the working-class movement *par excellence* is presented not as a supplement to the critique of Young Hegelianism, but as something opposed to it. It even declares in the Introduction that 'True humanism has no more dangerous enemy in Germany than spiritualism or speculative idealism, in which the actual individual human being is replaced by "self-consciousness" or "spirit."' In some important ways *The Holy Family* confirms, but with greater emphasis, Marx's theoretical standpoint as formulated in previous works, while in others it introduces new elements.

1. Communism as a historical trend. The class-consciousness of the proletariat

Marx expresses more plainly than hitherto the idea of the historical inevitability of the movement towards communism. Private property, by endeavouring to prolong itself indefinitely, creates its own antagonist, the proletariat. In the self-alienation which is strengthened by private property the possessing class enjoys the satisfaction procured by the outward show of humanity, while the working class is humiliated and impotent. Private property tends to destroy itself irrespective of the knowledge or will of the possessing class, since the proletariat which it creates is a dehumanization that is conscious of itself. The victorious proletariat does not simply turn the tables and substitute itself for the possessors, but puts an end to the situation by eliminating itself and its own opposition. It represents the maximum of dehumanization, but also awareness of that dehumanization and the inevitability of revolt. The misery of the proletariat obliges it to free itself, but it cannot do so without at the same time freeing the whole of society from inhuman conditions.

Marx's emphasis on the self-awareness of the proletariat in the process of emancipation is important in connection with the objection, sometimes put forward at a later date, that he appeared to believe that the revolution would come about as the result of an impersonal historical force, irrespective of the free activity of man. From his point of view there is no dilemma as between historical necessity and conscious action, since the class-consciousness of the proletariat is not only a condition of the revolution but is itself the historical process in which the revolution comes to maturity. For this reason the authors of *The Holy Family* join issue with any personification of history as an independent force. Bauer, Engels says, transforms history into a metaphysical being which manifests itself in individual men and women; but in actual fact 'history does nothing, has no "enormous wealth", wages no battles. It is not "history" but live human beings who own possessions, perform actions and fight battles. There is no independent entity called "history", using mankind to attain its ends: history is simply the purposeful activity of human beings.' These observations are the point of departure for the later controversy on Marx's alleged historical determinism. They leave room for differences of interpretation, as do subsequent statements of his: in particular, that men make their own history but do not make it irrespective of the conditions they are in. Are we to understand that man's ability to affect the historical process is limited, that existing conditions are not wholly obedient to human action but can to some extent be governed by the organized will of the community; or is it rather the case that the conditions in which a man acts are themselves the determinants of his consciousness and his action? These are key questions for the understanding of historical materialism, and we shall have occasion to revert to them in due course.

2. *Progress and the masses*

An essential topic of Marx's criticism of Bauer is the latter's opposition between the masses and progress, between the masses and the critical spirit. In Bauer's view the masses as such are an embodiment of conservatism, reaction, dogmatism, and mental inertia. Any ideas they assimilate, including revolutionary ones, are turned into conservatism; any doctrine absorbed by the masses becomes a religion. A creative idea is no sooner adopted by the masses than it loses its creativity. Ideas that need their support are foredoomed to distortion, degeneration, and defeat; all great historical enterprises that have come to grief have done so because the masses took possession of them. This analysis, in Marx's view, is an absurd attempt to condemn the course of history. Successful ideas, he contends, must be the expression of some mass interest ('The "idea" has always been a fiasco when divorced from "interest"'); but whenever 'interest' takes the form of an idea it goes beyond its real content and must present itself delusively as a general interest and not a particular one. By opposing progress to the conservatism of the masses, Bauer's criticism is condemned to remain a thing of the mind instead of an instrument of social transformation. In any case, Marx contends, the undifferentiated category of progress is itself without content. Socialist ideas originated in the historical observation that what is called progress has always come about in opposition to the majority of society and led to more and more inhuman conditions. This suggested that civilization was radically diseased; it pointed to a fundamental criticism of society, coinciding with a mass movement of social protest. We must not, then, be content with phrases about progress, since no absolute progress can be identified in history.

Marx here introduces for the first time a thought that recurs more than once in his later work. Instead of an incurable antagonism between the masses and the critical spirit—a parody, in his opinion, of the traditional opposition between 'spirit' and inert 'matter', the former being represented by the individual and the latter by the masses—he puts forward the idea of a fundamental antinomy that has pervaded history hitherto, whereby actual progress, especially in the technical field, has been effected at the expense of the great mass of toiling humanity. While Bauer's historiosophy is obliged by its nature to confine itself to purely theoretical ideas of liberation, socialist criticism is aimed at the material conditions which have produced a contradiction between the advance of civilization and the needs of the immediate creators of wealth. Ideas by themselves, Marx argues, can never burst the bonds of the old world; human beings, and the use of force, are necessary before ideas can be realized.

3. *The world of needs*

In *The Holy Family* Marx returns to the problem of the opposition between the true human community and the imaginary community of the state. Bauer holds that human beings are egoistic atoms which have to be welded into an organism by the state. To Marx this is a speculative fiction. An atom is self-sufficient and has no needs; a human individual may imagine himself to be an atom in this sense, but in fact he never can be, for the world of men is a world of needs and, despite all mystification, it is they which constitute the real links between members of the community. The social bond is not created by the state but by the fact that, although people may imagine themselves to be atoms, they are actually egoistic human beings. The state is a secondary product of the needs which constitute the social bond; this latter is not a product of the state. Only if the world of needs gives rise to a conflict, if needs are satisfied by means of a struggle between egoisms, and if the social bond assumes the aspect of social discord—only then does

the question arise as to the possibility of a real human community. Bauer, however, is content to maintain the Hegelian
opposition between the state as a community and civil society
as a tangle of egoisms, and regards this opposition as an eternal
principle of life.

4. The tradition of materialism

In *The Holy Family* Marx also expresses for the first time his
awareness of the link between socialist ideas and the tradition
of philosophical materialism. He distinguishes two trends in the
history of French materialism: the first, which goes back to
Descartes, is naturalistic in inspiration and evolves in the
direction of modern natural science. The second, that of Lockean
empiricism, represents the direct tradition of socialism, whose
ideological premises derive from the anti-metaphysical critique
of the eighteenth-century materialists and their attacks on the
dogmatism of the previous century. Locke's sensationalism
implied the doctrine of human equality: every man who comes
into the world is a *tabula rasa*, and mental or spiritual
differences are acquired and not innate. Since all men are by
nature egoists and morality can only be rationalized egoism,
the problem is to devise a form of social organization that will
reconcile the selfish interests of each with the needs of all. As
human beings are entirely the product of their education and
conditions of life, they can only be changed by changing the
social institutions that fashion them. Fourier's doctrine is the
fruit of the French materialism of the Enlightenment, while
Owen's socialist ideas are rooted in Bentham and, through him,
in Helvétius. The principles of empiricism and utilitarianism,
which lay down that human beings are neither good nor bad
by nature but only by upbringing, that interest is the mainspring
of morality, and so forth, naturally lead us to inquire what
social conditions are necessary to make the community of mankind a reality.

In this way Marx invokes the materialist tradition against
Bauer, who, following Hegel, makes self-consciousness into a
substantive entity (whereas it is in fact only an attribute of man
and not a separate form of Being) and imagines that he has
thus ensured the spirit's independence of nature. By the same
token Bauer reduces human life to intellectual activity and turns

all history into the history of thought, whereas it is, first and foremost, the history of material production.

The Holy Family thus contains, though as yet only in laconic and general formulas, the seminal ideas of the materialist interpretation of history: that of the mystification that befalls human interests when they are expressed in ideological form, and that of the genetic dependence of the history of ideas on the history of production. We find here the application to a new historiosophy of the classic schema of Hegel's dialectic, the negation of a negation. As private property develops it necessarily creates its own antagonist; this negative force is itself dehumanized, and as its dehumanization progresses it becomes the precondition of a synthesis that will abolish the existing opposition together with both its terms—private property and the proletariat—and will thus make it possible for man to become himself again.

The basis of the materialist interpretation of history was expounded in the next joint work by Marx and Engels, *The German Ideology*. Marx remained in Paris until the beginning of 1845, taking an active part in the meetings of socialist organizations and especially the League of the Just, while in Germany Engels spread the word of communism in speeches and writings and endeavoured to weld scattered socialist groups into a single organization. In February 1845 Marx was deported from Paris at the instance of the Prussian government and took up his abode in Brussels, where Engels joined him in the spring. In summer they visited England, where they made contact with the Chartists and took steps to establish a centre of co-operation of the revolutionary movements of different countries. Returning to Brussels, they continued to work for the unification of revolutionary associations and to carry on polemics with German philosophers.

CHAPTER VIII

The German Ideology

MARX and Engels finished *The German Ideology* in 1846, but were not able to publish it. Parts of the manuscript were lost; the remainder was published in an incomplete form by Bernstein in 1903, and in its entirety in the *MEGA* edition in 1932. The work was primarily an attack on Feuerbach, Max Stirner, and so-called 'true socialism'; Bruno Bauer is only referred to incidentally. From the philosophical point of view the most important sections are those criticizing Feuerbach's 'species-man' and Stirner's 'existential' conception of man. These also contain the most positive expression of the authors' own views; Feuerbach is in fact criticized indirectly, by the exposition of their own standpoint. To Feuerbach's anthropology they oppose the idea of humanity as a historical category; to Stirner's absolute of the individual self-consciousness, the idea of man actualizing his social nature in his own unique and individual character. The central ideas of *The German Ideology*, or at any rate those which gave rise to the liveliest discussion in the later development of Marxism, are those concerning the relationship between human thought and living conditions; these contain the basis of the materialist interpretation of history, which was developed later in fuller detail.

1. *The concept of ideology*

The term 'ideology' dates from the end of the eighteenth century, when it was introduced by Destutt de Tracy to denote the study of the origin and laws of operation of 'ideas' in Condillac's sense, i.e. psychic facts of all kinds, and their relation to language. The name 'idéologues' was given to the scholars and public men (Destutt, Cabanis, Volney, Daunou) who carried on the tradition of the *Encyclopédistes*; Napoleon applied the expression to them in the pejorative sense of 'political dreamers'. The Hegelians occasionally used 'ideology' to denote the subjective aspect of the cognitive process.

In the work of Marx and Engels 'ideology' is used in a peculiar sense which was later generalized: they do not define it expressly, but it is clear that they give it the meaning later expounded by Engels in *Ludwig Feuerbach* (1888) and in a letter to Mehring dated 14 July 1893. 'Ideology' in this sense is a false consciousness or an obfuscated mental process in which men do not understand the forces that actually guide their thinking, but imagine it to be wholly governed by logic and intellectual influences. When thus deluded, the thinker is unaware that all thought, and particularly his own, is subject in its course and outcome to extra-intellectual social conditions, which it expresses in a form distorted by the interests and preferences of some collectivity or other. Ideology is the sum total of ideas (views, convictions, *partis pris*) relating, first and foremost, to social life —opinions on philosophy, religion, economics, history, law, utopias of all kinds, political and economic programmes— which appear to exist in their own right in the minds of those who hold them. These ideas are in fact governed by laws of their own; they are characterized by the subject's unawareness of their origin in social conditions and of the part they play in maintaining or altering those conditions. The fact that human thought is determined by the conflicts of material life is not consciously reflected in ideological constructions, or they would not truly deserve the name of ideology. The ideologist is the intellectual exponent of a certain situation of social conflict; he is unaware of this fact and of the genetic and functional relationship between the situation and his ideas. All philosophers are ideologists in this sense; so are religious thinkers and reformers, jurists, the creators of political programmes, etc. It was not until much later, in Stalin's time, that Marxists came to use 'ideology' to denote all forms of social consciousness, including those that were supposed to present a scientific account of the world, free from mystification and distortion. In this sense it was possible to speak of 'scientific' of 'Marxist' ideology, which Marx and Engels, given their use of the term, could never have done.

The original Marxist concept was the basis of the twentieth-century theory of ideology and, more generally, the sociology of knowledge (Mannheim), i.e. the study of ideas irrespective of whether they are true or false—for ideological mystification is not the same as error in the cognitive sense; to define a product of the mind as ideology does not involve any judgement as to its truth or falsehood. Instead, this science considers ideas as manifestations of certain group interests, practical instruments whereby social classes and other sections of the community uphold their own interests and values. The study of ideology investigates social conflicts and structures from the standpoint of their intellectual expression; it considers ideas, theories, beliefs, programmes, and doctrines in the light of their dependence on the social situations that give rise to them, thought being a disguised version of reality. As Mannheim observed, this idea goes back beyond Marx; the hypocrisy of moral ideals, religious beliefs, and philosophical doctrines was pointed out by Machiavelli, and between Marx and Mannheim we can find similar ideas in Nietzsche and Sorel. In the modern analysis of ideas it is generally accepted that the ideological content must be distinguished from the cognitive value, that the functional-genetic conditioning of thought is one thing and its scientific legitimacy another. Marx was the pioneer of this distinction; he was concerned, however, not only with pointing out the dependence between thought and interests but also with identifying the particular type of interest that exerts the strongest influence on the construction of ideologies, namely that connected with the division of society into classes.

Marx begins by dealing with the central delusion of German ideologists who believe that while humanity is governed by false ideas and imaginations and men are enslaved to the creations of their own minds ('gods' in Feuerbach's sense), it is within the power of philosophy to expose and destroy these wrong ideas and revolutionize the society based upon them. The basic position of Marx and Engels is, on the contrary, that the authority of delusions over human minds is not a result of mental distortion that can be cured by working upon the consciousness, but is rooted in social conditions and is only the intellectual expression of social servitude.

2. Social being and consciousness

In this way, taking up a theme already sketched in their previous writings, Marx and Engels set out to overthrow the view of the Young Hegelians and Feuerbach that mental aberrations and distortions were the cause of social servitude and human misfortune, and not the other way about. They attempted to analyse the origin of ideas, not in Condillac's sense but by investigating the social conditioning of consciousness. The Hegelians, in their delusions, had not confined themselves to believing in the omnipotence of thought in social history. Holding as they did that the relations between human beings are the result of wrong ideas about the world and themselves— whereas in fact the contrary is the case—the Hegelians from Strauss to Stirner had reduced all human ideas on politics, law, morals, or metaphysics to the denominator of theology, making all social consciousness a religious consciousness, and seeing in the critique of religion a panacea for every human ailment.

The contention of Marx and Engels was that the distinguishing mark of humanity, that which primarily characterizes men as opposed to beasts, is not that they think but that they make tools. This is what first made man a separate species; then, in the course of history, men were distinguished by their way of reproducing their own life, and hence by their way of thinking. Human beings are what their behaviour shows them to be: they are, first and foremost, the totality of the actions whereby they reproduce their own material existence. ('As individuals express their lives, so they are. What they are, therefore, coincides with their production, both with what they produce and how they produce it. The nature of individuals thus depends on the material conditions of their production'.) The level of production determined by the productive forces, i.e. by the quality of tools and technical skill, itself determines the social structure. This latter consists primarily in the division of labour, and the historical development of humanity is divided into phases by the different forms that the division of labour assumes. Each of these forms in turn creates a new form of property. The tribal ownership of primitive times, the ancient world with communal and state property, feudalism with its estates, crafts and landed property, and finally capitalism—all these are forms of society owing their origin to the type of productive capacity available to the human race at each period. We cannot consider rationally any conscious human life except as a component of the whole of life as defined in the first place by the method of satisfying elementary needs, the widening of the range of needs, the method of reproducing the species in family life, and also the system of co-operation which is itself to be reckoned as a productive force. Consciousness is nothing but human existence made conscious; but the self-delusion of consciousness which imagines that it only defines itself in its own work is in fact conditioned by the division of labour. It is only when the level of production makes it possible to separate physical and intellectual labour that consciousness can imagine itself to be other than awareness of practical life and can devise pure, abstract forms of mental activity such as philosophy, theology, and ethics. In addition, the ruling thoughts of a particular era gradually become separate from the ruling individuals, i.e. intellectual labour becomes a distinct occupation and the profession of the ideologist is born. This encourages the idea that it is thought which governs history and that it is possible to deduce human relationships, as Hegel did, from the concept of humanity itself.

The imaginary creations of the human brain are the inevitable sublimations of the material process of existence, which can be observed empirically and which depends on material causes. Morality, religion, metaphysics and all other forms of ideology and the related forms of consciousness thus lose the independence they appeared to have. They have no history or development of their own; it is only people, developing their material production and mutual material relations, who as a result come to think different thoughts and create different intellectual systems. It is not consciousness that determines life, but life that determines consciousness ... and all consciousness is that of live individuals.

These first, somewhat crude formulations of the materialist interpretation of history foreshadow subsequent debates on the sense in which Marx regarded thought as dependent on social conditions. If such aspects of social life as religion, morality, and law have no history of their own, it would seem that for Marx human ideas are no more than a natural secretion of social life embodying no active principle, a mere by-product of the true history which consists in the material productive processes and the property relationships that correspond to them; or, as critics of Marxism later put it, that the life of the mind is an epiphenomenon of the conditions of production. There has been a controversy in this sphere between economic materialism and the version of Marxism which ascribes an active and independent historical function to 'subjective' factors, i.e. the workings of the intellect and freely directed political activity.

Clearly Marx cannot be saddled with the view that all history is the effect of 'historical laws', that it makes no difference what people think of their lives, and that the creations of thought are merely foam on the surface of history and not truly part of it. Marx speaks of the active function of ideas as an indispensable means of maintaining and transforming social life, and he includes human skill and technology among 'productive forces'. He does not, it is true, regard humanity as constituted by self-consciousness: the latter is 'given' as a product of life, not in a pure form but as articulated in language—i.e. as communicative self-knowledge, its form determined by the means of collective communication. In this sense consciousness is always a social product. But, Marx says, 'Circumstances create people in the same degree as people create circumstances.' Both social servitude and the movement towards its abolition have as their condition certain subjective factors. Material subjugation requires spiritual subjugation; the ideas of the dominant class are dominant ideas; the class which commands material force also commands the means of intellectual coercion, as it produces and propagates the ideas that express its own supremacy.

Marx, then, cannot be regarded as maintaining that history is an anonymous process in which conscious intentions and thoughts are a mere by-product or casual accretion. Yet there is room for controversy over his theory even if we accept that thoughts, feelings, intentions, and the human will are a necessary condition of the historical process. For this view is compatible with strict determination on the basis that although 'subjective' factors are necessary causal links, they are themselves entirely due to non-subjective factors; thoughts and feelings, on this assumption, have an auxiliary role in history but not an originating one. In short, even if we do not interpret Marx's position as one of economic determination, there is still room for argument as to the role of free action in the historical process. This controversy has in fact made its appearance in various forms of Marxism in the present century, and cannot by any means be regarded as settled.

3. *The division of labour, and its abolition*

In Marx's view the division of labour is, genetically speaking, the primary source of social conflict. It brings about inevitable disharmony between three aspects of life: productive forces, human relations, and consciousness. It leads to inequality, private property, and the opposition between individual interests and the general interest arising from the mutual dependence of human beings. As long as the division of labour runs riot and is outside human control, its social effects will be an alien force dominating individuals like an independent, superhuman power.

Marx, it will be seen, generalizes the concept of alienation, extending its operation to the whole historical process. Not only the imaginations of religion, as Feuerbach maintained, but the whole of history is alienated from mankind, since human beings cannot control its course; their actions result in a mysterious, impersonal process which tyrannizes over those who have brought it about. To remove this alienation man must be given power once more to shape the effects of his own actions—to turn history into something human, something controlled by man.

As the division of labour is the primary source of social inequality and private property, the chief purpose of communism must be to abolish the division of labour. Communism requires conditions in which men are not restricted to a particular type of work, but can take part successively in all types and thus achieve all-round development. The reification of human products, by which they come to dominate the individual, is one of the chief factors in the historical process; it also means that the 'general interest' assumes independent existence in the form of the state, which is at present necessary to enable the bourgeoisie to hold on to its own property. Political struggles within the state are an expression of the class conflict; every class aspiring to power must present its own interest as that of the whole community, and the purpose of its ideology is to confirm this mystification.

Marx later compared the situation of humanity faced with

the alienation of history to that of the sorcerer's apprentice in Goethe's poem, who called up magic powers which he could no longer control and which turned into a threat to himself. But to abolish alienation, two conditions are required. Firstly, the state of servitude must become intolerable, the masses must be deprived of possessions and totally opposed to the existing order. Secondly, technical development must have reached an advanced stage: communism in a premature state would only be generalized poverty. Moreover, this development must be worldwide: communism can only come about when the world is a single market and all countries are economically inter-dependent. It must be brought about by simultaneous revolution in the most advanced and dominant countries; a proletariat capable of effecting the revolution must be a class that exists on a world scale. (This last point, which is basic to Marx's theory of revolution, was hotly debated at the beginning of the Stalin era, when the possibility of building 'socialism in one country' was mooted.)

But the social conditions that make communism possible also mean that there will be an irresistible movement towards it. 'Communism is not merely a state to be brought about or an ideal to which reality should conform; what we call communism is an actual movement which is sweeping away the present state of things.' This view of Marx's, which he afterwards repeated in various forms, has given rise to another essential controversy. Should the communist movement await the spontaneous develop-ment of mass opposition and then impose a form on it, or should it organize that opposition from outside and not wait for the masses to become aware of their predicament? Should current political activity be geared to the achievement of a certain final state, or, as the reformists would have it, should the working-class movement be content with such piecemeal gains as can be extracted from particular situations? These problems were developed in later polemics. At the time of *The German Ideology* Marx and Engels were chiefly concerned to argue that communism is not an arbitrarily constructed ideal of a better world, but a natural part of the historic process. Until such time as the social preconditions of an upheaval are fully realized it is of no consequence how, and how often, the idea of that up-heaval is proclaimed. But the communist revolution is fundamentally different from all that have gone before. Previous revolutions have altered the division of labour and the distribution of social activity; but the communist revolution will abolish the division of labour and the class division, and will abolish classes and nations as divisions of the human race. Communism will for the first time bring about a universal transformation of the terms of production and exchange; it will treat all previous forms of social development as the work of man, and will subject them to the authority of united individuals.

4. *Individuality and freedom*

The restoration of man's full humanity, removing the tension between individual aspirations and the collective interest, does not imply a denial on Marx's part of the life and freedom of the individual. It has been a common misinterpretation by both Marxists and anti-Marxists to suppose that he regarded human beings merely as specimens of social classes, and that the 'restoration of their species-essence' meant the annihilation of individuality or its reduction to a common social nature. On this view, individuality has no place in Marxist doctrine except as an obstacle in the way of society attaining to homogeneous unity. No such doctrine, however, can be derived from *The German Ideology*, in which Marx distinguishes, as a fact of history, between the individual and the contingent nature of life. The opposition between the individual and the system of human relations is a continuation of the opposition between productive forces and productive relationships. As long as this contradiction does not exist, the conditions in which the individual operates do not appear to him as an external reality but as part of his individuality. Up to the present time the social relationships in which individuals of this or that class were involved were such that people stood to them not as individuals but as specimens of a class. At the same time, as the products of their activity escaped their control, the conditions of life were subordinated to a reified, extra-human power and the individual became a victim of absolute contingency, to which was given the name of freedom. Personal ties were transformed into material ones; people confronted one another as representatives of the impersonal forces that ruled the world—goods, money, or civil authority— while the individual's 'freedom' meant a lack of control over the conditions of his own life, a state of impotence *vis-à-vis* the external world. To reverse this reification and restore man's power over things is likewise to restore his individual life, the possibility of all-round development of his personal aptitudes and talents. In such a community people will for the first time be truly individuals and not merely specimens of their class.

While it is certain, therefore, that Marx does not follow the Cartesian tradition of conceiving man in terms of self-consciousness (which he regards as secondary both to physical and to social existence), it is also certain that he seeks to preserve the principle of individuality—not, however, as something antagonistic to the general interest, but as completely coincident with it. This should not be mistaken for a new version of the theory of 'enlightened self-interest', which holds that a properly organized system of laws can obviate the conflict between the individual, conceived as essentially selfish, and the collectivity, by so arranging matters that anti-social acts turn against their perpetrators, so that the true self-interest is to behave in a socially constructive manner. Marx for his part rejects the notion of 'innate egoism', and in this respect is closer to Fichte than to the Enlightenment. He believes that the abolition of dependence on alienated forces will restore to man his social nature, i.e. the individual will accept the community as his own interiorized nature. But this community, consciously present in each of its members, is not intended to be a merging of personality in an anonymous, homogeneous whole. There is no question of uniformity being either imposed or voluntarily accepted; this idea, in Marx's view, belongs to primitive utopian communism—not a state in which private property has been abolished, but one in which it has not yet developed. True communism, on the other hand, will enable everyone to deploy his abilities to the maximum: it will do away with the obstacles created by the power of things over human beings, the contingency of personal life, and the alienation of labour which reduces individuals to a dead level of mediocrity. At the same time, it was Marx's view that under communism men's in-dividual possibilities would display themselves only in socially constructive ways, so that conflicts among individuals would lose all *raison d'être*.

5. *Stirner and the philosophy of egocentrism*

Questions of personality and personal freedom are treated in *The German Ideology* in the form of a polemic with Max Stirner (1806–56; real name Johann Kaspar Schmidt). Stirner was one of the Berlin Young Hegelians, but his work *Der Einzige und sein Eigentum* (*The Ego and His Own*, 1844) belongs to the period of the dissolution of Left Hegelian views and reinterprets the cult of humanity in terms of extreme egocentrism. Prior to this, in 1841–2, Stirner wrote articles, reviews, and letters to various journals, especially the *Rheinische Zeitung* and the *Leipziger Allgemeine Zeitung*. He failed to obtain a post in the state educational system, and for some time taught in a private boarding-school for girls. Later he made a rich marriage and embarked on commercial speculation, which led to bankruptcy and imprisonment for debt. By what may seem a malicious irony of fate, the apostle of the absolute sovereignty of the Ego died of a gnat-bite. Subsequently to his main work he wrote some short articles and polemics and a compilation entitled *The History of Reaction* (1852). *The Ego* was celebrated for a short time in Germany and then forgotten till the 1890s, when it was the subject of extensive commentary and became a classic of anarchist literature. Some branches at least of the anarchist movement adopted Stirner as their chief ideologist, and today he is often thought of as an existentialist *avant la lettre*: his basic principle that personal self-consciousness cannot be reduced to anything other than itself may indeed be regarded as the keynote of existentialism in its earliest version. This is a matter of coincidence rather than historical continuity; there is, however, a link between Stirner and modern existentialism through Nietzsche, who had read Stirner's work though he nowhere expressly refers to it.

Stirner's book is a proclamation of absolute egoism, a philosophical affirmation of the Ego considered not as a distinct individual, body or soul, but as pure self-consciousness, an ego in which existence and the awareness of existence are identical. '*Der Einzige*'—'the unique one'—is deliberately opposed to '*der Einzelne*', the 'individual' of liberal philosophy. Stirner's apologia for the uniqueness of personality is an extreme reaction to Hegel's reduction of individuals to the role of instruments of the universal Idea; but it also stands in opposition to Feuerbach's cult of humanity as a species, to Christianity which subordinates mankind to values imposed by God, to liberalism with its democratic faith in the common nature of man, to socialism, and to some extent even to Marx, whom Stirner once quotes as the author of *Introduction to a Critique of Hegel's Philosophy of Right.*

Stirner maintains that the whole effort of philosophy has been, in one way or another, to subject the authentic human individual to some form of impersonal general Being. Hegel deprived human individuals of reality by treating them as manifestations of universal spirit. Feuerbach liberated man from religious alienation only to replace the tyranny of God by that of the species, man in his universal aspect. As Feuerbach opposed species-man to God, so Stirner sets up against Man the irreducible Ego, uniquely and solely present to itself in each particular case. All religions, philosophies, and political doctrines require me to fix my attention on outside things—God, man, society, the state, humanity, truth—and never simply on myself; yet my self is all that matters to me, and it requires no justification, precisely because it is mine. Hence Stirner adopted as his motto the line from Goethe 'Ich hab' mein Sach auf Nichts gestellt' ('I have put my trust in Nothing'). The Ego is not describable in words that are used to describe other things; it is absolutely irreducible, the self-sufficient plenitude of subjectivity, a perfect self-contained universe. In affirming my Ego I am simply myself; it is for me the only reality and the only value. My Ego is sovereign, it recognizes no authority or constraint such as humanity, truth, the state, or any other impersonal abstraction. All general values are foreign to myself and do not concern me. From this point of view the differences between moral or philosophical doctrines are insignificant. Christianity condemned self-love, egoism, and self-indulgence; so does liberalism, although on a different principle, and the result is the same. The idea of equality is no less destructive of the sovereign Ego than is the despotism of God. By reducing individuals to the level at which they share equally in the impersonal nature of humanity, I am circumscribing human personality and destroying it by turning it into a mere instance of a species. Socialism does the same when it seeks to reduce the unique Ego to the anonymity of social Being, subordinating its own values to those of the community. From the fundamental standpoint of the emancipation of the Ego, it is much the same whether I am enslaved to impersonal Hegelian Reason or to Humanity, to a divine being or to the mass of my fellow creatures. All these purport to reduce subjective human existence to some kind of universal essence, and to resolve the conflict between the thinking subject and society by destroying the former. The true way to put an end to human alienation is to abolish whatever subjects the Ego to universal, impersonal values. Stirner's philosophy is thus an affirmation of total egoism and egocentrism, in which the whole universe is only taken into account as a means to the realization of the individual's private values.

Is any community life possible on this basis? Yes, says Stirner, but the relations between individuals must be personal, i.e. not mediated by society or by institutions, and free from reified forms. The proper business of education, accordingly, is not to train people to render services to society. The kind of education which seeks to make 'good citizens', as in liberal doctrine, is an enslavement of the Ego, a triumph of generality over true existence. From this point of view liberalism is a continuation of Christianity, and communism of liberalism. The human individual is alienated, according to Stirner, whenever he is subjected to anything outside himself, including 'goodness' or 'truth' considered as values binding on everyone. There is no general good and no moral law that can be imposed on me as a duty; even the rules of logic are a tyranny over my unique existence. Language itself is a threat, being a reification of life. It is hard to see, indeed, how Stirner's programme of total egoism can be realized in practice. The whole of civilization, in his view, is a system of manifold pressure on the Ego, and a man's self-affirmation involves rejecting the *mores* and the scientific and cultural achievements of the community, which are all instruments of servitude as far as he is concerned. Apparently, therefore, a return from alienation to authenticity would mean the denial of civilization and a return to animality and the un-bridled sway of individual passions. Since specifically human behaviour is the outcome of a collective civilization, the whole-sale rejection of the norms of that civilization must mean a regression to a pre-human state. Stirner does not spell out this conclusion, but merely speaks of the Ego's need to rebel against enslavement. This it does, not by endeavouring to alter external conditions in any way, but by the emancipation of its personal self-consciousness independently of the outside world. My act of rebellion is a self-affirmation in which I oppose my Ego to every form of generality; it neither expects nor requires any external success. (Raskolnikov, in Dostoevsky's *Crime and Punishment*, may be taken as an embodiment of the Ego as conceived by Stirner.) The theory thus implies that in the last resort the source of each man's servitude lies within him: he is fettered by his own false imagination and deference to universals, and can accordingly liberate himself by a purely spiritual act.

In Stirner's system the Ego is always unique. This means not only that it possesses qualities peculiar to itself and to be found nowhere else, but that it is actually inexpressible in words. Its specific, irreducible subjectivity cannot be defined or conceptually understood, since language consists of signs denoting what is common to two or more objects. Subjectivity is beyond the reach of human utterance. The life of the Ego consists of recognizing oneself and one's thoughts simply as one's own and not as impersonal general truths. Man becomes exclusively himself, self-rooted and self-justified—not an individual in a community, but an Ego living its own life. The Ego's values are in complete opposition to such 'universal' notions as law or the public good. My freedom is an enemy to the general freedom; my Ego apprehends itself as a negation of the rest of the universe. The Ego's desires or whims are its own law; it is not bound by any state ordinances or 'rights of man'. It seeks no justification from society and acknowledges no obligation towards it; it has a right to everything it can lay hands on. If a criminal can get away with his crime, he is in the right as far as he is concerned; if he is punished, he has no call to blame anyone; what happens is proper in either case. 'Crime' is a politico–legal notion expressing the viewpoint of the generality, but the real crime is to violate the Ego. For the egoist in Stirner's sense, community life is worth while in so far as it increases his own strength. A community of egoists is conceivable, but it is not a stable polity founded on institutions, merely a constant process of uniting and disuniting. The Ego refuses to be measured by the yardstick of humanity: it asserts its own uniqueness and recognizes nothing outside itself, not even thought; my own thoughts are myself and acknowledge no master, no standard to which they must conform. In a community, or assembly, of egoists there are no ties between any man and his fellows, and hence no conflicts, since a conflict is itself a kind of tie.

Stirner's work represents a final breach between Young Hegelianism and the doctrine of Hegel himself; the criticism of Hegel is pushed to absurd lengths by the condemnation of human society and culture in the name of the monadic sovereignty of the subject. In his violent attack on Hegel Stirner invokes a theme that we also find in Marx, the protest against reducing individual human beings to instruments of the Absolute; but they apply this protest in quite different ways. Marx too denies that there is such a thing as 'humanity' over and above individuals, but he regards individuality as the product of civilization. To Stirner, on the other hand, individuality is the same thing as the experience of subjectivity; to exist is no more or less than to be aware that one exists. To this extent he is rightly to be regarded as a forerunner of existentialism. At the same time his philosophy is an attack on the value of all ties among human beings and the whole historical process of collective development. As recent studies by Helms have shown, Stirner's doctrine inspired not only anarchists but various German groups who were the immediate precursors of fascism. At first sight, Nazi totalitarianism may seem the opposite of Stirner's radical individualism. But fascism was above all an attempt to dissolve the social ties created by history and replace them by artificial bonds among individuals who were expected to render implicit obedience to the state on grounds of absolute egoism. Fascist education combined the tenets of asocial egoism and unquestioning conformism, the latter being the means by which the individual secured his own niche in the system. Stirner's philosophy has nothing to say against conformism, it only objects to the Ego being subordinated to any higher principle: the egoist is free to adjust to the world if it appears that he will better himself by doing so. His 'rebellion' may take the form of utter servility if it will further his interest; what he must not do is to be bound by 'general' values or myths of humanity. The totalitarian ideal of a barrack-like society from which all real, historical ties have been eliminated is perfectly consistent with Stirner's principles: the egoist, by his very nature, must be prepared to fight under any flag that suits his convenience.

6. *Critique of Stirner. The individual and the community*

In *The German Ideology* Marx and Engels criticize Stirner unmercifully, contrasting the sterility and hopelessness of the egoist's inward 'rebellion' with the act of revolution in which the individual participates with the community and liberates himself by so doing. This argument is in some respects an anticipation of the quarrel in our own day between Marxists and existentialists. Apart from its bitter sarcasm, the polemic of Marx and Engels contains some passages of key importance to the understanding of Marxism. Marx does not attack Stirner from the Hegelian point of view, or combat his doctrine of the sovereign Ego by subordinating the individual to any form of universal reason, society, or the state. Instead, he advances the outline of a theory in which true individuality (and not merely a fictitious, self-contained, and self-sufficient subject) is enabled to find a place in the community without sacrificing the uniqueness of its own essence.

Marx denounces as unreal the notion of a human being whose whole life is only a kaleidoscope of self-consciousness and who can be indifferent or insensitive to the physical and social changes which in fact condition mental ones. Stirner's 'Ego' is beyond understanding, and his acts are barren by definition. In Marx's opinion, Stirner expresses no more than the impotent, sentimental discontent of the *Philister* who rebels against the sanctities of his time, but keeps his thoughts to himself and does not attempt to turn them into reality. Stirner imagines that he can destroy the state by an intellectual act, when he is really only displaying his inability to criticize it in a material fashion. The difference between revolution and a revolt *à la* Stirner is not that one is a political act and the other an egoistic one, but that the latter is a mere state of mind and not an act of any sort. Stirner imagines that he can divest himself of human ties and that the state will collapse of its own accord when its members secede from it; he sets out to overcome the world by an attack in the realm of ideas. He seeks to liberate himself from all communal institutions as the embodiment of a 'general will', whereas the 'general will' is in fact the expression of the social compulsion which requires the governing class to invest its rule with an ideological aura of universality, although its position does not depend on its own preference in any way. Stirner's programme of liberation through egoism comes simply to this, that the egoist would like to do away with the world in so far as it hinders him, but has no objection to using it to further his career.

It is a pious illusion, Marx argues, to expect individuals to live together without the aid of the community and its institutions. It is not in the power of the individual to decide whether his relations with others are to be personal or institutional; the division of labour means that personal relations are bound to transform themselves into class relations, and the superiority of one individual over another is expressed in the social relationship of privilege. Whatever individuals may intend, the nature and level of needs and productive forces determine the social character of their mutual relations.

Individuals have always and in all circumstances stood on their own feet, but they were not 'unique' [*einzig*] in the sense of not needing one another: their needs (sex, trade, the division of labour) are such as to make them mutually dependent, and so they have been obliged to enter into relationships. This they did not as pure egos but as individuals at a particular stage of development of their productive forces and needs, which were in turn determined by their mutual intercourse. In this way their personal, individual behaviour towards one another has created their existing relationships and renews them day by day ... The history of an individual cannot be detached from that of his predecessors or contemporaries, but is determined by them.

For Marx, then, the intentions of individuals are of little account in determining the effect and social significance of their behaviour in a situation in which it is not individuals that regulate social ties, but the ties they have created become an independent, alien force regulating the lives of individuals. In the present age individuality is overwhelmed by material forms or by 'contingency'; this constraint has reached an extreme form and has thereby imposed on humanity the necessity of bringing about a revolution which will destroy the element of con-

tingency and give individuals the power once again to control
their mutual relations. That is what communism means:
restoring the control of individuals over the material, reified
forms in which their mutual ties are expressed. In the last
analysis, the task facing humanity consists of abolishing the division of labour; and this presupposes the attainment of a stage
of technological development at which the system of private
property and division of labour presents itself as a hindrance,
so that technology itself requires their abolition. 'Private
property can only be abolished on condition of an all-round
development of individuals, since the existing forms of exchange
and productive forces are universal and only individuals developing in a universal manner can assimilate them, that is to say
transform them into free vital activity.' In a communist society
the universal development of individuals is no empty phrase,
but it does not mean that the individual is to seek self-
affirmation independently of others (which is in any case
impossible), in monadic isolation and in the assertion of his
rights against the community. On the contrary, 'This development is conditioned by the existing link between them—a link
constituted partly by economic premises, partly by the necessary
solidarity of the free development of all, and finally by the
universal nature of the activity of individuals on the basis of
the productive forces existing at a given time.'

For this reason the idea of individual liberation based on
Stirner's category of the unique Ego is an idle fantasy. If
'uniqueness' is merely the consciousness of uniqueness, this can
of course be realized in any conditions, as an act of pure
thought, without any change of external reality. If it merely
signifies the obvious fact that everyone is different from everyone
else in some respect or other, then it cannot be a programme,
since, for what it is worth, it is already the case. As Leibniz
observed, there are no two identical things; even the passports
of no two men are alike, and in this way even officialdom or
the police ensure the identity and uniqueness of every human
being. But we are not concerned with such commonplace matters.
For the notion of 'uniqueness' to be any use it should denote
originality, a particular skill or ability; but these can only
display themselves as social values, within the community.
'Uniqueness in the sense of originality implies that the indivi-

dual's activity in a particular sphere is unlike that of other individuals of the same kind. La Persiani is an "incomparable" singer precisely because, as a singer, we compare her with others.'

In the light of this analysis we can easily perceive the error of those totalitarian interpretations of Marx, less frequent now than formerly, which represent his ideal of communism as a society in which the individual is identified with the species by the extinction of all creative initiative and all qualities that might distinguish him from his fellows. On the other hand, Marx does not believe that individuals can determine or assert their true personality by a mere act of self-knowledge. Self-affirmation of this kind can take place in any conditions, it calls for no change in the world of social ties, and therefore it cannot eradicate human servitude or the process by which human beings eternally forge and re-forge the bonds of their own alienation. In Marx's view, the affirmation of one's own individuality involves the restoration of man's 'social character' or 'species-nature' as distinct from, and opposed to, the state of 'contingency', i.e. enslavement to alienated forces. Under communism, the disappearance of the antagonism between personal aspirations and the species is not a matter of identification, whether forced or voluntary, between the two, and thus of generalized mediocrity and uniformity. What it means is that conditions will be such that individuals can develop their aptitudes fully, not in conflict with one another but in a socially valuable way, instead of superiority turning itself, as now, into privilege or the subjugation of others. 'Depersonalization', if we may introduce this modern term, derives from the subjection of individuals to the work of their own hands and brains; it cannot be cured by a mere reform of ideas, but by reasserting control over inanimate forces which have gained the upper hand over their creators.

However, to say that Marx did not intend the totalitarian version of his theory is not to say that that version is a mistake and nothing more. We shall have to consider in due course whether Marx's vision of social unity did not contain elements contrary to his own intention, and whether he is not to some extent responsible for the totalitarian form of Marxism. Can that unity in fact be imagined in any other way than that of a totalitarian state, however little Marx himself supposed this to be the case?

7. *Alienation and the division of labour*

In *The German Ideology* and subsequent writings Marx uses the
term 'alienation' less frequently, and some critics infer from
this that he no longer thought of society in the same categories as before. This, however, appears to be a mistake.
According to the Paris Manuscripts the process that engenders
all other forms of servitude is that of alienated labour, to which
private property is secondary; Marx does not inquire, however,
what gives rise to alienated labour itself. In *The German Ideology*
the root of all evil is the division of labour, private property
being once again a secondary phenomenon. It should not be
supposed, however, that the 'division of labour' is only a more
precise formulation of the rather vague term 'alienation'. Marx's
view is that the division of labour consequent on the improvement of tools is the first source of the alienating process and,
through it, of private property. This happens because the division
of labour leads necessarily to commerce, i.e. the transformation
of objects produced by man into vehicles of abstract exchange-
value. When things become commodities, the basic premiss of
alienation already exists. Inequality, private property, alienated
political institutions for the protection of privilege—all these
are a continuation of the same process. The phenomenon of
'alienated labour' continues to operate and to be created in
production. A particular form of alienation ensues when physical
and mental work are separated from each other. This leads to
the self-delusion of ideologists who believe that their thoughts
are not dictated by social needs but derive their power from
immanent sources; the very existence of ideologists as a group
increases support for the notion that ideas have an inherent
validity of their own.

A note appended to Part I of *The German Ideology* provides
evidence that Marx did not abandon the category of alienation
and did regard the division of labour as its primary source. It
runs:

Individuals have always regarded themselves as the point of departure;
their relations are part of the real process of their lives. How can it
be, then, that their relationships become independent of them, that
the forces of their own lives gain control over them? The answer,

in a word, is—the division of labour, the degree of which depends on the extent to which productive forces have developed.

Although the word 'alienation' occurs less often, the theory is present in Marx's social philosophy until the end of his life; 'commodity fetishism' in *Capital* is nothing but a particularization of it. When Marx writes that commodities produced for the market take on an independent form, that social relations in the commercial process appear to the participants as relations among things over which they have no control (exchange value being falsely represented as inherent in the object and not as an embodiment of labour), and that the supreme type of this fetishism is money as a standard of value and means of exchange—in all this Marx is reproducing the theory of self-alienation that he had formulated in 1844. That social relationships and the whole of history are the work of human beings, which escapes from their control and takes on a more and more autonomous aspect—this, to the very end, was a fundamental determinant of Marx's ideas on the degradation of mankind under capitalism and the social function of the proletarian revolution.

8. The liberation of man and the class struggle

There is another point which some critics have taken as signifying a change of attitude in *The German Ideology*, viz. that whereas in the Manuscripts and earlier writings Marx spoke of the emancipation of mankind in general, this idea is now replaced by that of the class struggle between the proletariat and the bourgeoisie. But here too there is no real alteration. Marx continued throughout his life to regard communism as the liberation of the whole of mankind; the proletariat was to be the conscious instrument of that liberation, as being the class which had suffered the extreme degree of dehumanization. It is generally recognized as essentially Marx's view that communism meant the abolition of the class system, not merely the substitution of one ruling class for another; and this view is in complete accordance with his early idea of liberation. Dehumanization cannot affect one class alone; it applies to all classes, though in different degrees, and although the possessing class turns it into a source of pride. It is true, indeed, that the aspect of universal liberation is less prominent in Marx's later works than that of the revolution inspired by the class interest of the proletariat. This is already the case in *The German Ideology*, and is easily explained by the polemical context and particularly the critique of 'true socialism'. According to this doctrine the socialist Utopia, which involved the general liberation of mankind, could and ought to be attained by a universal moral appeal to all social classes without distinction. In other words, 'true socialism' meant socialism without the class struggle and without a revolution inspired by class interests. Marx was convinced, however, that the particular interest of the proletariat and its struggle against the possessing classes was the motive force of the socialist revolution, and that while the revolution would bring about the final disappearance of classes and social antagonism, there must be a transitional period during which the proletariat would continue to oppose its exploiters. As Marx became more closely acquainted with political realities he took more interest in organizing the revolution than in portraying the ideal society, let alone planning the details of communism in action after the manner of Fourier and others; he was more interested, therefore, in the class struggle than in social eschatology. Nevertheless, the whole theory of the class struggle made no sense without that eschatology, and Marx adhered throughout his life to the basic premisses of communism as he had formulated them in 1844. He believed that in the class struggle it was no good appealing to general human interests, but only to those of the oppressed. Later, and especially in the *Critique of the Gotha Programme*, he expressly distinguished the first, negative, post-revolutionary phase from the universal community of the future. But the prospect of that community was continuously in his mind, as we may see, for example, from the third volume of *Capital*, and it is not inconsistent either with the class struggle or with the belief that the proletariat, by defending its own class-interest, will be the liberator of the whole human race.

9. *The epistemological meaning of the theory of false consciousness*

'False consciousness' is not regarded by Marx as 'error' in the cognitive sense, just as the emancipation of consciousness is not a matter of rediscovering 'truth' in the ordinary sense. In *The*

German Ideology as in the Paris Manuscripts, Marx refuses to concern himself with epistemological questions. For him there is no problem of the world being 'reflected' in the mind, except in the sense of his repeated statement that consciousness signifies people's awareness of the nature of their lives. Questions of the correspondence between thought and reality-in-itself are meaningless, as is the opposition of subject and object considered as two independent entities, one absorbing images produced by the other. As Marx says in the *Theses on Feuerbach*, the question of the reality of the world as distinct from practical human interests is a 'purely scholastic' one and is the result of ideological mystification. ('The whole problem of the transition from thought to reality, and thus from language to life, exists only as a philosophical illusion: it is justified only to the philosophic mind, puzzling over the origin and nature of its supposed detachment from real life.') Since extra-human nature is nothing to man, who knows nature only as the objectification of his own activity (which does not mean, of course, that he has physically created it), and since cognition signifies imparting a human sense to things, the difference between false and liberated consciousness is not that between error and truth but is a functional difference related to the purpose served by thought in the collective life of mankind. 'Wrong' thinking is that which confirms the state of human servitude and is unaware of its own proper function; emancipated thought is the affirmation of humanity, enabling man to develop his native abilities. Consciousness is the mental aspect of human life, a social process (for consciousness is realized only in speech) whereby men communicate with one another and assimilate nature in a humanized form. It can either intensify the slavery of man, imprisoned and dominated by material objects, or help towards his liberation. Consciousness determines things, but does not make them objective. As Marx puts it in his critique of Hegel, 'From the point of view of self-knowledge, what is unacceptable in alienation is not that the object is definite, but that it is objective.' Or as he says in one of his early articles: 'The character of things is a product of reason. Every object, in order to be something, must distinguish itself and remain distinguished. By imposing a definite form on every object of discourse and, as it were, giving permanent shape to flowing reality, reason creates the manifoldness of the world, which would not be universal without many one-sidednesses.'

For Marx, then, there is no question of knowledge having an epistemological value distinct from its value as an organ of human self-affirmation. The restoration of a sound consciousness is one aspect, and not merely a result, of the de-alienation of labour. Marx's epistemology is part of his social utopia. Communism does away with false consciousness, not by substituting a correct image of the world for an incorrect one, but by dispelling the illusion that thought is or can be anything other than the expression of a state of life. It is not a matter of providing new answers to questions of metaphysics and epistemology, but of denying their validity—whether the question be that of God's creation of the world, or that of 'being-in-itself' and the relation to it of subjective data. When we understand the genesis and function of human thought, purely epistemological questions fall to the ground. Thought is always an articulation of its own time in history, but whether it is 'good' or 'bad' does not depend merely on whether it is helpful to the governing class (those who govern materially and therefore intellectually) at that time—for if this were so, we should have to regard bourgeois thought as 'good' at the present day. Thought can and must be judged from an absolute standpoint—not, however, as related to a reality separate from man, but as related to the emancipated consciousness, affirming in an absolute manner the 'species-essence' of man. Consciousness may thus be false even when it correctly expresses the historical situation in which it arises; and we can only speak of false consciousness, or ideology, with reference to the absolute state of emancipation. Having in mind Marx's conception of reason as a practical organ of collective existence, and of the object as something defined though not objectivized by reason, we may describe his epistemology as one of generic subjectivism.

CHAPTER IX

Recapitulation

We may attempt at this stage to recapitulate Marx's thought in the form it had assumed by 1846. From 1843 onwards he developed his ideas with extreme consistency, and all his later work may be regarded as a continuation and elaboration of the body of thought which was already constituted by the time of *The German Ideology*.

1. Marx's point of departure is the eschatological question derived from Hegel: how is man to be reconciled with himself and with the world? According to Hegel this comes about when Mind, having passed through the travail of history, finally comes to understand the world as an exteriorization of itself; it assimilates and ratifies the world as its own truth, divests it of its objective character, and actualizes everything in it that was originally only potential. Marx, following Feuerbach, places in the centre of his picture the 'earthly reality' of Man, as opposed to the Hegelian Spirit developing through empirical individuals or using them as its instrument. 'For man, the root is man himself'—the basic reality, self-derived and self-justified.

2. Marx, like Hegel, looks forward to man's final reconciliation with the world, himself, and others. Again following Feuerbach against Hegel, he does not see this in terms of the recognition of being as a product of self-knowledge, but in the recognition of sources of alienation in man's terrestrial lot and in the overcoming of this state of affairs. Rejecting the Young Hegelian 'critical principle', he refuses to accept the eternal conflict between negative self-knowledge and the resistance of an unresponsive world, but envisages a de-alienated state in which man will affirm himself in a world of his own creation. On the other hand, he disagrees with Feuerbach's view that alienation results from the mythopoeic consciousness which makes God the concentration of human values; instead, he regards this conscious-

ness as itself the product of the alienation of labour.

3. Alienated labour is a consequence of the division of labour, which in its turn is due to technological progress, and is therefore an inevitable feature of history. Marx agrees with Hegel against Feuerbach in seeing alienation not merely as something destructive and inhuman but as a condition of the future all-round development of mankind. But he dissents from Hegel in regarding history up to the present time not as the progressive conquest of freedom but as a process of degradation that has reached its nadir in the maturity of capitalist society. However, it is necessary for man's future liberation that he should undergo the extremes of affliction and dehumanization, since we are not concerned with regaining a lost paradise, but with the reconquest of humanity.

4. Alienation means the subjugation of man by his own works, which have assumed the guise of independent things. The commodity character of products and their expression in money form (cf. Hess) has the effect that the social process of exchange is regulated by factors operating independently of human will, after the fashion of natural laws. Alienation gives rise to private property and to political institutions. The state creates a fictitious community to replace the lack of real community in civil society, where human relations inevitably take the form of a conflict of egoisms. The enslavement of the collectivity to its own products entails the mutual isolation of individuals.

5. Alienation is thus not to be cured by thinking about it, but by removing its causes. Man is a practical being, and his thoughts are the conscious aspect of his practical life, although this fact is obscured by false consciousness. Thought is governed by practical needs, and the image of the world in a human mind is regulated not by the intrinsic quality of objects but by the practical task in hand. Once we realize this we perceive the nullity of questions which have only arisen because philosophers did not understand the conditions that gave rise to them, namely the separation of intellectual from practical activity. We deny the validity of metaphysical and epistemological problems engendered by the false hope of attaining to some absolute reality beyond the practical horizon of human beings.

6. The transcendence of alienation is another name for communism—a total transformation of human existence, the

recovery by man of his species-essence. Communism puts an end to the division of life into public and private spheres, and to the difference between civil society and the state; it does away with the need for political institutions, political authority and governments, private property and its source in the division of labour. It destroys the class system and exploitation; it heals the split in man's nature and the crippled, one-sided development of the individual. Contrary to Hegel's view, the distinction between the state and civil society is not eternal. Contrary to the views of the liberal Enlightenment, social harmony is to be sought not by a legislative reform that will reconcile the egoism of each individual with the collective interest, but by removing the causes of antagonism. The individual will absorb society into himself: thanks to de-alienation, he will recognize humanity as his own internalized nature. Voluntary solidarity, not compulsion or the legal regulation of interests, will ensure the smooth harmony of human relations. The species (cf. Fichte) can then realize itself in the individual. Communism destroys the power of objectified relations over human beings, gives him control again over his own works, restores the social operation of his mind and senses, and bridges the gulf between humanity and nature. It is the fulfilment of the human calling, the reconciliation of essence and existence in human life. It also stands for the consciousness of the practical, humane and social character that belongs to all intellectual activity, and repudiates the false independence of existing forms of social thought: philosophy, law, religion. Communism turns philosophy into reality, and by so doing abolishes it.

7. Communism does not deprive man of individuality or reduce personal aspirations and abilities to a dead level of mediocrity. On the contrary, the powers of the individual can only flourish when he regards them as social forces, valuable and effective within a human community and not in isolation. Communism alone makes possible the proper use of human abilities: thanks to the variety of technical progress it ensures that specifically human activity is freed from the constraint of physical need and the pressure of hunger and is thus truly creative. It is the realization of freedom, not only from exploitation and political power but from immediate bodily needs. It is the solution to the problem of history and is also the end of

history as we have known it, in which individual and collective life are subject to contingency. Henceforth man can determine his own development in freedom, instead of being enslaved by material forces which he has created but can no longer control. Man, under communism, is not a prey to chance but is the captain of his fate, the conscious moulder of his own destiny.

8. Contrary to what the utopian socialists claim, communism is not an ideal in opposition to the real world, a theory which might have been invented and put into practice at any time in history. It is itself a trend in contemporary history, which is evolving the premisses of communism and moving unconsciously towards it. This is because the present age stands for the maximum of dehumanization: on the one hand it degrades the worker by turning him into a commodity, on the other it reduces the capitalist to the status of an entry in a ledger. The proletariat, being the epitome of dehumanization and the pure negation of civil society, is destined to bring about an upheaval that will put an end to all social classes, including itself. The interest of the proletariat, and that of no other class, coincides with the needs of humanity as a whole. The proletariat, therefore, is not a mere agglomeration of suffering, degradation, and misery, but also the historical instrument by which man is to recover his heritage. The alienation of labour has operated through the ages to create the working class, the agent of its destruction.

9. But the proletariat is more than the instrument of an impersonal historical process: it fulfils its destiny by being conscious of that destiny and of its own exceptional situation. The consciousness of the proletariat is not mere passive awareness of the part assigned to it by history, but a free consciousness and a fount of revolutionary initiative. Here the opposition of freedom and necessity disappears, for what is in fact the inevitability of history takes the form of a free initiative in the proletariat consciousness. By understanding its own position the proletariat not only understands the world but *ipso facto* sets about changing it. This consciousness is not a mere Hegelian acknowledgement and assimilation of past history; it is turned towards the future, in an active impulse of transformation. At the same time it is not, as Fichte and the Young Hegelians would have it, a mere negation of the existing order, but an urge to

create a movement that is already potentially there—an innate trend of history, but one that can only be set in motion by the free initiative of human beings. In this way the situation of the proletariat combines historical necessity and freedom.

10. While communism is the final transformation of all spheres of life and human consciousness, the motive force of the revolution that brings it about must be the class-interest of the exploited and destitute proletariat. The revolution has a negative task to perform, and this devolves on the proletariat as long as it is necessary to carry on the struggle with the possessing classes. Communism is not established merely by abolishing private property; it requires a long period of social convulsion, which is bound to result in the consummation demanded by history and by the improvement of instruments of production. Communism has as its precondition advanced technical development and a world market, and will itself result in more intensive technical development; this, however, will not turn against its creators as in the past, but will help them to full self-realization as human beings.

These are the fundamental principles of Marx's theory, from which he never departed. The whole of his work, down to the last page of *Capital*, was a confirmation and elaboration of these ideas. Engels, from a more empirical point of view, gave expression to the same vision of a classless communist society, to be brought about by the initiative of the working class activating the natural trend of history. On the other hand, Engels adopted a different standpoint as regards the cognitive and ontological link between man and nature. In his later works the idea of the 'philosophy of praxis' as we have discussed it gives place to a theory which subjects humanity to the general laws of nature and makes human history a particularization of those laws, thus departing from the conception of man as 'the root' (in Marx's phrase) and of the 'humanization' of nature. In so doing Engels created a new version of Marxist philosophy, differing as much from its original as did post-Darwinian European culture from the age that preceded it.

CHAPTER X

Socialist Ideas in the First Half of the Nineteenth Century as Compared with Marxian Socialism

1. The rise of the socialist idea

FROM 1847 onwards Marx occasionally reverted to philosophic speculations of the kind that dominate his early writings. The instances of this are important, as they confirm the essential continuity of his thought and enable us to relate his political and economic ideas to the trends of his earliest thinking. However, his mature writings are directly focused on an increasingly precise analysis, of which *Capital* provides the most finished version, of the functioning of the capitalist economy, together with polemics against various socialist doctrines and programmes which, in his opinion, misinterpreted the historical and economic facts and impeded the development of the workers' revolutionary movement. Having joined issue with German 'true socialism' he proceeded to challenge Proudhon, utopian socialism, Bakunin, and Lassalle. All these quarrels and controversies were of great importance to the history of the workers' movement, but not all of them involved new departures in the realm of theory.

At the time when Marx came into the field as a theoretician of the proletarian revolution, socialist ideas already had a long life behind them. If we sought to provide a general definition of socialism in historical as opposed to normative terms, i.e. to identify the common features of the ideas that went under that name in the first half of the nineteenth century, we should find the result extremely jejune and imprecise. The mainspring of the socialist ideas that arose under the combined influence of the Industrial and the French Revolution was the conviction that the uncontrolled concentration of wealth and unbridled competition were bound to lead to increasing misery and crises, and that the system must be replaced by one in which the organization of production and exchange would do away with poverty and oppression and bring about a redistribution of the world's goods on a basis of equality. This might imply the complete equalization of wealth, or the principle of 'to each according to his labour', or, eventually, 'to each according to his needs'. Beyond the general conception of equality, socialist programmes and ideas differed in every respect. Not all of them even proposed to abolish private ownership of the means of production. Some advocates of socialism regarded it as essentially the cause of the working class, while others saw it as a universal human ideal which all classes should help to bring about. Some proclaimed the necessity of a political revolution, others relied on the force of propaganda or example. Some thought that all forms of state organization would soon be done away with, others that they were indispensible. Some regarded freedom as the supreme good, others were prepared to limit it drastically in the name of equality or efficient production. Some appealed to the international interest of the oppressed classes, while others did not look beyond the national horizon. Some, finally, were content to imagine a perfect society, while others studied the course of historical evolution in order to identify the natural laws which would ensure the advent of socialism.

The invention of the term 'socialism' was claimed by Pierre Leroux, a follower of Saint-Simon, who used it in the journal *Le Globe* in 1832; it was also used in Britain in the 1830s by the disciples of Robert Owen. As the name and the concept became widely known, theorists and adherents of the new doctrine naturally turned their attention to its antecedents in Plato's Republic, the communist ideas of medieval sectarians and the Renaissance utopianists, especially Thomas More and Campanella. In these writers and their imitators in the seventeenth and eighteenth centuries it was possible to discern a continuity of ideas despite their very different philosophies. Plato's hierarchic society was a long way from the egalitarian tenets of most modern socialists, and the ascetic ideals of medieval doctrinaires were specifically religious in character. But More's Utopia owed its origin to reflection on the first symptoms of capitalist accumulation, and the advocates of socialism found much to sympathize with in its ideals: the abolition of private property, the universal obligation to work, the equalization of rights and wealth, the organization of production by the state, and the eradication of poverty and exploitation. From the sixteenth to the eighteenth century socialist ideas were generally inspired not merely by reflection on the sufferings of the downtrodden classes, but by a philosophical or religious belief that antagonisms and conflicts of interst, inequality and oppression were contrary to God's plan or Nature's, which intended men to live in a state of peace and harmony. Some exponents of these ideas went so far as to maintain that the perfect society required all its members to be completely uniform in all respects—not only their rights and duties, but their way of life and thought, food and clothing, and even (according to Dom Deschamps and others) their physical appearance. In some cases the ideal of static perfection excluded any notion of creativity or progress. Campanella was an exception: his *Civitas Solis*, unlike More's *Utopia*, left plenty of room for scientific and technical discovery.

2. Babouvism

The first active manifestation of socialism after the Revolution of 1789 was the conspiracy of Gracchus Babeuf. Filippo Buonarroti, who took part in the conspiracy, published an account of it in 1828, thanks to which its ideas became for the first time generally known. Babeuf and the Babouvists took their philosophy in the main from Rousseau and the utopianists of the Enlightenment, and regarded themselves as the successors of Robespierre. Their basic premiss was the idea of equality: as Buonarroti wrote, 'the perpetual cause of the enslavement of peoples is nothing but inequality, and as long as it exists the assertion of national rights will be illusory as far as the masses are concerned, sunk as they are beneath the level of human dignity' (*Conspiration pour l'égalité dite Babeuf*, i. 100). As all men have by nature the same right to all earthly goods, the source of inequality is private property and this must be done away with. In the future society wealth will be distributed equally to all, irrespective of the work they do; there will be no right of inheritance, no large cities; all will be compelled to do physical work and to live in the same manner. In addition to laying down the principles of the new society, the Babouvists planned the way to it by organizing, under the Directory, a conspiracy to overthrow the existing order. Since the masses were not yet liberated from the spiritual influence of the exploiters they could not at once exercise power, which would be wielded for them by the conspirators. Later, when education became universal, the populace would govern itself through elected bodies. Babeuf's conspiracy was detected in 1796, and he was tried and guillotined. His ideas were to some extent carried on by Louis Blanqui. The Babouvist programme was not expressed in specific class-categories, but merely distinguished rich and poor, or peoples and tyrants; its egalitarian rhetoric, however, was one of the first attempts at an economic criticism of private property as the foundation of society.

The Babouvist movement is also important because it reflected for the first time a conscious conflict between the revolutionary ideal of freedom and that of equality. Freedom meant not only the right of assembly and the abolition of legal differences between estates of the realm, but also the right of every man to carry on economic activity without hindrance and to defend his property; freedom, therefore, meant inequality, exploitation, and misery. Babeuf's conspiracy was in its immediate origin a reaction of the Jacobin Left to the *coup* of Thermidor, but ideologically it went far beyond the Jacobin tradition. The Babouvists took over the Jacobin conception of society in terms of political power acquired by force, and bequeathed this to the French socialist movement. (British socialism, originating as it did not from a political revolution but from the process of industrialization, was dominated from the outset by a reformist tendency.) The *Manifeste des égaux* drawn up in 1796 by Sylvain Maréchal described the French Revolution as the prelude to another, much greater and final revolution. The leaders would not allow this document to be published, as they drew the line at two of its statements. The first was 'Let all the arts perish, if need be, so that we may have true equality'; the second called for the abolition of all differences not only between rich and poor, masters and servants, but also between rulers and the ruled. The former statement reveals a tendency that was often to recur in communist movements. Equality is the supreme value, and in particular equality in the enjoyment of material goods. Taken to an extreme, this means that it matters less whether people have much or little so long as they all have the same. If there is a choice between improving the lot of the poor but allowing inequality to subsist, or leaving the poor as they are and depressing everyone to their level, it is the second alternative that must be chosen. The various communist and socialist groups did not actually envisage the matter in these terms, since they were all certain that the equalization of wealth would produce, if not abundance, at any rate a sufficiency for all. Most of them also naïvely assumed that the deprivation of the workers was due to the conspicuous consumption of the rich, and that if all the goods enjoyed by the privileged classes were distributed among the people, the result would be general prosperity. In the first stage of socialist ideas, however, moral indignation at poverty and inequality was not distinguished from economic analysis of capitalist production, but rather took the place of such analysis. As with the utopianists of the Enlightenment, Morelly or Mably, the principle of the community of goods was deduced from a normative theory according to which human beings, simply as such, have an identical right to whatever the earth provides. Whether this view was defended by quotations from the New Testament (as in many socialist writings) or by the materialist tradition of the Enlightenment, the conclusion was the same: inequality of consumption is contrary to human nature, and so is rent, interest, and any unearned income.

As to abolishing the difference between the rulers and the ruled, this, as an immediate revolutionary aim, belongs rather to the tradition of anarchism. The Babouvists rejected it, as they envisaged a period of dictatorship in the general interest for as long as might be necessary to destroy or disarm the enemies of equality.

Altogether, the Babouvist movement marks the point at which liberal democracy and communism began to part company, as it came to be seen that equality was not a completion of liberty but a limitation of it. This does not mean, however, that the dilemma was at once obvious to all. For some time, liberal democracy and socialism were present in mixed and inter-mediate forms; only 1848 drew a clear line between them.

Similarly, the terms 'communist' and 'socialist' were for a long time not clearly distinguished. By the 1830s, however, the former name was in general used by those radical reformers and utopians who demanded the abolition of private property (at first chiefly the ownership of land, then also factories) and absolute equality of consumption, and who did not rely on the goodwill of governments or possessors, but on the use of force by the exploited.

After 1830, in both France and England—the parent countries of socialism—socialist ideas and the embryonic workers' movement appeared in combination in various ways. Even before this, however, ideas of a radical reform of society on socialist though not communist (i.e. not Babouvist) lines were ventilated in both countries in the form of theoretical reflections on the development of industry. This type of socialism, in which the chief names are Saint-Simon, Fourier, and Robert Owen, had an important influence on Marx's thought, both positively and negatively. It was not itself a protest by the deprived classes, but sprang from the observation and analysis of social misery, exploitation, and unemployment.

3. Saint-Simonism

Claude Henri, comte de Saint-Simon (1760–1825), was the real founder of modern theoretical socialism, conceived not merely as an ideal but as the outcome of a historical process. He was a descendant of the famous Duke, fought in the American War of Independence and, after the Revolution, engaged in trade operations which led to bankruptcy. He had a lifelong interest in philosophical subjects and in the possibility of reforming society by reforming the method of studying it. He also formulated the idea, taken up afterwards by Auguste Comte, of reducing every branch of knowledge to a positive state, having first liberated it from its theological and metaphysical phases. In his early works, including *Lettres d'un habitant de Genève* (1803) and *Introduction aux travaux scientifiques du XIXe siècle* (1807), he called for a form of political science that would be as positive and reliable as the physical sciences. Another Newton was needed to impose unity on the body of knowledge accumulated since his day; scholars would in time lead the nations on their path towards happiness. In 1814–18, aided by the future historian

Augustin Thierry, he drew up plans of political reform on a European scale (*De la réorganisation de la société européenne*, 1814); these included parliamentary government on British lines and a supranational European assembly to ensure peace, co-operation, and unity on the medieval lines, but inspired by liberalism instead of theocracy. As time went on he took an increasing interest in broad problems of economic organization. He reached the conclusion (*'Industrie*, 1817) that the proper function of the state was to look after productivity and that it should apply methods of industrial management to all social questions. Developing this subject with the assistance of Auguste Comte, who was his secretary from 1818 to 1822, he finally abandoned economic liberalism and formulated the principle of a future 'organic' social community, which gained many adherents and was the basis of his fame.

Saint-Simon believed that the future of humanity was to be discerned in the light of past historical changes and trends. The conclusion he came to, although he did not work it out systematically, was similar to that of historical materialism, viz. that all political change has been due to the evolution of the instruments of production, and today's technology calls for corresponding political change. Poverty and crises are caused by free competition and the resulting anarchy of production and exchange. This anarchy, however, subjects those who contribute to production—manufacturers, merchants, industrial and agricultural workers—to the authority of incompetent drones and idlers. The most important dividing line, in Saint-Simon's opinion, was between producers and those who merely consumed the fruits of others' labour. The future society, to which industrial concentration was leading, would be one in which industry was managed by the producers of wealth; production would be planned and measured by social needs, and private property, while still permitted, would change its character, as its use would be subordinated to the general good and not left to the owner's whim; inheritance would be abolished, so that property would be enjoyed only by those who had earned it by their abilities and application. Competition would give way to emulation; private interest would become an instrument of self-improvement, devoted to serving the community instead of opposing it. The social hierarchy would be preserved but would no longer be hereditary; the highest positions would be held by bankers allocating investment resources and wise men supervising the general development of society. The new industrial order would put an end to the poverty and humiliation of the most afflicted class of society, the proletariat; however, Saint-Simon did not look to the oppressed workers to carry out his plans, but believed that society would be transformed for their benefit by manufacturers, bankers, scholars, and artists, once they had been convinced by the new doctrine. Political power would undergo a complete change: it would not be a matter of governing people but of administering things, i.e. ensuring that human beings made the best possible use of Nature's gifts. To bring about this change, nothing more was needed than peaceful reforms such as the acquisition of parliamentary power by industrialists; from time to time Saint-Simon also appeared to the governing class to support his plan. In his last major work (*Le Nouveau Christianisme*, 1825) he declared that political science must be based on still more fundamental principles, namely religious ones. Far from spelling the ruin of Christian civilization, the industrial society would fulfil its essential meaning and especially the precept to 'love one another'. Self-interest was not enough as a basis for social organization; sentiment and religion were necessary, and religious life was a permanent feature of human existence and could never become obsolete.

The religious strain in Saint-Simon's programme was emphasized by his immediate followers, who systematized his thought and contributed elements of their own. In the *Exposition de la doctrine de Saint-Simon* by Enfantin and Bazard, of which the first volume appeared in 1830, we see clearly the process by which his social philosophy transformed itself into a dogma and his adherents into a sect; we also find there a detailed exposition of ideas which in some cases were no more than outlined by Saint-Simon himself.

The Saint-Simonist school regards history as a continual progress in which, however, two phases alternate: the organic and the critical. 'Organic' periods are those in which certain principles of thought are generally accepted, there is a clear-cut social hierarchy and an unbroken unity of faith. 'Critical' periods are necessary transitional phases of disharmony and disunity, in which the sense of community is lost and the bonds of society are relaxed. Europe has been in this state since the Reformation but is now moving into a new organic period, which will be permanent and not succeeded by one of anarchy. It will be a kind of return to medieval theocracy, but without its contempt for the body and for temporal needs. Instead, the new Christianity will be imbued with the spirit of science and technical progress and will regard productive work as essentially valuable. Belief in God and a future life will be maintained, as will the priesthood, but the whole system of religion will be harmonized with man's concern for his earthly welfare.

This prospect, according to the Saint-Simonians, is not an arbitrary one but can be deduced from the whole of history, in which we may trace the gradual development of co-operative principles, The growth of industry and its increasing centralization call for a fundamental change in the organization of production. Idlers are sharing less and less in the fruits of labour, as we see from the falling rates of interest in industrial countries. But the seeds of future development must be encouraged to sprout. At present competition and anarchy are widening the gap between classes, since manufacturers reduce wages in order to cut prices. Thanks to the hereditary principle the means of production are controlled by incompetent persons and the irrational privilege of birth has replaced that of estates of the realm. In the new society, instead of man exploiting man the earth will be made to fructify by co-operating producers, and their output will not be consumed by idlers. This will be achieved by doing away with the right of inheritance, especially as regards the means of production, abolishing interest on capital, and organizing productivity on a state-centralized basis. The state will allocate investment credits and all the means of production to manufacturers in accordance with their abilities and with social needs. The right to use the means of production will depend on ability, and the exercise of that right under state supervision will be the sole form of property. Men will not be governed by selfish interests alone but by sentiment and enthusiasm, willingness to work for others, morality and religion. Incomes will not be equal, since the principle is 'To each according to his labour', but this inequality will not be due to exploitation and will therefore not be injurious to the community or tend to revive classes and class antagonisms. The illusory freedom which means nothing to the starving, and the equality before the law which is nullified by the privilege of wealth, will be superseded by the universal brotherhood of working man. Industrialists, artists, and men of learning will work harmoniously to improve the human race and satisfy its material, moral, and intellectual needs, while preserving the invaluable link with Divinity that alone enables man to be happy and to love and help his neighbour.

Like other moral and philosophical doctrines, Saint-Simonism showed itself capable of evolving in opposite directions. Its authoritarian elements—the emphasis on social hierarchy, and a theocratic strain—contributed, partly through Comte, to a conservative school of thought which stressed everything connecting Saint-Simon with de Maistre and other traditionalist critics of the post-Revolutionary order. But, on the other hand, Louis Blanc was a disciple of Saint-Simon and so, through him, was Lassalle. Socialist ideas, wherein the state was expected to play an important part in resolving class antagonisms, were largely an intellectual legacy of Saint-Simon. As far as Marxian socialism is concerned, the most important features of his doctrine may be listed as follows: the firm belief in the regularity of history and its inexorable march towards socialism; the ruinous consequences of anarchic competition and the necessity of state economic planning; the replacement of political government by economic administration; science as the instrument of social progress; and the internationalist approach to politico-economic problems. What is contrary to Marxism, on the other hand, is the idea that the state as it now exists can be used to bring about a socialist transformation; likewise Saint-Simon's appeal for co-operation between classes, and the religious overtones of his 'industrial order'. The formula 'From each according to his ability, to each according to his needs' was taken over by Marxist socialism from Louis Blanc, who modified Saint-Simon's doctrine on this point.

Like early Marxism, Saint-Simon's doctrine requires to be judged within the framework of the Romantic movement or rather as an attempt to overcome Romanticism from within. His critique of post-Revolutionary society reflected not only sympathy with the downtrodden masses, but also alarm at the dissolution of the bonds which had held the old society together. The Romantics, Saint-Simon, and the young Marx condemned industrial civilization not only for its social injustice but because it replaced almost every link between human beings by the negative principle of private interest. The new world was one in which everything was for sale and was worth just what it would fetch in the market, while selfish motives took the place of human solidarity and fellow-feeling. The Romantics for the most part blamed this state of affairs on technical progress, and idealized the rural or chivalrous communities of pre-industrial times. The Saint-Simonists agreed with the Romantics in disliking the new industrial order, or rather disorder, but they saw the answer not in calling back the past but in a rational organization of production. They also believed, like Marx, that technical progress would cure its own destructive effects and restore to humanity—by which they meant mainly Europe—an organic unity based on scientific development instead of, as in former times, on the stagnation of a primitive agricultural community.

The later fortunes and extravagances of the Saint-Simonists—a priestly hierarchy, sexual mysticism, the Near Eastern quest for a female Messiah—are irrelevant to the history of socialism. However, some manufacturers were attracted to the doctrine by the cult of industrial organization, technical efficiency, and the entrepreneurial spirit. In France, unlike Britain, the dawn of industrialization was associated with a semi-Romantic ideology in which engineers and businessmen figured as the knights-errant and explorers of the new world of applied science. The 'Père Enfantin' ended his career as the manager of a railway line, and another of Saint-Simon's disciples, Ferdinand de Lesseps, built the Suez Canal.

Of all pre-Marxist doctrines Saint-Simonism had the strongest effect in diffusing socialist ideas among the educated classes. Two or three generations grew up on the novels of George Sand, who was among the converted. It was chiefly due to the Saint-Simonists that a belief in socialism spread to the intellectuals of the great European countries, including German Romantics, British utilitarians, and Russian and Polish radicals.

4. Owen

Unlike most of the socialist thinkers of his day, Robert Owen (1771–1858) had been an industrialist and in direct contact with working-class life for many years before he put pen to paper. As compared with the French socialists, moreover, he lived in a country which suffered much more grievously from the ill-effects of industrialization and mechanization.

The son of a poor craftsman, Owen began to earn his living at an early age. By dint of great energy and ingenuity he set up a workshop of his own in Manchester. He later became the manager of a large cotton mill, married a manufacturer's daughter, and became manager and co-owner of a large textile factory at New Lanark in Scotland. There, from 1800 onwards, he carried on social and educational experiments designed to rescue workers and their families from poverty, degradation, and debauchery. His career as a manufacturer and philanthropist continued for many years. He reduced working hours to ten and a half, employed no children under 10 years of age, introduced free primary education and relatively hygienic working conditions, and eliminated drunkenness and theft by persuasion instead of punishment. To the general surprise he showed that on this basis he could achieve better results in production and trade than employers in whose factories adult and child workers were decimated by cruel and inhuman conditions, while disease, starvation, drunkenness, crime, and slave-driving methods degraded the labouring class to the level of animals.

Owen described his experiments and their philosophical basis in *A New View of Society, or Essays on the Principle of the Formation of the Human Character* (1813–14). In this work he sought to convince manufacturers and the aristocracy of the need for a reform of the industrial and monetary system, wages and education, in the interest not only of workers but of capitalists and the whole of society. In numerous subsequent pamphlets, periodicals, articles, memorandums and appeals to Parliament he continued to advocate his reformist ideas, exposing the horrors of industrialization and urging the adoption of social and educational measures which would remedy abuses without hindering technical progress. Above all he sought to relieve the cruelty of the system which obliged children of 6 to work fourteen or sixteen hours a day in spinning mills. With great difficulty he succeeded in obtaining the passage of the Factory Act of 1819, the first law in England to limit child working hours in the textile industry. In speeches and writings from 1817 onwards he attacked the Established Church for keeping the masses in a state of poverty and superstition; the most erroneous and harmful of its doctrines, in his opinion, was that of the individual's responsibility for his character and actions. In later years Owen turned from philanthropy to organizing trade unions and co-operatives and planning a new type of society based on voluntary mutual aid without exploitation or antagonism. Pilloried for his attacks on private property and religion, he went to America in 1824 and attempted unsuccessfully to set up communist settlements there. He returned to England in 1829 and spent the rest of his life promoting the trade union and co-operative movement, being thus the first outstanding organizer of the British proletariat in its economic struggle. He advocated a 'labour currency' to enable the price of products to be fixed at their real value, i.e. the average labour-time required for their manufacture, and organized a 'labour exchange' for the direct marketing of goods. Although the British trade unions and co-operatives subsequently changed the basis of their activity, they had in Owen not only a champion and theoretician but also their first large-scale organizer.

The aims for which Owen fought were the practical ones of eliminating poverty, unemployment, crime, and exploitation. In this he was inspired by a few simple principles, the recognition of which would, in his opinion, suffice to cure all the ills of humanity. Above all, he took over from the eighteenth-century utilitarians the view that man does not form his own character, feelings, opinions, or beliefs but is irresistibly influenced by environment, family, and education. It is a fatal error to hold, as all religions do, that a man's will has any effect on his opinions or that the individual is responsible for his character and habits; experience shows that people are conditioned by up-bringing and circumstances, and criminals, no less than judges, are the product of their environment. Man has an innate desire for happiness, he has intellectual powers and animal instincts, and everyone comes into the world with different abilities and inclinations. But knowledge and convictions are wholly the work of education, and man's prosperity and adversity depend on the knowledge he receives. The only source of the evil and unhappiness that have beset mankind through the ages is ignorance, particularly ignorance of human nature, and knowledge is the cure for all human ills. From all this it follows that a man cannot achieve happiness by acting against his neighbour, but only by means of actions directed to the happiness of all.

The idea that man can be moulded at will, and that there can be a social harmony which does not do away with private interests but reconciles them through education, is part of the stock-in-trade of the Enlightenment; but Owen derived from it practical conclusions which were intended to revolutionize the social system. The essential need, in his view, was to transform the educational milieu. Children, if properly taught, would imbibe lifelong instincts of co-operation and charity towards their fellows; but for this they must be given training at an early age and not driven to work in factories where they were physically degraded and kept in ignorance.

Children are, without exception, passive and wonderfully contrived compounds; which, by an accurate previous and subsequent attention, founded on a correct knowledge of the subject, may be formed collectively to have any human character. And although these compounds, like all the other works of nature, possess endless varieties, yet they partake of that plastic quality which, by perseverance under judicious management, may be ultimately moulded into the very image of rational wishes and desires. (*New View of Society*, Second Essay)

The reform of education must be accompanied by a reform of labour conditions. It is in the manufacturers' own interest to improve the lot of the workers, since they provide a mass demand for the goods they themselves produce. Poverty and low wages lead to crises of overproduction, in which goods remain on the market and employers are ruined. Owen at first hoped that by convincing capitalists of this he could enlist their support for his reforms. He finally decided, however, that the workers would have to rely on their own efforts for any improvement in their lot, although he never ceased to believe that reform was in the interest of the whole of society and could be brought about without revolution, by gradual change and peaceful propaganda.

In his later years Owen put his trust in communist settlements engaged in agriculture and industry, the nuclei of a future harmonious society. Here, thanks to good organization and loyal co-operation, people would produce more willingly, in greater quantity, and at a cheaper rate than elsewhere. Education would inculcate love of the community from the child's earliest years, and would eliminate religious intolerance and sectarian strife. The desire to help one's neighbour would be a sufficient incentive to work, without the stimulus of competition or public honours. Value would be measured by labour; the currency in circulation would correspond to the amount produced, and the economy would thus be immune to crises, overproduction, depression, or inflation. There would be no crime, drunkenness, or debauchery, no punishments, prisons, or executions. It was not true, as Malthus contended, that food supplies could not keep up with natural increase and that part of the population was therefore condemned to undernourishment and starvation. People could produce far more than they consumed; there was no known limit to the fertility of the soil, and production was increasing at an ever-faster rate.

Owen believed that if these simple truths were not universally accepted, it was only because people's minds were not ready for them: thanks to ignorance, mankind had for centuries been in a kind of conspiracy to work its own undoing. Now that the moment of clarity had come, the whole of life could be reformed easily and quickly. In time the reform would spread all over the world, since it applied to the whole human species. National prejudices and enmities, belief in the inequality of man and the class system—all this was the fruit of superstition, and would disappear when superstition was eradicated.

Owen's belief that human nature was unchangeable did not conflict with his theory that character can be moulded, since he maintained that the permanent factor in humanity was its liability to change and also the desire for happiness. He often uses the term 'human nature' in a normative rather than a descriptive sense, signifying man's duty to live in harmony and concord despite individual differences.

Although it originated in practical experience, Owen's

doctrine, like that of the French socialists, centred round the conviction that socialism was a unique and heaven-sent discovery, so manifestly right that it was bound to be accepted by all classes as soon as proclaimed. Since Owen is never tired of repeating that innate determinism puts men at the mercy of inherited beliefs and prejudices, it is not clear how some of them, like Owen himself, are able suddenly to break free and to show others the way to social reform. These defiers of omnipotent tradition are, it would seem, endowed by the spontaneity of genius with the power to inaugurate a new era. Owen himself did not discuss this problem; he was only interested in philosophy in so far as it related directly to his plans for society, and even then contented himself with general formulas drawn from the Enlightenment tradition. He does not discuss the function of class-consciousness, and is inclined, like most of the system-builders of socialism, to ascribe to himself the rôle of a demiurge in the historical process. This is the chief point of difference between Owenite socialism and Marxism, and is the source of such other important differences as that concerning the respective role of economic and political reforms. Marx shared the view of Owen and others that in a socialist society the power of the state over men would in the end be superseded by the administration of things, i.e. of the productive process, but he held that this could only come about after a political upheaval. Owen, on the other hand, thought that a radical economic reform in a socialist spirit could be effected by appealing to universal human interests and with the aid of the existing state power. The British trade union movement is still marked by this outlook, which directly subordinates the political struggle to economic interests. The social democratic theories which treated workers' political parties as organs of the trade unions are a continuation of the same doctrine. In a more developed form the question was to become a source of polemics at the time of the Second International.

Owen's doctrine initiated a new phase of the British workers' movement, in which it ceased to be merely an outburst of despair and became a systematic force which in the end brought about immense social changes. Moreover, his attack on capitalism and his plans for a new society contained enduring features, although some of his ideas—for example, that of a labour currency, developed by his followers John Gray and John Francis Bray—were soon discarded as they proved to be based on entirely false economic diagnoses.

Meanwhile, at the end of the 1830s a political workers' movement made its appearance in England in the form of Chartism, which remained in the public view for the next ten years. Engels wrote for its newspaper the *Northern Star*, founded in 1838 by Feargus O'Connor. The main Chartist demand was for equal and universal male suffrage; this they did not achieve, but their agitation led to the passage of further legislation against exploitation in industry.

5. *Fourier*

Charles Fourier (1772–1837), who enjoys the deserved reputation of a visionary and crank of the first order, described the future socialist paradise in more grandiose detail than any of the utopians who preceded him throughout history. Nevertheless, he was the first to make certain observations that proved of importance in the evolution of socialist ideas. He was an eyewitness and to some extent a victim of the economic crises, destitution, and speculation of the Revolutionary and Napoleonic era; these experiences formed the background to his system, which he regarded as the most important event in the history of the human race.

Born at Besançon, the son of a rich merchant, Fourier was destined against his will for a business career. He became a commercial agent at Lyons in 1791, and in this capacity travelled extensively in France, Germany, and Holland. Eventually he founded a firm of his own, but was ruined by the events of the Revolution and thereafter held its ideas in abhorrence. Conscripted into the army, he was discharged in 1796, and became once more an agent at Lyons and subsequently a broker. After some years he moved to Paris, then returned to Lyons as a bank cashier, and finally settled in Paris, first as a trade official and then as a modest *rentier*. The last four decades of his life were spent in elaborating and publicizing his ideal of a perfect society: nearly all his spare time was devoted to writing, and only a small fraction of it to reading. He sought incessantly for a patron who would invest a few million francs in the first 'phalanstery' or cell of the new society; it it were once set up, he was convinced that the example would prove irresistible in four years at the longest. Though embittered by failure he continued his efforts and managed to recruit a small band of disciples, the chief of whom was Victor Considérant (1808–93). Fourier began writing in 1800, and in 1808 expounded his system in the anonymous *Théorie des quatre mouvements et des destinées générales*. In 1822 he published his *Traité de l'association domestique et agricole*, and in 1829 *Le Nouveau Monde industriel et sociétaire*. He left a number of manuscripts, some of which were published by his followers, while others have only recently seen the light of day.

Fourier's extraordinary cast of mind is well illustrated by the account he gives of the manner in which he hit upon the basic principle of his system. Travelling from Rouen to Paris in 1798, he noticed a wide difference in the price of apples from one place to another, although the climate was the same. This brought home to him the harmful and destructive effect of middlemen, and thus inspired the whole conception of the new society. Fourier goes on to observe that in the history of the world there have been two pernicious apples, those of Adam and Paris (the apple of discord), and two beneficial ones, Newton's and his own; the latter is more salutary than all previous human inventions put together. The world might, he adds, have been organized on his system at almost any time in the past, for instance in the age of Pericles, and this would have saved much suffering and unhappiness. Fourier was not the only theoretician of his time to see himself in the role of a saviour, but he was more open about it than most.

Fourier's doctrine was inspired by the phenomena of crisis, speculation, exploitation, and the misery of the workers. All this, he thought, was not an inevitable consequence of human nature but was due to a wrongful system of labour and exchange. Human needs and passions were ineradicable, but they only led to unhappiness because society was badly organized; the problem was to order matters in such a way that they conduced to the general good instead of to antagonism. Modern civilization was contrary to the natural order as established by God; we must rediscover Nature's demands and organize public life accordingly. The society of the future would be composed of settlements called 'phalansteries', in which all passions would be satisfied and would serve constructive ends. Altogether twelve passions were common to human beings, though in varying proportion: four related to sentiment (friendship, love, ambition, and family feeling), one to each of the five senses, and the remaining three were 'distributive': the desire for change, love of intrigue, and the tendency to unite in competing groups. By means of elaborate calculation Fourier showed that the combinations of these passions produced 810 types of character, and the basic unit of his society, which he called the 'phalanx', should consist, for maximum variety, of twice this number of individuals plus a reserve, making a total of 2,000. Production was to be organized in such a way that everyone had an occupation congenial to his character. Work would not be a form of drudgery but a stimulus and a source of pleasure. No one would be obliged to stick at the same job; everyone would have at least forty different aptitudes, and could change employment several times a day if he felt like it. Unpleasant jobs like killing animals or cleaning sewers and drains could be performed by children, who like playing in the dirt. The phalanstery was to be an agricultural and industrial unit. Life would be communal, but privacy would not be sacrificed; the dwellings would be hotels rather than barracks; everyone would be completely free to follow his own bent. Women would enjoy full equality with men; family life would be abolished, and children brought up communally at public expense; burdensome domestic cares would cease, and all restrictions on sexual life would be removed. This was a basic feature of the new society: people could live monogamously if they wanted to, but love was to be absolutely free and brothels would be among the most respected institutions of the new order.

Private property, inheritance, and economic inequality would not be done away with, but would lose their antagonistic character. The phalanx would provide minimum subsistence for all, even if they did not wish to work (but everyone would wish to, for all work would be pleasurable). Production would be by co-operatives whose share in the general wealth would be determined by the usefulness of their output, the enjoyment of creating it etc. Each individual would work in several groups and be paid differently in each according to his ability. There would be inequality but no envy, only zeal and healthy competition. All were entitled to share in the capital of the co-operative, but this could not give rise to exploitation. The education of all children free of charge would ensure that they engaged in useful work at an early age. The organs of political authority would become superfluous; public affairs would be decided on democratic principles and government would be reduced to economic administration; however, for the sake of variety and emulation, the new order would maintain a system of titles, dignities, and representative functions. Fourier calculated with precision how many phalanxes, combined in units of increasing size, would be necessary to comprise the world state of 'omnarchy' (*sic*, i.e. rule of all). Since the evils of the present system had affected the animal and vegetable kingdoms as well, the new order would see a transformation of these and the assertion of man's dominion over them. The seas would turn into orangeade, deserts would blossom and glaciers melt, spring would be eternal, and wild beasts would die out or become friends of man, 'anti-lions' and 'anti-whales' to do his bidding. There would be a single language for all mankind; all would live life to the full, developing their personality in all directions, in a happy and harmonious community embracing every kind of sentiment and avocation.

The extravagance of Fourier's description and the naïvety with which he attributed his own tastes to other (sexual promiscuity, gluttony, love of flowers and cats, etc.) caused him to be regarded as a hopeless crank, with the result that some of his acuter observations were overlooked. His whole theory was enveloped in a speculative cosmology and theology which sought to explain human affairs by universal laws. The pursuit of knowledge was to him a form of worship, and the laws of nature were divine decrees. Newton's law of gravity applied to souls as well; all human passions were instances of 'attraction', all were natural, therefore divine and deserving of satisfaction. The universe was a kind of phalanstery composed of heavenly bodies in a hierarchical order: the planets copulated, the stars had souls, and so on. Fourier adopted Schelling's view that the world was a unity, and he believed that the human soul and the universe were constructed according to an identical schema.

Despite these absurdities, Fourier's critique of 'civilization' (a term he always uses in a pejorative sense) and his ideas of a future harmonious state contain many elements that became part of the socialist tradition. His view that exploitation and poverty are due to a discrepancy between social conditions and the developed instruments of production appears in a more precise form in Marx's writings. Fourier pointed out the parasitic nature of trade in conditions of economic anarchy, and also the harm done by tiny land-holdings. He showed that technical progress increased the poverty of the proletariat (the remedy being not to halt progress but to alter the property system), and that wages gravitated to a minimum subsistence level. His ideal was a unified economic system which would prevent human energies being wasted in intermediary occupations and would eliminate the chaos of unplanned production leading to a glut of merchandise and pauperizing the workers. Fourier criticized the republican doctrines that extolled political freedom, which, he pointed out, was of little use without social freedom, i.e. liberty to develop one's own inclinations. He argued that hired labour is a form of slavery, that humanity aspires to freedom based on conformity between the individual's desires and the work he does, and that the aim is a voluntary society of harmonious co-operation. All these ideas are close to those of Marx. As to Fourier's conception of an all-round man liberated from occupational one-sidedness, capable of performing a variety of tasks, and living in a system which enables him to do so, this too can be found many times in Marx, from the Paris Manuscripts to *Capital*. Again, Fourier was one of the first to advocate the emancipation of women: he believed that human progress depended on the liberation of the sex, and, like the Marxists, he condemned the element of prostitution in bourgeois marriage. His Utopia was the antithesis of the monastic imaginations of the Renaissance and Enlightenment; he held that asceticism was contrary to nature and that the liberation of man signified, not least, the liberation of his passions. In this respect he appears to have more in common with Rabelais than with the classical utopianists. He is close to Marxian socialism, again, in the important place assigned in his ideal world to aesthetic experience and artistic creation.

Fantastic though some of his answers were, Fourier posed a real and important problem: as men are endowed with different desires and with aggressive and selfish impulses, how are these natural rivalries to be turned into constructive channels instead of leading to social antagonism? Unlike most utopianists, Fourier saw the remedy in terms of a new social order and not of transforming human nature. He believed that the conflict of interests was a universal law and that it was no use trying to prevent it, but that society must be so organized that conflict invariably led to harmony. He thought it useless to contemplate a general levelling and the equalization of man, and from this point of view he disagreed with both Saint-Simon and Owen; the idea of complete equality and community of goods seemed to him chimerical. He was convinced, however, that partial reforms of civilization were no good. Society must be transformed root and branch, or nothing would be changed; yet he believed that this transformation could be brought about by the mere force of example.

Fourier's disciples were not interested in the religious and cosmological trappings of his system, but they upheld the view that political struggles could not lead to any change and that it was only social reform that counted. They tried in various ways to modify Fourier's ideas in the direction of realism. Workers' consumer co-operatives were an outcome of his system, as were attempts to establish producer co-operatives in which the workers were shareholders.

Victor Considérant published Fourierist journals (*Le Phalans-tère*, 1832–4, and *La Phalange*, 1836–49) and tried to start model colonies in Texas (many utopians sought to put their theories into practice in the New World, including Owen, Cabet, and Weitling). Another disciple of Fourier's was Flora Tristan (1803–44), an early feminist known for the amorous adventures described in her autobiography.

6. *Proudhon*

Pierre Joseph Proudhon (1809–65) is noteworthy among the early socialists for the many directions in which his influence extended, a fact largely due to the incoherence of his writings and the contradictions they contain. His lifelong passion for social justice was not equalled by his education (he was largely self-taught) or powers of historical analysis. Born at Besançon, the son of a brewery workman, he was sent to school by benefactors and became a printer. Subsequently he received a scholarship and migrated to Paris. In 1840 he published the pamphlet *Qu'est-ce que la propriété?*, which aroused fury and admiration in equal measure. Henceforth, to his pride, he was identified with the slogan 'Property is theft', though these exact words had in fact been used by Brissot before the Revolution. He was tried and acquitted, and soon published two further pamphlets on the same subject (*Lettre à M. Blanqui sur la propriété*, 1841; *Avertissement aux propriétaires*, 1842), for which he was again tried and acquitted. Until 1847 he earned his living as the agent of a transport firm, and in these years published two important books: *De la création de l'ordre dans l'humanité, ou Principes d'organisation politique* (1843) and the lengthy *Système des contradictions économiques, ou Philosophie de la misère* (1846). The latter work provoked a crushing reply from Marx entitled *Misère de la philosophie* (1847). Marx had met Proudhon and, in the course of long conversations, imparted to him, or so he claimed, the ideas of Hegelian philosophy. Proudhon did not know German, but he may also have heard about Hegel from the lectures and books of Heinrich Ahrens, who was then teaching in Paris.

After the Revolution of 1848 Proudhon went into politics in the hope of persuading the republican government to enact his programme of social reform. In June he was elected to the constituent assembly, where he became the chief representative of the Left; however, he was shortly sentenced to three years' imprisonment for articles criticizing Louis Napoleon. He continued to work in prison, and in 1851 published *L'Idée générale de la révolution au XIXe siècle*. After the *coup d'état* in December of that year he hoped for a while that he might use the Prince-President to carry out his socialist plans. Undeterred by failure, poverty, and obloquy he continued to agitate and published numerous writings. In 1858 he was again sentenced to three years' imprisonment for his large work *De la justice dans la révolution et dans l'église*, but escaped by fleeing to Belgium. Four years later he was expelled from Belgium and returned to France, where he again attempted unsuccessfully to found a party and a literary organ. He died at Passy.

Proudhon, as he admitted, never reread his own works and did not seem aware of their contradictions. His plan belongs to the category of socialist utopias in so far as it is purely normative and invokes ideals of justice and equality, but he sought to found it on an analysis of contemporary economic life and to assess the possibility of change in practical terms. It was he who coined the expression 'scientific socialism'.

Proudhon believed in a 'natural' social harmony and in the inalienable rights of man, which were violated by the existing economic system: the right to freedom, equality, and the sovereignty of the individual. These were part of man's destiny as prescribed by the will of God (though elsewhere Proudhon represents himself as God's enemy). The system of competition, inequality, and exploitation is incompatible with human rights, and economists who confine themselves to describing it are ratifying a state of chaos. However, the contradictions of the system cannot be simply removed by an act of synthesis. Proudhon, in his limited acquaintance with Hegel's dialectic, was especially attracted by the well-known schema of thesis, anti-thesis, and synthesis; this in fact plays quite a secondary role in Hegel's philosophy, but has always appealed to the imagination of those who know little of it. In Proudhon's opinion, the Hegelian 'synthesis' by which the terms of a contradiction are assimilated is thought of in such a way as logically to precede those terms. The belief that all contradictions are resolved by the synthesizing movement of progress is the foundation of Hegel's cult of the state and of the absolutism which subordinates the value and dignity of the human personality to the state apparatus. To this logic Proudhon opposes his own negative dialectic based on the view that antagonistic terms are not dissolved in synthesis, but balance each other without ceasing to be distinct; such balance, moreover, is not an inevitable law of progress but only a possibility of which people may or may not succeed in taking advantage. Men and women are not instruments of progress, working itself out independently of their will; if progress occurs, it is the result of human effort.

Despite Marx's scornful criticism, it is not the case that Proudhon regarded actual social conditions and economic forces as the embodiment of abstract philosophical categories antecedent to social reality. On the contrary, he is at pains to state that the intellectual organization of social reality in abstract categories is secondary to that reality. The first determinant of human existence is productive work, while intellectual activity is the outcome of such work. If spiritual life has become alienated from its true origins, and if ideas are not aware that their source lies not in themselves but in the world of labour, this is a symptom of an illness in society that must be cured.

However, 'labour' in Proudhon is a normative as well as a descriptive category. His criticism of property is based on the moral indignation aroused by unearned income. 'Property is theft' may sound like a call to do away with all private property, but Proudhon was far from actually being a communist. When he sets out to prove in his pamphlet that 'property is a physical and mathematical impossibility', what he really has in mind is that the system which permits the enjoyment of unearned income is immoral and leads to social contradictions. To draw dividends, interest, rent, etc. on the mere ground that one possesses capital is as though one were creating something out of nothing. It is irrelevant whether the property-owner performs productive work or not; if he does, he is entitled to a proper reward, but anything he enjoys over and above this, merely as an owner of wealth, represents a theft from other workers. Property in a monopolistic form, i.e. the privilege of unearned income, is a source of inequality and wrong and destroys personal life; it owes its origin to violence, of which it is the crystallization. The antithesis of a system based on property, however, is not communism but the abolition of incomes not justified by labour, i.e. a society in which goods are exchanged among producers at a rate determined by the amount of labour that has gone into them.

In this respect Proudhon claims to modify the theories of Ricardo and Adam Smith. Ricardo held that labour was the only measure of value, and the market value of any product was a crystallization of the man-hours required to manufacture it; the proceeds were then shared out between capitalists (in the form of return on capital), landowners (as rent), and workers (as wages). This led British socialist reformers of the 1820s and 1830s to point out that the immediate producer of goods was at the same time the only creator of value, and that he was entitled to the whole of the value he created; it was equally unjust that goods were manifestly not exchanged according to their value and that some people enjoyed what they had not created. Proudhon for his part did not entirely accept this naïve interpretation of Ricardo, but he accepted its ultimate consequence. His view was that none of the three factors of production—tools, land, and labour—created value by itself, but only all three together. Tools and land had no productive force without labour, but the mere expenditure of energy was unproductive too unless it were used to change the face of nature by means of tools. The sea, the fisherman, and his net are all needed before we can have fish to eat. Present-day economy, however, was based on the false premiss that capital (tools and machinery) or land are in themselves productive forces, so that the owners of land, capital, or buildings are entitled to charge for their use. In a just economy this could not be, nor could it happen that goods were bought and sold in accordance with the fluctuations of supply and demand instead of at their true value. As to what that value is, Proudhon is not altogether clear. On the one hand he says it depends on utility, on the other that it derives from all three factors of production, or again from labour alone. But the guiding principle of his economic Utopia is clear, even if its theoretical foundation is shaky. What it requires is that each person should receive, from the products of others' labour, the exact equivalent of what he himself produces, and this equivalence must be measured in hours of work. Unearned income must be abolished, and a system of exchange created on the basis of the number of work-hours embodied in a given product, so that each producer receives an income sufficient to buy what he himself produces.

Thus property in the sense of monopoly is done away with, but not in the sense of the producer's right to use the means of production as he wishes—a right which is the condition of personal freedom and individual sovereignty. The concentration of wealth in the hands of a few, and the resulting pauperization of the working masses, can only be remedied by the abolition of monopoly income. The Malthusians are mistaken in regarding overpopulation as the cause of poverty, for overpopulation is relative to the quantity of resources shared among the non-property-owning classes. It cannot be cured so long as goods are not exchanged on a basis of equivalence and the working man's wage will only buy part of what he produces. In such conditions, no matter how many people emigrate from a country it will still be overpopulated in the sense that the masses will be in a state of impoverishment.

It thus appears that Proudhon (like Fourier, though his moral and philosophical reasons are quite different) does not really wish to abolish property but to generalize it. Communism, he believed (having chiefly Cabet and Blanc in mind), would never be compatible with the dignity of the individual and the values of family life; its outcome would be universal poverty and the suffocating mediocrity of a regimented existence. The advocates of communism were power-thirsty fanatics who aimed to set up an omnipotent state on the basis of public property. Far from abolishing the harmful effects of property, the communists would carry it to an absurd extreme: the individual, in their system, would have no property, but the whole lawlessness of its use would be conferred upon the state, which would own the country's wealth and the bodies of its citizens as well. The lives, talents, and aspirations of human beings would, at a stroke, become state property, and the monopoly principle, the source of all social evil, would be intensified to the utmost. Communism, in short, had nothing to offer but the extremity of police despotism.

In order to ensure 'equivalent exchange' and eliminate competition, the first necessity was to reorganize the credit system and do away with interest, which was an especial cause of injustice. Proudhon proposed to create a people's exchange bank which would make interest-free loans to small producers and thus turn the whole of society into property-owners, assuring them of freedom, equality, and a fair share in the fruits of their own activity. The bank would issue bonds or coupons which would serve as a means of exchange between producers on the principle 'To each according to his labour'. From some of Proudhon's writings it might be inferred that his ideal was a petty-bourgeois community of small individual producers, as the only way to ensure social justice. However, it appears elsewhere that he did not contemplate a return from mechanized industry to craftsmanship. He was concerned rather with what he called 'industrial democracy', i.e. that the workers should retain control over the means of production. Productive units must be the collective property of all those employed in them, and the whole of society would consist of a federation of producers, both industrial and agricultural. This, among other things, would resolve the contradiction inherent in machinery, which on the one hand was a triumph of the human spirit over matter, but on the other hand spelt unemployment, low wages, overproduction, and the ruin of the working class. The plan would also resolve the contradiction in the division of labour, which was an instrument of progress yet which degraded human beings into mere parts of themselves.

The new 'mutualistic' society would thus reconcile, for the first time in history, property with equality and freedom with co-operation. Proudhon made light of purely political problems, regarding the social issue as the only important one. In his early writings he takes an anarchistic view of the state as an instrument of the possessing classes, to be replaced by a system of free agreements among economic co-operatives. Later he came to acknowledge the need for state power, not as the weapon of a class but as the organizer of production for the common good. His ideal, however, continued to be decentralized production and a state consisting of a loose federation of communities.

For the translation of his dreams into reality Proudhon relied neither on political nor on economic action by the proletariat. He was opposed to revolutions and even strikes, on the ground that violent action against the 'haves' would lead to disorder and despotism and would only exacerbate class hostility. He believed that, as his ideals were rooted in human nature and their realization would be no more than the fulfilment of human destiny, he could reasonably direct his appeal to all classes without distinction. In several places he invites the bourgeoisie to take the lead in bringing about the desired reform, and he also relied from time to time on the state as an auxiliary factor. He continued for many years to believe in co-operation among different classes. However, in his posthumous work *De la capacité politique des classes ouvrières* (1865) he reverted to the idea of the uniqueness of the proletariat and called for a combination of the economic and the political struggle (and, as before, a boycott of state institutions). On the other hand, his theories show no trace of internationalism: his plans for reform are geared to French conditions, he had no quarrel with French national values, and in one work (*La Guerre et la paix*, 1861) he even glorified war as a strengthener of moral fibre and developer of the highest virtues.

Proudhon's work as a whole presents a chaotic and incoherent aspect, and its inconsistencies were fully reflected in its subsequent influence. Marx, who greeted his first publication as a political event comparable with Sieyès's *Qu'est-ce que le Tiers Etat?*, was mercilessly sarcastic at the expense of *La Philosophie de la misère*, reproaching Proudhon with ignorance of economics, the fanciful use of misunderstood Hegelian schemas, a moralistic conception of socialism, and a reactionary petty-bourgeois Utopia.

Proudhon regarded this attack as a farrago of coarse slander, misrepresentation and plagiarism, but he did not join issue with Marx in public. There was clearly a wide difference between them as regards the interpretation of economic life, their ideas of the future of socialism, and their choice of political tactics.

While Marx's criticism was unjust and dishonest in some respects, he was intellectually far superior to Proudhon, who had all the faults of the clever autodidact: self-assurance, unawareness of the limitations of his knowledge, incomplete or desultory reading, lack of skill in selecting and organizing material, and the hasty condemnation of authors whom, for the most part, he did not properly understand. Nevertheless, his influence was of considerable duration. It was clearly felt in the French syndicalist movement of the 1860s, which rejected political action and hoped to liberate the workers by organizing co-operatives and credit on a mutual basis. Most of the French members of the First International, notably Tolain and Fribourg, were Proudhonists and upheld the principle of 'mutuality' in preference to strikes, let alone political revolution. Proudhon also exerted a strong influence over Bakunin, particularly from the anarcho-syndical point of view, and many of his followers were active in the Paris Commune; he was also looked up to by later anarchists such as Kropotkin. In the years before the First World War his teaching was acknowledged by the Action Française monarchists under Charles Maurras, who perceived in it the spirit of the first counter-revolutionary ideologists, de Maistre and Antoine de Rivarol: the defence of individual and family property, French patriotism and the praise of war, domestic virtues and a patriarchal system (together with the natural inferiority of women), the decentralization of power, hostility to the unification of Germany and Italy (Proudhon was also opposed to the cause of Polish independence), and finally racism and anti-Semitism. Georges Sorel, the advocate of revolutionary syndicalism, invoked the authority of Proudhon, who opposed strikes on principle.

After the Paris Commune there was no 'Proudhonism' properly so called in the workers' movement itself, but particular ideas and proposals featured in French socialism for a considerable time. Anti-centralist and anti-etatiste tendencies are part of Proudhon's heritage; the objection to communism as a system of extreme political and economic centralization is a theme that he implanted in the French workers' movement and that has retained its actuality from his time onwards. He originated the idea of 'industrial democracy' and also what is called *ouvriérisme*—the tendency to disparage purely political and parliamentary action, to mistrust intellectuals in the workers' movement, and to look with suspicion on all ideologies that do not serve the immediate material interests of the proletariat.

7. *Weitling*

The works of Wilhelm Weitling (1808–71) stand out amid the communist utopias of the 1840s not because he was in any sense a forerunner of Marx, but because he was himself a member of the working class and therefore a better exponent of its attitude at the time than were theorists belonging to the privileged classes. His form of communism was less close to Babouvism than to the German anabaptists of the early sixteenth century.

After an impoverished childhood, Weitling left his native Magdeburg at an early age and earned his living as an itinerant tailor. His travels took him to Vienna, Paris, and Switzerland. Paris was at that time the home of thousands of working-class German *émigrés*, and Weitling made contact with two clandestine communist organizations, the League of Outlaws (Bund der Geächteten) and its offshoot the League of the Just (Bund der Gerechten). In 1838 he published in Paris a pamphlet in German on *Humanity as it is and as it ought to be* (*Die Menschheit wie sie ist und wie sie sein sollte*). Fearing prosecution he fled to Switzerland, where he published *Guarantees of Harmony and Freedom* (*Garantien der Harmonie und Freiheit*; 1842) and *The Gospel of a Poor Sinner* (*Das Evangelium eines armen Sünders*; 1843); the latter earned him some months' imprisonment at Zurich. He later went to London and collaborated for a time with Karl Schapper, the leading spirit in German *émigré* worker organizations there. By this time his writings were known throughout Europe, but their religious and prophetic strain was equally uncongenial to the more down-to-earth workers' leaders and to sophisticated theorists. In the spring of 1846, on his return to the Continent, Weitling encountered Marx, who was at Brussels organizing a liaison centre for European communist groups. The meeting was a disaster, as Marx attacked the self-taught worker with the arrogance of an intellectual, accusing him of ignorance and naïvety; Weitling, for his part, thought that having shared the sufferings of the proletariat he was able to understand its position and prospects better than a doctrinaire scholar. After a short visit to America Weitling returned in time to take part in the 1848 Revolution in Berlin, after which he emigrated to America for good.

Weitling's works are a typical example of primitive evangelical communism, in the form of sermons on justice and the need to rebel against tyranny. They make ample use of everything in the Gospels that can be turned against the rich and the oppressor, and present a picture of Christ as a communist urging the destruction of the system of exploitation and injustice. The world is governed by the selfishness of the rich, while the workers who create their wealth live in poverty and insecurity. It is not machines that are to blame: in a just society technical progress would be a blessing, but as things are it makes the poor worse off than before. The real cause of social misery is the unequal distribution of goods and obligations and the craving for luxury. When wealth is held in common and all are obliged to work, all evil will disappear in a twinkling; working hours will be greatly shortened, and work will be a delight instead of a curse. There will be no money or accumulation of wealth; class differences will vanish, all benefits of body and soul will be available to everyone. This is the true message of Christianity. Not surprisingly, the Gospel teaching has been distorted and falsified by kings and priests who have used it to defend their own privileges; but the time has come to unmask their imposture and to build a new world of freedom, equality, and Christian love. We must not expect governments and capitalists, however, to recognize this ideal and bring it about of their own accord; the workers can rely only on themselves and their own strength. From the medieval preachers of the millennium, Weitling takes over the division of history into three ages: the ancient times of primitive communism, the present era of private property, and the communism of the future. He describes in some detail the earthly paradise in which there will be no hatred or envy, no crime or evil desires. Men will be brothers again, and the national languages that divide them will die out within three generations. As all will have equal obligations, wealth and luxury will be accessible to all. Anyone, for instance, who wishes to wear different clothes from those provided by the community will be able to afford them by working overtime, especially as the compulsory working day will soon be no longer than three hours.

In this way Weitling naïvely reflected the current notions and day-dreams of the poor. Marx, inevitably, was irritated by his lay-preacher's tone. Yet Weitling imparted to the German working class something of the ethos of medieval chiliasm, and, while he could contribute nothing to the scientific analysis of capitalism, undoubtedly helped to awaken the rudimentary class-consciousness of the proletariat in his country.

8. Cabet

If Weitling embodies the traditions of sectarian revolutionism of the pre-capitalist era, Étienne Cabet (1788–1856) provided the early industrial age with a specimen of a classic literary genre in his utopian description of a communist island.

Cabet, who was trained as a lawyer, took part in the Revolution of 1830, and his political and literary activity belongs almost entirely to the period of the July Monarchy. In 1839–40 he published a four-volume *Histoire populaire de la Révolution française*. In 1840 his best-known work, *Voyage en Icarie*, appeared under a pseudonym in England, where he had emigrated for fear of prosecution and where he was influenced by Owen's ideas. On returning to France he resumed publication of the journal *Le Populaire*, advocating non-revolutionary communism as the teaching of Christ. He emigrated to America at the beginning of 1849 and established communist settlements in Texas and later in Illinois; one of these lasted for several decades. He died at Saint Louis.

Cabet's 'Icaria' is an egalitarian community with some totalitarian features, like many utopias of the Renaissance and Enlightenment. Since inequality is the cause of all social evils and can only be remedied by holding goods in common, and since the equality of rights and duties is commanded by 'true' human nature and by the Christian faith, it follows that in the ideal society there is no private property and no monetary system. All social production is the work of a single organism of which individuals are parts. All are equally obliged to work according to their powers and to share in the general revenue according to their needs. The community must do its best to see that everyone eats the same food, wears the same clothes, and lives in the same kind of dwelling; obligatory living standards are laid down by the authorities, and all towns look alike. The people as a whole is the sovereign power in its territory, and elects for a limited period administrators to look after production. There are no parties or political clubs (there would be nothing for them to do), and the written word is strictly supervised to prevent any danger to morale. All this is brought about without violence and without a revolution. Cabet expressly dissents from Babeuf and believes that revolutions, conspiracies, and *coups* have brought mankind more unhappiness than gain. Since the perfect society is based on the dictates of natural law and all men are equally partakers in it, it would be a fatal mistake to inaugurate it with force, oppression, and hatred. Rich men and oppressors are the victims of a faulty social system, and their prejudices must be cured by education and not repression. The better world is not to be brought about by violence and conspiracy, but by gradual reform and through a transitional system which will merge into the ideal society of the future.

Among Cabet's other works are *L'Ouvrier, ses misères actuelles, leur cause et leur remède* (1845); *Comment je suis communiste* (1845); and *Le Vrai Christianisme suivant Jésus-Christ* (1846). Together with 'Icaria', they have all the attributes of utopianism in the pejorative sense generally given to this term in Marxist literature. However, as a widely read writer in a popular style, he did much to spread communist ideals: he had no influence whatever on Marx, but helped to acquaint French readers with the basic values of communism.

9. Blanqui

In the history of socialism Blanqui is of importance not so much as a theoretician but because he transmitted the heritage of Babouvism to the generation of 1848 and its successors, thus providing a link between the Jacobin Left and the nineteenth-century radicals and introducing the idea of revolutionary conspiracy into the workers' movement. He was also the originator of the idea (though not the phrase) of a 'dictatorship of the proletariat', to be exercised in its name by an organized minority. The son of a Girondin, Louis-Auguste Blanqui (1805–81) studied law and medicine in Paris. He became acquainted with the various socialist doctrines that were going about, and took an active part in the July Revolution. In the thirties he organized clandestine societies of a radical-democratic nature, inclining more and more to socialism. He was put on trial in January 1832 and made a celebrated speech of accusation rather than defence, proclaiming the just war of the proletariat against the rich and the oppressor. He was imprisoned for a year, after which he resumed conspiratorial activity and led an unsuccessful revolt against the monarchy in May 1839. Sentence of death was passed, but commuted to life imprisonment. Liberated by the 1848 Revolution, he became one of the chief leaders of the Paris working class, but was soon behind bars again. He was released in 1859 for a brief period, but spent most of the sixties in gaol. Under Thiers's regime he was released, and again arrested in March 1871; he was elected *in absentia* to the leadership of the Paris Commune, in whose ranks his followers were the most active and resolute faction. He remained in prison till 1879, and thereafter continued to agitate for the remaining two years of his life.

Those of Blanqui's writings that appeared during his lifetime were of a more propagandist than theoretical character, except for the philosophical work *L'Éternité par les astres* (1872). This was based on the mechanistic materialism of the Enlightenment and put forward the Stoic notion of the unceasing repetition of worlds—the state of the universe is entirely determined by the arrangement of its material particles, and, as the number of such arrangements is finite, each one must repeat itself an infinite number of times in the course of history. In 1885 the two-volume *Critique sociale* appeared posthumously. Blanqui's critique of capitalism does not go beyond the usual rhetoric of his day and is fairly simplistic on the economic side. He shared the view that inequality and exploitation occur because goods are not exchanged at their 'true' value as determined by labour-content; as to the future communist society, he has no more than generalities to offer. His chief role in the history of socialist movements is that he inculcated the importance of revolutionary organization and helped to improve the technique of conspiracy. The term 'Blanquism' in socialist parlance came to mean much the same as 'revolutionary voluntarism'—that is to say, the belief that the success of a communist movement does not depend on 'objective' economic circumstances, that a properly organized conspiratorial group may seize power if the political situation is favourable, and that it may then proceed to exercise a dictatorship on behalf of the working masses and establish a communist system regardless of other social conditions. 'Blanquism' in this sense was a pejorative label affixed by reformists to revolutionaries, notably in Russia after the split in the social democratic party in 1903, when the Mensheviks accused Lenin of following a non-Marxist, conspiratorial strategy of revolution.

10. Blanc

Blanqui and Blanc were the nineteenth-century protagonists of two sharply opposed tendencies in the socialist movement, both of which are contrary to Marxism. Blanqui believed in the all-conquering force of the revolutionary will embodied in an armed conspiracy, while Blanc trusted that gradual reform by the state would abolish inequality, exploitation, crises and unemployment. The former doctrine is derived from Babouvism; the latter from Saint-Simon, with some attenuation as regards democracy and the take-over of all means of production by the state. Blanqui's ideas were adopted by Tkachev and afterwards by Lenin; those of Blanc by Lassalle and the modern social democrats. The former was a conspirator, the latter a reformer and scholar. Lenin was accused of 'Blanquism' by Plekhanov and Martov, and retorted on many occasions between the February and the October Revolution by comparing the attitude of his Menshevik opponents to that of Blanc in 1848, with his indecision, proneness to compromise, and lack of revolutionary will-power.

Louis Blanc (1811–82) studied in Paris under the Restoration and in 1839 founded the *Revue du progrès*, in which he published in instalments *L'Organisation du travail*, one of the most popular socialist texts of the 1840s. Besides major works on the revolutions of 1789 and 1848, the Empire, and the July Monarchy, he published *Le Socialisme. Droit au travail* (1848) and many articles on political and social questions. He was a member of the Provisional Government in 1848 and put forward an extensive programme of reforms and public works to overcome unemployment and poverty. After the savage repression of the June insurrection the right wing accused him of responsibility for the outbreak (although he had hoped to prevent riots by reforms); he fled the country and spent the next two decades in England, returning in 1870 after the collapse of the Second Empire. His attempts to reconcile the Commune and Versailles brought on him the obloquy of both sides. He was a deputy of the moderate republican Left from 1876 until his death, and in 1879 inspired the law granting amnesty to the Communards.

His classic work, *L'Organisation du travail*, argued that revolution was inevitable, but by this he meant radical social reform and not violent political change. Unlike the utopianists with their detailed plans for a perfect social order, Blanc set out to be a practical reformer and to indicate what steps might be taken on the basis of the existing state of affairs. He did not want to provoke a violent upheaval, but to prevent one; however, an explosion was bound to occur soon if the starving, desperate masses could not find jobs, and the most urgent need was to cure unemployment. The system based on unimpeded competition among entrepreneurs led infallibly to crises, poverty, ignorance and crime, the barbarous exploitation of children, and the decay of family life. Unless Malthus's doctrine was to be applied by simply killing off workers' children in excess of a certain number, the state must use all its power to carry out social reform, of which political reform was a necessary prerequisite. History had shown that violent revolutions whose leaders started with no definite plan but imagined that they could work one out after seizing power resulted merely in pointless slaughter: it sufficed to compare 1789 with 1793 and the following years. The proposals of Owen, Saint-Simon, and Fourier contained many useful ideas but were lacking in practical sense, and the changes they advocated could not be put into effect in a short time. What could be done was for the state to assume control over production immediately, and to put an end by degrees to unbridled competition. A grand design for industry, based on public property, should be set on foot with the aid of a national loan; workers whose earnings depended on productivity and the success of the concern they worked for would display far more energy than when driven by private capitalists. The competition between socialized and private enterprises would soon be resolved in favour of the former, which would produce better articles more cheaply. No more competition, no more crises, no more so-called overpopulation; technical progress, instead of harming the workers' interests, would lessen the weight of toil and shorten the working day. Free compulsory education would bring benefits to all. Wage rates would have to be differentiated for some time to come, as faulty education had so conditioned people that they had to be tempted to work harder. The administrative hierarchy would be elective, and the units of production would enjoy autonomy. The right to work would be universally recognized as the basic principle of social organization.

Blanc may justly be considered one of the chief precursors of the welfare state. He believed that it was possible, without violence or mass expropriation, to carry out peaceful economic reforms within a system of political and industrial democracy which would eliminate poverty and harmful competition and would gradually lead to social equality and to the socialization of means of production. Of all the writers discussed in this chapter he was certainly the least 'utopian' in the usual sense, and indeed the only one whose ideas proved to some extent workable—apart from the idea of political dictatorship, which became a reality but not for the purposes that its authors intended.

11. Marxism and 'utopian socialism'

As may be seen from this rapid survey, the socialist writers of the first half of the nineteenth century can be classified in various ways. We may oppose reformists to conspirators, novelists to theoreticians, democrats to advocates of revolutionary despotism, and working-class leaders to philanthropists. On the other hand, the division into those whose philosophy is based on eighteenth-century materialism and those, such as Weitling, Cabet, and Lamennais, who invoke Christian values, is not essential. In both cases their Utopia is founded on the premiss that all human beings possess the same dignity by virtue of their humanity, and that, whatever the innate differences among individuals, they are identical as far as their rights and duties are concerned. This conception of human nature is both descriptive and normative. We may deduce from it what a man needs, and is entitled to receive, in order to be truly a man, but we know in advance that the answer will be the same for every individual. The idea of

human nature presupposes equality, whatever its other implications may turn out to be.

The conception of human nature is at the same time a description of man's proper calling. Throughout utopian literature it is
assumed that men are intended to live in a state of equality and
mutual love, and that exploitation, oppression, and conflict of all
kinds are contrary to nature's ordinance. The question of course
arises: how can it be, in that case, that men have lived for centuries
in a manner at variance with their true destiny? This is the hardest
question to answer from the utopian point of view. Even if we
suppose that somebody at some time happened to devise the
system of private property, which would otherwise not have been
instituted, how are we to explain the fact that his crazy and
inhuman notion was unanimously adopted? If we lay the blame
on 'evil desires', how is it that such desires came to dominate
society? If it is man's nature to live in amity and equality with his
fellows, why is it that we seldom or never find him doing so? How
can a majority of mankind 'truly' want something which, as a
matter of experience, they do not want? On the utopian view, the
whole of human history is a monstrous calamity, and incomprehensible to boot. For traditional Christianity there is no
problem, on account of the doctrine of original sin and the corruption of humanity at its source. But the utopians of this period, even
when they called themselves Christians, did not believe in
original sin; they were thus deprived of this explanation, and had
no other to offer. They wanted the good, but evil was to them
inconceivable and inexplicable. They fell back, without exception, on a confused idea of human nature as something already
'given' and not a mere arbitrary norm (for in that case there
would be no reason to expect people to conform to it)—a kind of
reality or 'essence', dormant in every individual.

Thinking in this way, the utopians were naturally attracted to
the idea of communist despotism. If we know that human nature is
fulfilled by the communist system, it is of no importance, in
establishing this system, what proportion of humanity wants to
accept it. Jean-Jacques Pillot, at the end of his pamphlet *Ni
châteaux ni chaumières* (1840), puts the question 'What if people do
not want this?', and replies 'What if the inmates of Bicêtre [lunatic
asylum] refuse to have baths?' If people are out of their minds,
they must be cured by force. The utopians did not put the further

question, which calls to mind Poe's story of Professor Tarr and Dr. Fether—how do we decide which are the lunatics and which are the keepers? Is a man really entitled to claim that everyone is out of step except himself? To say that mankind should decide its own destiny may mean that history is to be left in the hands of lunatics, but if we disagree with our fellow men we must prove that we ourselves are sane. As long as it was possible to appeal to the divine will as an irrefragable authority, the matter was simple enough. The utopians do so appeal when it suits them; but, as we know, Scripture has been used for centuries to justify inequality and the hierarchical order of society.

The same objection could be put to all the utopians, not only the advocates of communist despotism, and it was in fact put to Owen by Marx: who is to educate the educators? In the answer to this question lies the principal difference between Marx's Utopia and those of all his predecessors, between the heir of Hegelian phenomenology and the heirs of French materialism.

It is not difficult to select from the works of the utopian socialists a series of propositions that seem to anticipate the most important ideas of Marx, though they are not set out in the same order or expounded in the same way. They comprise three main topics: historiosophical premises, the analysis of capitalist society, and the depiction of the future socialist order.

Under the first two headings we may list the following points:

No essential change is possible in the system of the distribution of wealth without a complete change in the system of production and property relations.

Throughout history, constitutional changes have been conditioned by technological ones.

Socialism is the outcome of inevitable historical laws.

The organization of capitalist society is in contradiction with the state of development of productive forces.

Wages, under capitalism, tend naturally to remain at the minimum level consistent with survival.

Competition and the anarchic system of production lead inevitably to exploitation, overproduction crises, poverty, and unemployment.

Technical progress leads to social disaster, not for inherent reasons but because of the property system.

The working class can only free itself by its own efforts.

Political freedom is of little value if the mass of society is enslaved by economic pressure.

As regards the socialist future—whether this goes by the name of Harmony, mutualism, or the industrial system—we may enumerate the following ideals:

The abolition of private ownership of the means of production.

A planned economy on a national or world scale, subordinated to social needs and eliminating competition, anarchy, and crises.

The right to work, as a basic human entitlement.

The abolition of class divisions and social antagonisms.

The whole-hearted, voluntary co-operation of associated producers.

Free education of children at the public expense, including technical training.

The abolition of the division of labour and the degrading consequences of specialization; instead, the all-round development of the individual, and free opportunity for the use of human skills in every direction.

Abolition of the difference between town and country, while permitting industry to concentrate as at present.

Political power to be replaced by economic administration; no more exploitation of man by man, or rule of one man over another.

Gradual effacement of national differences.

Complete equality of rights and opportunities as between men and women.

The arts and sciences to flourish in complete freedom.

Socialism as a boon to humanity as a whole; the exploitation of the proletariat as the chief factor tending to bring about socialism.

Impressive as these analogies are, there is a basic difference between Marx and all other socialist thinkers of the first half of the nineteenth century. Moreover, this difference affects the meaning of many ideas which, in themselves, show a striking similarity and no doubt testify to the utopians' influence on Marx's thought. It is often said that he and they were not at variance as regards the end to be attained but only as regards the means, i.e. revolution versus peaceful persuasion; but this is a superficial and misleading view. It is, in fact, incorrect, since Marx never adopts the ethical, normative point of view which first establishes an aim and then seeks the best means of achieving it. On the other hand, it is not the case that he regarded socialism as the inevitable result of historical determination and was not interested in whether it was desirable or not. It is an essential feature of Marx's thought that he avoided both the normative and the purely deterministic approach, and it is in this that he shows himself to be a Hegelian and not a member of the utopian school. The utopians, admittedly, did not always regard socialism as a 'free' ideal; we may find references to historical necessity in Owen, Fourier, and the Saint-Simonists; but they do not probe the question to any depth or indicate how their deterministic fancies are to be reconciled with the conception of socialism as an ideal or as a moral imperative. On the one hand they insist that socialism (or whatever name they give it) is bound, in the nature of things, to conquer the world, on the other they regard its discovery as a happy effect of intellectual genius; and they oscillate between these points of view without seeming to perceive their inconsistency. Again, the utopians are convinced that political changes cannot by themselves bring about the new economic order and the redistribution of wealth; they believe that economic reforms must be achieved by economic action, and in consequence they undervalue politics and reject the prospect of a revolution. The starting-point of their reflections is poverty, especially that of the proletariat, which they are bent on relieving.

Marx's starting-point, however, is not poverty but dehumanization—the fact that individuals are alienated from their own labour and its material, spiritual, and social consequences in the form of goods, ideas, and political institutions, and not only from these but from their fellow beings and, ultimately, from themselves. The germ of socialism in capitalist society consists in the working class's awareness of dehumanization, not of poverty. This comes about when dehumanization has reached its uttermost limit, and in that sense the proletariat's class-consciousness is an effect of historical development. But it is also a revolutionary consciousness, the awareness of the working class that its liberation must come from its own efforts. The proletariat cannot do away with the system of wage-labour and competition by peaceful persuasion, because the consciousness of the bourgeoisie, which is likewise determined by its part in the productive process, prevents it from abandoning its role voluntarily. Dehumanization, although in a different form, is also an attribute of the possessing class, but the privileges that class enjoys prevent it from being clearly aware of its own dehumanized condition, in which it rejoices instead of chafing at it. Socialism is the effect of history in the sense that history gives birth to the revolutionary consciousness of the proletariat, but it is the effect of freedom inasmuch as the act of revolution is free, so that, in the revolutionary workers' movement, historical necessity expresses itself in free action. Revolution, a political act, is the indispensable condition of socialism, for the institutions that purport to represent the community in fact embody the particular interest of the possessing classes and cannot be the instrument by which that interest is overthrown. Civil society, or the collectivity of actual individuals with their private interests, is destined to 'absorb' the ostensible community and turn it into a real one. Free human action cannot bring about a radical change of conditions if it is only a question of ideals and an attempt to reform society from outside; it is constructive only when it proceeds from that society's awareness of itself as a dehumanized society, and this awareness can only arise in the working class, which experiences the acme of dehumanization. It is a demystified consciousness, presenting itself from the outset as awareness of actual reality, and by the same token a revolutionary consciousness, that is to say a practical attempt to change the world by violently destroying the political institutions that protect the existing order. In that consciousness, but not otherwise, historical inevitability and freedom of action are the same: as we read in the *Theses on Feuerbach*, 'The coincidence of the changing of circumstances and of human activity can be conceived and rationally understood only as revolutionary praxis.'

The suggestion that Marx differs from the utopians in soteriology but not in eschatology, i.e. that he more or less shares their ideal of the future while not agreeing that it can be achieved by peaceful means, is thus seen to be erroneous. As a disciple of Hegel he knew that truth is not only a result but also a way. The picture of a harmonious community, a society without conflict in which all human needs are satisfied, and so forth—all this can be found in Marx in similar formulas to those of the utopians. But socialism means more to Marx than a welfare society, the abolition of competition and want, the removal of conditions that make man an enemy to man: it is also, and above all, the abolition of the estrangement between man and the world, the assimilation of the world by the human subject. In the class-consciousness of the proletariat society attains to a state in which there is no longer any opposition between subject and object, educator and pupil, for the act of revolution is one in which society transforms itself by being conscious of its own situation. There is no longer a difference between ideologists above the community, and the community itself; consciousness knows itself to be part of the conditions that have produced it, and it also knows that men's fetters are forged, and can only be broken, by themselves. Socialism is not a mere matter of consumer satisfaction, but the liberation of human forces—the forces of each and every individual, aware that his own energy is likewise social energy. The fact that productive forces determine productive relationships and, through them, political institutions does not mean, in Marx's view, that socialism can be brought about by direct action in the economic field: for political institutions are not simply the outcome of the system of production but are its means of self-defence, and they must be swept away before it can be altered. Socialism, therefore, can only result from a political revolution with a 'social soul'. As we have seen, it is neither an arbitrary goal nor the mere result of history working in the manner of a natural law, but is the outcome of the conscious struggle of dehumanized man to recover his humanity and to make the world a human place again. The proletariat, as the spearhead of that struggle, is not a mere tool of history but a conscious agent; nevertheless, it was necessary for the historical process to dehumanize it completely before the struggle was possible.

12. *Marx's critique of Proudhon*

Marx's critique of Proudhon in *La Misère de la philosophie* may be summed up under three heads.

In the first place, Proudhon fails to perceive the inevitable consequences of competition and, in his anxiety to eliminate its 'evil aspects', adopts a moralistic point of view at the expense of economic analysis. The same substitution of moral indignation for economic thought appears in the slogan 'Property is theft'— which, moreover, is not accurate in itself, for theft by definition presupposes property. It is a utopian fantasy to hope to establish the true value of commodities in accordance with a labour standard, while maintaining the system of individual production and exchange and therefore of competition. Proudhon constantly confuses labour-time as a standard of value with the value of labour itself. Since labour is itself a commodity (Marx at this stage still held the view that wage-labour is a sale of labour, and not of labour-power as in the final formulation of his theory of surplus value), it is not clear how it, more than any other commodity, can be a standard of value. The true standard of value is labour-time— not the time it actually takes to make a particular article, but the shortest time in which it could be made in present conditions of technology and the organization of production. Competition fixes the price of goods on the basis of socially necessary labour-time, and thus necessarily involves inequality among competing producers. As long as competition exists, there can be no equivalent exchange because, as Marx later argued in more detail, the movement of capital evens out the rate of profit while fixing prices at a level above or below the actual value (it is impossible to maintain prices corresponding to value and at the same time ensure equal profit rates in different branches of production). In competitive conditions, moreoever, the system of exchange serves the needs of production and not consumption, and industry does not await demand but creates it. To attempt to maintain private property and competition while abolishing their 'evil aspects' is a moralist's chimera.

Secondly, Marx accuses Proudhon of a reactionary and hopeless endeavour to revive medieval production methods based on individual craftsmanship. The ideal of individual exchange on a value basis is as utopian in the industrial age as is the ideal of abolishing the division of labour in conditions of small-scale production. Marx himself regards the division of labour in its present form as a source of physical and mental degradation, and envisages that it will somehow be abolished; but on Proudhon's view this can only happen if the worker carries out the entire process of producing a given article, i.e. if he reverts to being a craftsman. Industry dominated by competition entails ever-increasing division of labour for the sake of increased output, and one can only imagine its abolition if competition is done away with and production regulated by actual human needs. Proudhon's doctrine is a petty-bourgeois fantasy—a dream of preserving the bourgeoisie while eliminating the proletariat, in other words turning everybody into a bourgeois.

Thirdly, Proudhon attempts to apply Hegelian schemata in a fantastic and arbitrary manner. Having taken over from Hegelian idealism the notion that economic categories are independent historical factors, spiritual forces to which actual phenomena are secondary, he imagines that social reality can be transformed by the intellectual manipulation of categories. But the latter are no more than abstractions, the reflection in human minds of social conditions at a given moment in history; the only reality of social life are human beings, who form links determined by history and then convert them into mental 'categories'. Above all, it is quite wrong and contrary to the Hegelian dialectic to suppose that one can set out to abolish the 'evil aspect' of a particular category while preserving its positive values. The contradictions that belong to each historical era are not ordinary blemishes that can be removed by simply taking thought; they are indispensable conditions of social development and of society's evolution towards maturity.

Suppose that the economists of the feudal era, captivated by all that was good in feudalism—the virtues of chivalry, the harmony of rights and obligations, the patriarchal life of the cities, the flourishing of cottage industry in the villages, the development of production in guilds, corporations and fraternities—suppose they had decided to preserve all this and simply remove the blemishes of serfdom, privilege and anarchy, what would have been the result? They would have rooted out all the elements of conflict and stifled the bourgeoisie at its very origin. They would have set themselves the absurd task of eliminating history.

Marx here follows the Hegelian interpretation of progress as the result of internal conflict, a process incompatible with the simple elimination of defects. 'Since the dawn of civilization,' Marx writes, 'production has been based on the antagonism of groups, estates and classes, and finally the antagonism between accumulated labour and direct labour. Where there is no antagonism there is no progress. This has been the rule of civilization until our own time. Up to this very day, class antagonism has been the cause of the development of productive forces.' It was absurd for Proudhon to seek to turn everyone into a capitalist and thereby cure the defects of capitalism—inequality, exploitation, and the anarchy of production—since this amounted to 'removing' social antagonisms while retaining their basic cause, or abolishing the proletariat while preserving the bourgeoisie.

All three of Marx's chief criticisms are aspects of a single idea: the historical process has a dynamic of its own, which is governed by the level of technology ('The hand-mill gives you society with the feudal lord; the steam-mill, society with the industrial capitalist') and which works itself out by means of the class struggle. It follows that a social upheaval cannot be brought about by moralizing, that outworn structures cannot be revived, and that social conflicts cannot be resolved by eliminating one of the contenders. The struggle must be allowed to reach its final form, in which both antagonists will give place to a higher type of organization: the proletariat, in the revolution, will liquidate itself as a class and by so doing will destroy all class differences.

13. The Communist Manifesto

In 1847–8 events took place which decisively affected the communist movement and its propaganda on Marxist lines. A group of German communists in Brussels, with whom Marx collaborated, were in contact with similar bodies in other countries, including the Bund der Gerechten, which at the end of 1846 had transferred its headquarters from Paris to London. One of its leaders, Joseph Moll, invited Marx and Engels to join the League and draft a programme: the League at this time was operating on the basis of an eclectic mixture of socialist ideas and lacked a coherent theoretical basis. In June 1847 Engels attended the League's congress in London. On the advice of Marx and Engels its name was changed to The Communist League, and its motto 'All men are brothers' was replaced by the class-conscious slogan 'Proletarians of all countries, unite!' Marx and Engels organized branches in Brussels and Paris respectively, and Engels drew up a programme in question-and-answer form entitled 'Principles of Communism': this dealt with capitalist exploitation and the inevitability of crises and described the future society based on community of goods, political democracy, equal wages, and planned industrial production. The document also spoke of the necessity of a simultaneous political revolution in all civilized countries. At the end of November and beginning of December Marx and Engels both attended the League's second congress in London and were entrusted with the task of composing what became a fundamental text of scientific socialism—the *Manifesto of the Communist Party*. This masterpiece of propagandist literature was first published in February 1848, and in subsequent editions was entitled *The Communist Manifesto*.

The *Manifesto* deals in turn with relations between the bourgeoisie and the proletariat, between the communists and the proletariat, and between communism and existing socialist doctrines. The first section contains the classic sentence: 'The history of all hitherto existing society is the history of class struggles.' After the antagonisms in the ancient world between freeman and slave, patrician and plebeian, after the lords and serfs of the feudal era, the basic structure of the present age consisted in the opposition between the bourgeoisie and the proletariat. Modern society had simplified the class situation: the division into two basic classes was more and more evident and was becoming more and more widespread. The discovery of America and the rise of industry had created a world market and, after long struggles, had given the bourgeoisie a commanding role in political life. The bourgeoisie had accomplished a revolutionary task without precedent by destroying the patriarchal, so-called 'natural' ties between human beings and reducing their mutual relations to the level of unabashed self-interest. It had turned the working man's 'vocation' into wage labour and had impressed a cosmopolitan stamp on trade, industry, and the whole of civilization, breaking down national barriers and involving the world in a breathless rush of technical and cultural progress. 'The bourgeoisie ... has been the first to show what man's activity can bring about.' But, unlike the dominant classes of earlier times, the bourgeoisie is neither able to preserve the means of production unchanged nor desirous of doing so. It can only exist if technology, and therefore social relations, are being constantly revolutionized. More and more it subordinates agricultural production to itself, concentrates the means of production in general, and organizes, to serve its own

interest, national states with uniform legislative systems. But just as the victory of the bourgeoisie was due to the incompatibility of the social and legal institutions of feudal society with the productive forces that evolved in that society, so its downfall will be due to the contradiction between its own technology and the property relationships of capitalism. This contradiction manifests itself in periodic crises of overproduction which are overcome by the destruction of productive forces and the conquest of new markets, but these methods in turn lead to more and graver crises. 'Not only has the bourgeoisie forged the weapons that bring death to itself; it has also called into existence the men who are to wield those weapons—the modern working class, the proletariat.' The workers are obliged to sell themselves to the bourgeoisie at a price equal to the cost of reproducing their labour, i.e. the minumum that will keep them alive; they have become an appendage to the machine. Exploited by manufacturers, tenement owners, trades-men, and usurers, they rise in revolt, firstly against the new machines which throw them out of jobs and increase their insecurity, then against exploitation by their own employers, and finally against the capitalist system itself. At this stage their struggle becomes a political one, embracing ever-wider areas and uniting the proletariat on a national and then a worldwide basis. The proletariat is the only class that is truly revolutionary. The particular interests of the middle classes—peasants, craftsmen, small traders—are conservative; they would, if they could, arrest the inevitable process whereby capital is centralized and con-centrated and they themselves are forced down into the pro-letariat. They are in a state of gradual disappearance and can only be a revolutionary force in so far as they are proletarianized. The bourgeoisie, as industry develops, creates worse and worse con-ditions for the workers and thus drives them into solidary, united action. In this way it creates, unconsciously but inevitably, its own grave-digger. The bourgeoisie has proved that it cannot maintain itself as the ruling class and is doomed to destruction. The workers, for their part, can only gain control of productive forces by demolishing the whole system by which wealth has hitherto been acquired. 'The proletarians ... have nothing of their own to secure and fortify; their mission is to destroy all previous securities for, and insurances of, individual property.'
    Communists have no interests apart from those of the pro-

letariat, and they are distinguished from other proletarian parties by the fact that they stand for the proletariat's interest as a whole, irrespective of national differences. They are in advance of the proletarian masses owing to their theoretical understanding of the world in which the struggle is going on. Their aim is to lead the proletariat in the conquest of political power, to destroy the bourgeois property system which enables the capitalist to appropriate others' labour, and to abolish the bourgeoisie and the proletariat as social classes. In addition, the *Manifesto* replies as follows to the accusations most often levelled against communism:

1. 'The abolition of private property will lead to general idleness and the collapse of production.' But private property does not exist today for the masses, yet society exists and maintains itself.

2. 'Communism is a denial of individuality.' Yes—of such individuals as are enabled by the system to use their own property as an instrument for the enslavement of others.

3. 'Communism destroys the family.' It destroys the bourgeois family, based on property-ownership on the one hand and on prostitution and hypocrisy on the other. Big business has destroyed the family life of the proletariat.

4. 'Communism is against nationality.' But the working man has no fatherland, so how can he be deprived of one? In any case the world market is effacing national differences, and the victory of the proletariat will intensify this process. When the exploitation of man by man is abolished we shall also see an end of exploitation, oppression, and enmity among nations. National oppression is the outcome of social oppression.

5. 'Communism seeks to destroy the eternal truths and sublime ideas of religion, ethics, and philosophy.' But all the ideas bequeathed by history are absolute only in so far as exploitation and oppression have persisted despite all changes in political systems. The spiritual output of mankind is as changeable as the conditions of human existence; ideas are permanent in so far as particular social relations have hitherto been permanent. Communism overthrows 'eternal' ideas by destroying the class system which, by existing from time immemorial, gave them the appearance of eternity.

The socialist propaganda of the time is criticized in the *Manifesto* according to its class origin. In the first place there is feudal socialism, which opposes capitalism from the standpoint of the aristocracy ruined by the bourgeois property system (the French legitimists, 'Young England'): invoking the patriarchal bliss of olden times, it attacks the bourgeois for subverting the ancient order and, above all, for creating the revolutionary proletariat. The same may be said of Christian socialism, 'holy water with which the priest consecrates the heart-burnings of the aristocrat'. Petty-bourgeois socialism (Sismondi) reflects the small producers' fear that industry will drive them out of existence. It argues that increased mechanization, the concentration of capital, and the division of labour infallibly lead to crises, poverty, gross inequality, war, and moral disintegration; this is true, but the proposed remedy of a return to the pre-capitalist system of production and exchange, with guilds and a patriarchal peasant economy, is reactionary and useless. As for the 'true socialism' of Grün and other German writers, it is a sentimental tissue of speculation and generalities about mankind regardless of class divisions and the particular interests of the workers. Socialists of this school attract the approval of the feudal classes who still govern Germany by attacking the liberal bourgeoisie which, in that country, is the true vehicle of progress.

Such are the brands of reactionary socialism. Then there is the bourgeois socialism of Proudhon and others, which seeks to preserve existing conditions by eliminating everything that tends to revolutionize society—'to keep the bourgeoisie and get rid of the proletariat'. It relies on philanthropic slogans and administrative reforms, making no effort whatever to abolish the bourgeois property system.

Lastly, utopian socialism or communism as preached by Saint-Simon, Owen, and Fourier, while aware of the class struggle and the oppression of the proletariat, fails to perceive the latter's key historical role and makes it a mere passive object of reformist plans. These theorists reject the prospect of revolution and fix their sights on the community as a whole or on the privileged classes. They have played a useful part in criticizing bourgeois society and advocating reforms, but, having attempted to rise above the actual class struggle, their successors in later generations turn into reactionary sects whose aim is to extinguish class antagonisms and prevent independent political action by the proletariat.

Communists in different countries support various political

movements, but only those which aim at a radical transformation of existing conditions. Germany is especially important to them, as the imminent bourgeois revolution in that country will take place against the background of more advanced social conditions, in Europe and even in Germany herself, than did the bourgeois revolutions in France and England: so much so that the German bourgeois revolution 'can only be the direct prelude to a revolution of the proletariat'.

Marx and Engels saw little cause to revise subsequent editions of the *Manifesto* as far as its theoretical bases were concerned. Apart from their over-sanguine expectations of revolution in Europe and their failure to foresee developments which could not have been predicted at the time (the *Manifesto* does not mention either Russia or America as potentially revolutionary countries), their later prefaces or amendments only involve one important point of theory: the experience of the Paris Commune convinced them that the revolutionary proletariat cannot capture the state machine and use it for its own purposes, but must start by destroying it.

As regards the controversy with socialists of the earlier part of the century, Engels reverted to this in 1878 in the *Anti-Dühring*, which repeats the *Manifesto*'s main criticisms of utopian socialism. This doctrine, he says, is the product of a situation in which the working class has not yet matured to the point of taking a historical initiative of its own, and appears merely as an oppressed and suffering group and not as the vehicle of social revolution. Utopian socialism is precluded, by the very conditions of its origin, from envisaging socialism as a historical necessity of the present time rather than an ingenious theory, an intellectual windfall which might have occurred at any period. Whenever Marx and Engels, the creators of scientific socialism, revert to the subject of their utopian predecessors they repeat the three basic charges of philanthropism towards the working class, rejection of the prospect of revolution, and the conception of socialism as an accidental theory. To these errors they oppose their own view of socialist theory as the self-awareness of the actual revolutionary initiative of the working class, a free activity which is nevertheless historically necessary. Engels pays tribute to the utopians, however, for the sharpness and boldness of their attack on the contemporary world and the inventiveness of their predictions of the future; he does not look down on them from the height of a superior revelation, since he is aware of the historical conditions that restricted their field of vision.

With the appearance of *The Communist Manifesto*, we may say that Marx's theory of society and his precepts for action had attained completion in the form of a well-defined and permanent outline. His later works did not modify what he had written in any essential respect, but enriched it with specific analyses and transformed what were sometimes no more than aphorisms, slogans, or heads of argument into a massive theoretical structure. We may, therefore, after a short review of relevant historical events, abandon our chronological exposition for one based on subject-matter. Special attention is due, however, to Engels's theory of the dialectic of nature and his interpretation of philosophical materialism, since these may be regarded as a substantive change in Marxism as it existed in the years before 1848. Naturally, the principles that were established then and elaborated later were at no time so expressed as to preclude mutually inconsistent interpretations. As the socialist movement and socialist theory progressed, it often happened that Marx's views on this or that subject—historical determinism, the theory of classes, of the state, or of revolution—were understood differently by different people. This is the natural fate of all social theories without exception—at all events those that have been a real force in politics and social development, and from this point of view no modern theory can rival Marxism. However, the most important controversies as to the exact interpretation of Marx's theory took place after his own lifetime.

CHAPTER XI

The Writings and Struggles of Marx and Engels after 1847

1. Developments in the 1850s

THE publication of *The Communist Manifesto* coincided with the political convulsions of 1848. After the February Revolution in Paris the Belgian government adopted repressive measures against the *émigré* revolutionaries; Marx was expelled from Brussels and returned to Paris, where he worked for the German revolutionary cause on behalf of the Communist League. After the Vienna and Berlin revolutions in March many German *émigrés* made their way from France to Germany; Marx and Engels established themselves in Cologne, where communist propaganda was most active, and from June onwards published a newspaper, the *Neue Rheinische Zeitung*, with a programme based on a flysheet previously composed by them, entitled *Demands of the Communist Party in Germany*. These aims were not communistic as such, but radical-democratic and republican: they included the confiscation of large estates, free universal education, a progressive income tax, and the nationalization of railways. The paper, of which Marx was chief editor, condemned the pliant and irresolute attitude of the bourgeoisie and advocated a united Germany under a republican constitution with direct and universal suffrage; it championed the oppressed national minorities, especially the Poles, and called for war with Russia as the mainstay of reaction in Europe. The programme of alliance between the proletariat and the republican bourgeoisie for the sake of a democratic revolution was looked at askance by many German communists, who feared that if the working class did not maintain itself as a separate political entity it would be merely the instrument of a revolution in the bourgeois interest.

The victory of reaction in Europe and the collapse of the Frankfurt parliament put an end to Marx's revolutionary activity in Germany. The *Neue Rheinische Zeitung* closed down in May 1849; Marx was expelled from Prussia and made his way, not without difficulty, back to Paris, where he expected a fresh revolutionary outbreak at any moment. The French government, however, put obstacles in the way of his remaining, and in August, with no money and no means of livelihood, he embarked on a new life of exile in London. He was to spend the rest of his days there, wrestling with poverty, illness, and domestic troubles. Engels settled in Manchester in 1850 and remained there for twenty years, drawing an income from the cotton mill of which his father was co-owner. For many years he supported Marx financially, sacrificing his own literary work in order that his friend might be able to devote himself to academic writing.

Soon after their arrival in London Marx, Engels, and a few friends set about resuscitating the Communist League, which had been dissolved during the revolution. The manifesto they wrote for this purpose advocated a different programme from that of the *Neue Rheinische Zeitung*: it urged that the proletariat should organize itself independently of the republican bourgeoisie and, while supporting all democratic claims, should aim at a state of 'permanent revolution' which would enable it eventually to seize political power. Marx and Engels believed that the growing economic crisis was bound to touch off revolution in Europe, and especially in France, at an early date. When this hope proved vain, the League was condemned to an early demise; it was in fact wound up in 1852. The *Neue Rheinische Zeitung*, with the sub-title *Politisch-ökonomische Revue*, appeared in London for a few months only, in 1849. During the next two decades the European socialist movement subsisted in the margin of political life, but thanks to Marx's efforts it acquired a new theoretical basis which enabled it to spring vigorously into life when conditions changed. During the 1850s Marx reverted to economic studies and did not himself take part in any political organization, though he maintained some links with the Chartist leaders.

The first important work of Marx's to be published during his London period was *The Eighteenth Brumaire of Louis Bonaparte*, an essay on the *coup d'état* of December 1851: it constituted the first number of a New York journal, the *Revolution*, which had been started by Marx's friend Joseph Weydemeyer. The essay formed a sequel to *Class Struggles in France, 1848–50*, which had appeared in London in the *Neue Rheinische Zeitung*; in his new work Marx analysed the class situation which had enabled such a 'mediocrity' as Louis Napoleon to seize power. It is rich in general observations and contains some of Marx's most frequently quoted aphorisms.

The trial of a group of communists in Cologne in 1852, in which evidence was produced purporting to incriminate Marx, provoked him to expose the fabrications of the Prussian police; the most important document, *Revelations concerning the Communist Trial in Cologne*, appeared anonymously at Basle in 1853. From 1851 to 1862 Marx contributed articles on current affairs to the *New York Daily Tribune*, some of which were by Engels though they appeared over Marx's signature. This did not suffice to provide a livelihood, but it helped to mitigate the family's direst poverty. For years on end they were short of money for rent, paper, and footwear; Marx was notoriously incapable of keeping accounts, and Jenny was a regular customer of the London pawnbrokers. Marx at one stage tried for a job as a railway official, but was rejected on account of his execrable handwriting.

Marx's chief occupation during these years, however, was the elaboration of his critique of political economy which had begun with the Paris Manuscripts of 1844. Again and again be believed that he had come to the end of the work, but his restless thoroughness impelled him constantly to seek fresh data and new sources with which to improve the draft. The economic crisis of 1857 prompted him to compose a revised version, but this was never completed and was not published in his lifetime. The Introduction to this work was published in 1903 by Kautsky in *Die Neue Zeit* (Stuttgart), and is Marx's fullest and most important study of the problems of method in the social sciences. The whole work, entitled *Grundrisse der Kritik der politischen Ökonomie (Outline of a Critique of Political Economy)*, was first published in Moscow in 1939–41, an unpropitious time for academic study. It was republished in East Berlin in 1953, but was not subjected to thorough examination and discussion till the 1960s. It is of interest as showing the continuity of Marx's thought from the Paris Manuscripts to *Capital*; it contains, for instance, a new version of the theory of alienated labour which throws light on the significance of this category in Marx's later work.

In general the text of the *Grundrisse* shows that Marx had not abandoned his anthropological ideas of the 1840s but was attempting to translate them into economic terms. We also know from a letter of his that the method of the work was influenced by a rereading of Hegel's *Logic*, a copy of which had happened to come his way. The Introduction contained a general plan of the work he intended to write, and as this plan is only partially fulfilled in *Capital* there has been discussion as to whether, or how far, he subsequently changed his mind. However, the recent studies of McLellan and others have shown clearly that there is no reason to think he made any essential change. The three volumes of *Capital*, which deal with the theory of value, money, surplus value, and capital accumulation (Volume I), circulation and reproduction (Volume II), and profits, rent, and credit (Volume III), constitute a portion of the structure as originally planned, while the *Grundrisse* is the first sketch, and the only one, covering the whole ground, i.e. it provides the most comprehensive exposition of Marx's economic doctrine that we possess. It contains the first statement of some important ideas that appear in *Capital*—for example, the theory of the average rate of profit and the distinction between constant and variable capital—and also some themes that are not to be found in the later work. Among these—apart from the earliest portion, criticizing Carey and Bastiat—are the observations on foreign trade and the world market and the philosophical passages scattered through the work, in the style of the 1844 Manuscripts. The publication of the *Grundrisse* has not altered the general picture of Marxist doctrine in any important respect, but it has upheld the view of those who believed in the continuity of Marx's philosophical inspiration, and not of those who postulated a radical breach between the anthropological theories of his youth and the economic tenets of his maturer years.

Another economic work of Marx's did see the light of day at this time, viz. *Zur Kritik der politischen Ökonomie* (*Contribution to a Critique of Political Economy*), published with Lassalle's help at Berlin in 1859. Here Marx expressed for the first time his theory of value, different from Ricardo's, though he did not develop it to a conclusion. The Preface to this work is one of Marx's most-quoted texts, as it contains the most concise and general formulation of what was later called historical materialism.

In 1859–60 much of Marx's energy was devoted to a polemic with Karl Vogt, a German politician and naturalist who was then teaching at the University of Berne. The immediate cause of the quarrel was that Marx accused Vogt—without much evidence, but, as later events showed, correctly—of intriguing in support of Napoleon III at the time of the Franco-Austrian War. Apart from this, Vogt was the advocate of a crude and vulgar form of materialism ('Thought is a secretion of the brain just as bile is of the liver'). Marx's work *Herr Vogt*, published in 1860, denounced him for intrigue, slander, and double-dealing; it is now, however, of no more than biographical interest.

2. Lassalle

Apart from Proudhon, Marx's chief rival as a theoretician in the 1860s was Lassalle, who for many years outclassed him as far as ideological influence in Germany was concerned.

Ferdinand Lassalle (1825–64) was the son of a Jewish trader from Breslau. He studied philosophy and philology in Berlin and Breslau in 1843–6 and intended to embrace an academic career. He became a Hegelian (though not a Young Hegelian), read socialist literature, and decided at an early age that he was destined to be an eminent philosopher and to transform social conditions in Germany. However, his energies were for a long time absorbed by personal affairs. He fell in love with Countess Sophie von Hatzfeld, who was nearly twice his age, and for ten years chivalrously defended her financial interests against her estranged husband in innumerable German courts. In this connection he was arrested early in 1848 for complicity in the theft of certain documents. He was released six months later, but reimprisoned for some months in November for incendiary speeches in support of the revolution. From 1849 to 1857 he lived at Düsseldorf. During this time he corresponded with Marx (they had first met in 1848) and also wrote a large work on Heraclitus (*Die Philosophie Herakleitos des Dunklen von Ephesos*, 1857); Marx in a letter to Engels, dismissed this as a diluted version of the relevant part of Hegel's *History of Philosophy*. In 1859 Lassalle published a historical drama, *Franz von Sickingen*, on the subject of a sixteenth-century knight who headed a league to spread the Reformation in Germany; his tragic fate was apparently intended to symbolize the defeat of the 1848 Revolution. The work is full of patriotic sentiment and faith in the German mission. In 1860 Lassalle wrote articles on Fichte and Lessing, and in 1861 he published his most important work, *The System of Acquired Rights*—a philosophical, historical, and political treatise which was well received by the academic world. After reviewing the history of the Roman and Germanic laws of inheritance Lassalle discussed the question which had also been raised by Savigny: in what circumstances can acquired rights lose their validity? This had a clear bearing on current politics, as the defenders of privilege invoked the classic rule that a law cannot act retrospectively: from this they deduced that new laws could not extinguish rights acquired under earlier ones. Lassalle's counter-argument was on the following lines. Acquired rights are those created by the deliberate activity of an individual; but the law tacitly presupposes that they are valid only for so long as such rights are allowed in general by the legal system in force, and the legal system derives its legitimacy from the consciousness of the nation as a whole. If a certain type of right or privilege is forbidden by later laws, the individual cannot appeal to the formula *lex retro non agit* and claim, for instance, that he has a right to keep slaves or serfs or to be immune from taxation, simply because 'it has always been so'. In this way Lassalle defended the legality of social changes that involved the abolition of privilege.

Lassalle's activity as a politician and ideologist of the workers' movement began, properly speaking, in 1862 and lasted (owing to his early death) for little more than two years. He was now living in Berlin and took an active part in the Prussian constitutional controversy, attacking the liberals of the Progressive party (Deutsche Fortschrittspartei). In the spring of 1862 he published an address to the workers, later known as the *Arbeiterprogramm*, which became the classic exposition of his views; also a speech on the constitution and a lecture on Fichte.

The Progressive party had a strong following in the Prussian working class; one of its leaders, Schulze-Delitzsch, was a promoter of friendly societies, insurance funds, and consumer co-operatives as methods of improving the lot of the proletariat within the framework of co-operation between capital and labour. However, some groups were not content with the patronage of the liberal bourgeoisie, and one of these, in Leipzig, appealed to Lassalle to state his position in regard to the workers' movement. Lassalle responded in January 1863 with an Open Letter which became a kind of charter of the first German working-class socialist party, the Allgemeiner Deutscher Arbeiterverein, founded in May of that year.

At the same time, as became known afterwards. Lassalle entered into contact with Bismarck in the evident hope of contracting an alliance with the conservatives against the bourgeoisie. In a speech in the Reichstag in 1878, Bismarck said he had had several talks with Lassalle at the latter's request, but these had not been negotiations, for the simple reason that Lassalle represented no political force and had nothing to offer; he described Lassalle, however, as a man of intellect and a true patriot.

The Arbeiterverein had no special success during Lassalle's lifetime, but it grew to a membership of about a thousand and was the first independent political expression of the German working class. In August 1864 Lassalle was killed in Geneva in a duel over a girl of seventeen whom he wished to marry; her aristocratic family refused to receive him, she herself changed her mind and returned to a previous fiancé, and Lassalle wrote an insulting letter which resulted in a challenge and his death.

Marx and Lassalle met in Berlin in 1861 and in London in the following year. They were never on cordial terms; Marx distrusted Lassalle and criticized him repeatedly in letters to Engels and others, while his political disagreement was expressed most notably in 1875, years after Lassalle's death, in the *Critique of the Gotha Programme*. There were also personal grounds of dislike and irritation. Lassalle was a man of outstanding powers, but he was also an ostentatious parvenu and something of a play-actor. In 1860 he wrote, to a woman with whom he was in love at the time, a 'confession' which is an extraordinary specimen of naïve self-praise. He represents himself as a genius adored by the people, the leader of a revolutionary party (which then existed only in his imagination), a new Robespierre, the terror of his enemies; a man of thirty-five with the experience of a sage of ninety, and the possessor of an income of 4,000 talers a year.

However, Marx's conflicts with Lassalle were not mainly due to personal antipathy. They differed on almost every point of substance: economic doctrine, political tactics, their attitude to the state in general and the Prussian state in particular. In general it may be said that the points on which their views coincided had nothing specifically Marxist about them. Some of their dissensions were as follows.

Firstly, the economic diagnosis of the situation of the proletariat. Lassalle stated in his Open Letter of 1863 that the liberals were mistaken in thinking they could liberate the working class by means of insurance funds, co-operatives, and so on; this, of course, was in accordance with Marx's view. However, Lassalle went on to prove his point by the 'iron law of economics' that, when wages are determined by the supply of labour and the demand for it, they are bound to gravitate to the 'physiological minimum' necessary to keep the workers and their children alive. If wages rise for any reason, the working classes will have more children and the increased supply of labour will push down wages; if wages fall below the minimum the workers will have fewer children, the demand for labour will exceed supply and wages will go up. The vicious circle is inevitable as long as supply and demand govern the wage level.

Lassalle took this doctrine over more or less literally from Malthus and Ricardo. Marx never professed it in this form, and although he sometimes took the view (especially in his earlier works) that wages must tend towards the physiological minimum, he did not accept Lassalle's supporting argument, which gave sole weight to the demographic factor in determining labour supply and demand. It was clear, indeed, that supply and demand could not be measured absolutely but only in relation to the whole economic picture, including such matters as boom and slump, the state of world markets, technical progress, the proletarianization of the peasantry and the petty bourgeoisie, and finally the effect of working-class pressure on wages. According to circumstances these factors might collectively push wages up or down, but in any case it was a gross over-simplification to reduce the whole problem to that of the birth-rate of an existing proletariat. Moreover, Lassalle contradicted himself in the same document when he said that minimum needs increase as general progress increases, so that one cannot speak of an improvement in the workers' lot by comparing their present position with the past: workers may be earning more in absolute terms, yet be worse off in relation to their total needs. It follows that the minimum is not merely a physiological but also a social and cultural one. Thus understood, the theory of 'relative pauperization' is close to Marx's views as he expressed them in the fifties and sixties.

Secondly, Lassalle differs radically from Marx in inferring from the 'iron law of economics' that the right way to emancipate the workers is to develop producer co-operatives, in which they will be paid wages equal to the value of the goods they produce. As the proletarians cannot set these up by their own efforts, the state must help them with credit institutions. For this to happen the workers must be able to exert pressure on the state, which they can only do if there is universal, direct, and equal suffrage.

This programme was contrary to Marx's theory in at least three important respects. In his view the domination of the economy by producers' associations was simply a repetition of Proudhon's Utopia: units of this kind, even if they belonged to the workers, could only exist in a state of competition like that which now prevailed. The laws of the market would continue to operate; there would still be crises, bankruptcies, and the concentration of capital. In any case, wages could never be fully equal to the value of the goods produced, since part of that value must be devoted to public needs, necessary unproductive work, reserves, etc. Finally, the programme whereby the state was to be the agent of working-class emancipation under capitalist conditions was contrary to Marx's idea of the state as a defensive weapon of the privileged classes.

Lassalle criticized the liberal theory of the state from a Hegelian point of view: as he wrote in the *Arbeiterprogramm*, the state's only function according to the bourgeoisie was to protect the freedom and property of individuals, so that if there were no criminals it would have nothing to do. In reality, however, the state was the highest form of human organization, in which all human values were actualized, and its function was to lead the human race to freedom; it was a unity of individuals in a single moral entity, and the instrument whereby man is to fulfil his destiny. In writing this, Lassalle had in mind the Prussian state; unlike Marx, he was a German patriot and saw the events of his time, including wars, from a national viewpoint rather than a proletarian international one. He believed German unity to be an issue of supreme importance, and thought Bismarck's policies would bring more gain than loss; moreover, the true antagonist of the proletariat was the bourgeoisie, so an alliance with the conservatives might well be desirable. This was directly contrary to Marx's general line that when the claims of the liberal bourgeoisie conflict with the interests of conservative, feudal, or monarchist elements, the proper course for the proletariat is to ally itself with the former.

The philosophical basis of Lassalle's nationalism is seen most clearly in his lectures on Fichte, where he says that the latter's ideas embody the spiritual greatness of the German people. The endeavour of all German philosophy is to overcome the duality of subject and object, to reconcile the spirit with the world and to achieve the mastery of 'spiritual inwardness' (*die Innerlichkeit des Geistes*) over reality. Fichte had proclaimed the mission of the German people to march in the forefront of human progress and to vindicate the divine plan of creation by attaining national independence. Germany was not only a necessary aspect (*Moment*) of world history but was destined to be sole champion of the idea of liberty on which the future of mankind depends. Precisely because it had had no proper history for centuries, being a 'pure metaphysical inwardness' and not a state, it had become the birthplace of the philosophical idea which set out to reconcile thought and being.

The metaphysical nation, the German nation, has had bestowed on it, throughout its development and in the perfect accordance of its internal and external history, the supreme metaphysical destiny and the uttermost honour in world history—namely, that of creating a national territory out of the spiritual concept of a nation, and evolving its own being out of pure thought. To a metaphysical nation belongs a metaphysical task, an achievement no less than that of the divine creation. Pure spirit not only informs the reality presented to it but creates a territory, the very seat of its own existence. There has been nothing like this since the beginning of history. ('Die Philosophie Fichtes', in F. Lassalle, *Reden und Schriften*, ed. Hans Feigl, 1920, p. 362)

The Fichtean–Romantic conception of state and nation took precedence, in Lassalle's thinking, over his semi-Marxist vision of the proletariat as the liberator of the world. He appears to have felt his Jewish origin as a stigma, though he made no attempt to hide it—he used to say that he had always hated two kinds of people, Jews and literary men, and that unfortunately he himself was both—and he lost no opportunity of proclaiming his patriotic feelings. In his glorification of the state, the organic unity of the nation, and the spiritual leadership of Germany he was, like Fichte before him, a pioneer of national socialism. His inflated, prophetic style exasperated Marx no less than their theoretical disagreements. Yet his practical success is beyond dispute: his insistence on an independent proletarian movement laid the foundations of organized socialism in Germany. Among later orthodox Marxists opinions on him were divided. Mehring emphasized Marx's personal dislike of Lassalle and minimized the political and theoretical differences between the two, whereas Kautsky held that their ideas of socialism were completely different. At all events it was clear that Lasalle's theoretical horizon, unlike Marx's, was limited to Germany; so was his political influence, but in that country it was powerful and lasting. Even in later years, when German social democracy had finally abandoned Lassalle's programme, his spirit was still discernible in the party, both in the strain of nationalism which persisted beneath the surface and in the belief that the existing machinery of the state could be made to serve the interest of the proletariat.

3. The First International. Bakunin

From the mid-sixties onwards Marx was less involved in combating Lassalles's views than in polemics against other schools of thought within the International, especially those of Proudhon and Bakunin.

The International Working Men's Association, to give it its full title, was established at a public meeting in London in September 1864. A year earlier the first organizational links had been formed between British and French trade unionists on the occasion of demonstrations in support of the Polish insurrection against Russia. The meeting in 1864 was attended by German, Polish, and Italian *émigrés* as well as the British and French, and it was decided to create an international body to co-ordinate the working class struggle in different countries. A General Council of thirty-four members was elected, with George Odger, a London trade unionist, as its president. Marx was elected to the Council and was made corresponding secretary for Germany; he also played a leading part in drafting the Rules and Inaugural Address. The latter described briefly the fortunes of the European proletariat since 1848. It pointed out that the working class was increasingly impoverished and property more concentrated, that there had nevertheless been successes in reducing working hours and in the co-operative movement, but that the emancipation of the proletariat depended on its conquering political power. This could only be achieved by the international action of the workers, who were a class with common interests independent of country or nationality. They were not fighting to replace existing privileges by others, but to put an end to class domination. However, the approved texts contained no express revolutionary demands.

During the next few years the International endeavoured with moderate success to organize sections in various European countries: outside Britain, these were formed in various towns in France, Belgium, and Switzerland, generally on the basis of existing organizations. Lassalle's party remained outside the International, largely owing to disagreement over its attitude to Bismarck and to German bourgeois democracy. The British unions, some of which joined the International, pursued a separate policy of their own. The French were mostly Proudhonists, and expressed their differences from Marx at the congresses at Geneva (September 1866) and Lausanne (September 1867). Among other things they objected to the Polish question being discussed at meetings or mentioned in manifestos; Marx, on the other hand, believed that Polish independence was inseparable from the cause of the European workers and that the most urgent task was to break the reactionary power of the Tsardom. The Proudhonists, like their master, were mistrustful of political action in general and held to the belief in 'mutualism', which in Marx's eyes was purely utopian.

The Rules of the International were loose enough to permit the membership of a large variety of groups. Besides the British unions and the Proudhonists it included, for some years, French radicals and the partisans of Mazzini. The federation was a loose one, and the General Council had no executive powers over its members. Throughout its existence Marx devoted the greater part of his time to its affairs, with three chief objectives that were especially evident in the later years. He wanted the International to become a centralized body that could impose a uniform policy on its sections; he strove to make the whole movement accept the ideological bases he had himself worked out; and he hoped to turn the International into a weapon against Russia. Despite his prestige he failed in all three of these aims, and his policies led to a breach within the International which was a major cause, if not the main cause, of its collapse. Marx himself attended only one congress of the International, the final one, held at The Hague in 1872.

The economic crisis of 1867 and the wave of strikes in many European countries were propitious to the International: new sections were created in Spain, Italy, Holland, and Austria, while in Germany a new social democratic party was formed, alongside the Lassallists, by Liebknecht and Bebel; this did not formally join the International, but was closer to Marx on the main issues. The influence of the Proudhonists grew weaker; at the Brussels Congress in September 1868 the International called for the collective ownership of arable land, forests, roads, canals, and mines, and declared itself in favour of the strike weapon.

The year 1869 marked the zenith of the International's activity and influence, but also witnessed the beginning of the fatal split between the outstanding figures of the nineteenth-century revolutionary movement, Marx and Bakunin. These two leaders held diametrically opposite views on strategy and on the subject of the working class, revolution, the state, and socialism.

Mikhail Alexandrovich Bakunin (1814–76) had a long and adventurous political past behind him when he joined the International in 1869. Born in the province of Tver, of aristocratic family, he began his education at a military school but left it after a short time. He spent some years in Moscow, where he frequented intellectuals who discussed the future of Russia and the world in the light of Hegel's philosophy of history. For a time he was a Hegelian conservative, believing in the rationality of actual history and holding that the individual has no right to assert his accidental subjectivity against the decrees of universal reason. Soon, however, he passed to the opposite extreme, which was certainly more suited to his temperament. He went to Berlin in 1840, met the Young Hegelians and was infected by their ideas. In further travels to Switzerland, Belgium, and France he met the chief socialist writers of the time: Cabet, Weitling, Proudhon, and finally Marx and Engels. He also met many Poles of the post-1830 emigration, and from then onwards devoted much attention in his writings to the cause of Polish independence. In the 1840s he agitated for a Slav federation, an idea which he later rejected as ineffective or reactionary. He never abandoned his hatred of Germany, however, which was as violent as Marx's hatred of Russia.

The two men clashed for the first time during the 1848 Revolution, when an article in the *Neue Rheinische Zeitung* accused Bakunin of being a Tsarist agent—a libel which the newspaper was compelled to withdraw. Bakunin took an active part in the revolutionary struggle in Prague and Dresden; he was twice condemned to death and finally expelled to Russia, where he spent the next twelve years in prison and exile. From one of his prisons he addressed an extraordinary Confession to Tsar Nicholas I, (first published after the October Revolution), expressing repentance for his subversive activity but warning that the fearful conditions in Russia might lead to revolution. In 1862 he escaped from Siberia to Japan and made his way via America to London. His career as a theoretical and practical anarchist dates from 1864, when he founded a clandestine group known as Fraternité Internationale—a loose organization of his friends and adherents, chiefly in Spain, Italy, and Switzerland. In September 1868 he established an overt anarchist association, the Alliance Internationale de la Démocratie Socialiste, which applied to join the International. The latter's Council refused to accept the Alliance as such, but in 1869 it agreed that individual sections might join, including the Geneva one to which Bakunin belonged and which was the only properly organized group. From then on Marx and Bakunin were engaged in a conflict in which it is hard to distinguish political from personal animosities. Marx did his best to persuade everybody that Bakunin was only using the International for his private ends, and in March 1870 he circulated a confidential letter to this effect. He also saw the hand of Bakunin (whom he never met after 1864) on every occasion when his own policies were opposed in the International. Bakunin, for his part, not only combated Marx's political programme but, as he often wrote, regarded Marx as a disloyal, revengeful man, obsessed with power and determined to impose his own despotic authority on the whole revolutionary movement. Marx, he said, had all the merits and defects of the Jewish character; he was highly intelligent and deeply read, but an inveterate doctrinaire and fantastically vain, an intriguer and morbidly envious of all who, like Lassalle, had cut a more important figure than himself in public life.

Politics apart, the history of Marx's relations with Bakunin does not show the former in a favourable light. His charge that Bakunin was using the International for personal advantage was groundless, and his efforts to have Bakunin expelled were finally successful (in 1872) thanks in the main to the Nechayev letter, for which Marx must have known that Bakunin bore no responsibility. Bakunin, of course, worked for the victory of his own ideas in the International, just as Marx did. At the Basle Congress in 1869 the Bakuninists secured the adoption (contrary to Marx's standpoint) of a proposal declaring that the abolition of the right of inheritance was a basic feature of the social revolution. From 1870 there was increasing dissension within sections of the International, and in Switzerland, Italy, and Spain Bakunin's adherents predominated over those of Marx. Bakunin's last years were mainly devoted to writing. In 1870 he published *L'Empire knouto-germanique et la révolution sociale*, and in 1873, in Russian, his only work of any size, *Statehood and Anarchy* (translated as *Etatisme et anarchie*). This was intended as the introduction to a larger work (which he never wrote), and contains all the important ideas of his anarchist period. It is an unsystematic collection of remarks on the most varied subjects: European and world politics, Russia, Germany, Poland, France, China, the 1848 Revolution, the Paris Commune, attacks on communism, and various philosophical observations.

Bakunin had not the gifts of a theoretician or a founder of systems. He was full of inexhaustible revolutionary energy, bent on destructive aims and inspired by anarchistic Messianism. He could not endure situations which required long-term political calculations, tactical manoeuvres, and temporary alliances. He expressed, as he was well aware, all the spirit of revolt which grew among the most deprived elements of the working class, the lumpenproletariat and the peasantry. According to him, 'state communism', i.e. Marx's variety, was supported by the better-off, relatively secure workers who had acquired bourgeois habits, while he himself appealed to the ragged paupers who were still uncorrupted and had nothing to lose. He referred repeatedly to the rebellions of Pugachev and Stenka Razin in Russia—elemental, instinctive uprisings of the desperate peasantry led by 'bandits' (his own expression). Marx's adherents, he declared, despised the people; had not Lassalle written that the suppression of the peasant revolt in sixteenth-century Germany had been a major contribution to historical progress? Marx and Lassalle, who were divided by nothing but Marx's personal jealousy, were the upholders of a new state despotism which was bound to develop out of their 'scientific socialism'.

Bakunin's whole doctrine centred in the word 'freedom', while the term 'state' epitomized all the evil which must be banished from the world. He accepted to some extent the theory of historical materialism, in the sense that human history depends on 'economic facts' and that men's ideas are a reflection of the material conditions in which they live. He also espoused philosophical materialism (under this name), based on atheism and the rejection of any notion of 'another world'. But he believed that the Marxists absolutized the principle, in itself correct, of historical materialism into a kind of fatalism which left no room for the individual will, for rebellion, or for moral factors in history.

Maintaining the primacy of 'life' over 'ideas', Bakunin rejected the doctrine of 'scientific socialism' which assumed that it was possible to organize social life on the basis of schemata devised by intellectuals and imposed on the people. Political or moral propaganda could only convince the masses in so far as it accorded with what was in their minds and hearts but had not yet found expression. It was no use hoping to enlighten the Russian people by means of academic theories; they would only accept what they already knew after a fashion but had not been able to articulate. In general, science was no more than a function of life and could not claim supremacy over its other manifestations. It was necessary and should be respected, but it could not grasp phenomena in their fullness: it reduced them to abstractions and ignored individuality and human freedom. Life was creative; science was uncreative and was no more than a facet of reality. The social sciences in particular, which were still in their infancy, could not claim to foretell the future or impose ideals on mankind. History is a process of spontaneous creation, not the working-out of scientific schemes; it develops like life itself, instinctively and in an unrationalized manner.

Bakunin's idea of the revolt of life against science, though hedged with reservations concerning the value of knowledge, was to serve as the basis for versions of anarchism which regarded all academic thought as a crafty invention of the intelligentsia to maintain their privileges under the cloak of mental superiority. Bakunin did not go so far as this, but he inveighed against universities as the abodes of élitism and seminaries of a privileged caste; he also warned that Marxist socialism would lead to a tyranny of intellectuals that would be worse than any yet known to man.

'Life', in Bakunin's sense, is an endless, indefatigable endeavour towards freedom for every individual, every community, and the whole human race. Freedom in turn presupposes equality, not merely before the law but in reality, that is to say economic equality. Freedom and equality are opposed by the system of privileges and private property safeguarded by state power. The state is a historically necessary form of communal life, but it is not eternal and is not merely a superstructure imposed on 'economic facts'; on the contrary, it is an essential factor in maintaining privilege, exploitation, and all forms of slavery. The state by its very nature signifies the enslavement of the masses by a despotic, privileged minority, whether priestly, feudal, bourgeois, or 'scientific'. 'Any state, even the most republican and the most democratic, even the pseudo-popular state imagined by Marx, is essentially nothing but the government of the masses by an educated and therefore privileged minority, which is supposed to understand the people's needs better than they do themselves' (*Statehood and Anarchy*, pp. 34–5). The task of the revolution, accordingly, is not to transform the state but to abolish it. The state is not to be confused with society: the former is an artificial means of oppression, the latter a natural extension of the instinctive ties that bind human beings together. To abolish the state does not mean abolishing all forms of co-operation and organization; it means that every social organization must be built up entirely from below, without authoritarian institutions. Bakunin does not accept Stirner's doctrine that in the society of the future everyone will pursue his own private interest; on the contrary, human beings have a natural, instinctive solidarity which makes them capable of self-sacrifice and concern for others. The state not only does not foster this solidarity but opposes it: at most, it organizes the solidarity of the privileged classes in so far as they have a common interest in maintaining exploitation. When the machinery of the state is destroyed, society will be organized in small autonomous communes which will allow their members absolute freedom. Any larger units will be formed on a completely voluntary basis, and every commune will be able to withdraw from the federation whenever it wishes. No administrative functions will be permanently assigned to any individuals; all social hierarchies will be abolished, and the functions of government will be completely merged in the community. There will be no law or codes, no judges, no family as a legal unit; no citizens, only human beings. Children will not be the property of their parents or of society, but of their own selves as they are destined to be: society will take care of them and remove them from their parents if they are in danger of being depraved or hampered in their development. There will be absolute freedom to maintain any views, even false ones, including religious beliefs; freedom, too, to form associations to propagate one's views or for any other purpose. Crime, if any there still is, will be regarded as a symptom of disease and treated accordingly.

Since it is clear that all privilege is connected with the right to bequeath one's property and that the state serves to perpetuate this unjust arrangement, the first step towards destroying the present system must be to abolish the right of inheritance. This is the road towards equality, which is unthinkable without freedom; and freedom is indivisible.

In the light of these principles the state communism of the German doctrinaires—Marx, Engels, Lassalle, and Liebknecht —is revealed as the threat of a new tyranny of self-styled 'scientists' in a new form of state organization. 'If there is a state, there is bound to be domination and therefore slavery. A state without slavery, open or disguised, is unthinkable—that is why we are enemies of the state.' (*Statehood and Anarchy*. p. 280.) In one way or another, the minority will govern the majority.

But, the Marxists say, this minority will consist of the workers. Yes, no doubt—of former workers, who, as soon as they become governors or representatives of the people, cease to be workers and start looking down on the working masses from the heights of state authority, so that they represent not the people but themselves and their own claim to rule over others. Anyone who can doubt this knows nothing of human nature ... The terms 'scientific socialist' and 'scientific socialism', which we meet incessantly in the works and speeches of the Lassallists and Marxists, are sufficient to prove that the so-called people's state will be nothing but a despotism over the masses, exercised by a new and quite small aristocracy of real or bogus 'scientists'. The people, being unlearned, will be completely exempted from the task of governing and will be forced into the herd of those who are governed. A fine sort of emancipation! ... They [the Marxists] claim that only a dictatorship, their own of course, can bring the people freedom; we reply that a dictatorship can have no other aim than to perpetuate itself, and that it can engender and foster nothing but slavery in the people subjected to it. Freedom can be created only by freedom, that is by a rising of the whole people and by the free organization of the working masses from below. (*Statehood and Anarchy*, pp. 280-1.)

In short, the object of the revolutionary movement cannot be to gain control of the existing state or to create a new one, for in that case the outcome would defeat the idea. For the same reason, the movement cannot pin its faith on a political struggle within the framework of existing state and parliamentary institutions. Liberation can only be attained by a single apocalyptic upheaval sweeping away the whole apparatus of the state, law, and private property. From this point of view the coming social revolution differs fundamentally from all its predecessors and especially the French Revolution, which turned into a despotism inspired by the sick mind of Rousseau. Bakunin speaks of Rousseau and Robespierre in tones of abhorrence; nor has he much good to say of any socialist thinker except for Proudhon, who knew the value of freedom.

Must there not, however, be a state organization and means of compulsion or restriction so as to limit conflicts and keep human egoism within bounds? No, Bakunin replies: it is precisely because the state exists that even the best individuals, emerging from the mass of humanity, become tyrants and executioners. In a society based on freedom even the most selfish and ill-disposed will be cured of their vices; for a society without a state and without privileges is not only better but is the only mode of life compatible with human nature, spontaneous, creative, and unrestricted. Anarchy is more than an ideal, it is the realization of man as he was meant to be. This does not signify, however, that it is guaranteed by the laws of history or part of a destined plan: it is essentially the work of human purpose, but there is every reason to think that purpose will prevail. Bakunin believed strongly in the natural revolutionary instinct of the working masses, and he considered the problem chiefly as it affected Russia. Revolution required as a prior condition extreme poverty and desperation, plus the ideal of a new society: this ideal could not be imposed on the people from outside, but must already be dormant within them. What the people needed was not teachers to invent ideals, but revolutionaries to arouse them from their slumbers. The Russian people, i.e. the peasantry, had a deeply rooted sense of anarchy: they felt that the land belonged to everyone and that the village commune, the *mir*, should be completely autonomous, and they were naturally hostile to the state. This feeling, however, was overlaid by the patriarchal tradition, by their faith in the Tsar, and by the fact that the *mir* absorbed human personality and hampered its development, while the opium of religion kept the peasants in spiritual bondage. Consequently, the village communes were inert and isolated from one another; but there might arise from among the people rebels who would stir it up and awaken its natural revolutionary tendencies. Moreover, the same natural ideals were dormant among the poor of other countries, as was most clearly seen in Italy, where the anarchist revolution became more imminent every day. The great exception was Germany, where there were always plenty of theoreticians chattering about revolution but not enough people working for it. The Germans were natural state-worshippers, delighting to obey as much as to command, and it was not surprising that they could rise no higher than the state socialism of Marx and Lassalle, or that Bismarck's Germany was now the bastion of world reaction. The Tsardom, whatever Marx might say, did not compare with Germany in this respect: it was certainly always trying to meddle in European affairs, but with very little success.

Bakunin's statements about Russia do not form a consistent whole. On the one hand, he says that the Slavs are incapable of forming states and that all their political systems have been created by foreigners. But, on the other hand, he maintains that Russia is not only a military state (as opposed to a commercial one like Britain) but has evolved a system in which the interests of all classes, and the whole of industrial and agricultural activity, are subordinated to the central power, so that the nation's wealth is no more than a means to aggrandize the state. On this point Bakunin repeats an observation that had often been made in the nineteenth century: the primacy of the Russian state *vis-à-vis* the civil community was so absolute that even the distinction of classes was secondary to it. But it is hard to see how this can be reconciled with the view that the Slavs have no state-forming abilities.

From this brief review of Bakunin's ideas it will be clear that he differed widely from Marx as regards both theory and tactics. Apart from their dispute over the leadership of the International, each accusing the other of dictatorial aims, and apart from the question whether Russia (as Marx insisted) or Prussia (as Bakunin maintained) was the headquarters of world reaction, they disagreed on several points of key importance to the socialist movement.

In the first place, Marx regarded the call for the immediate abolition of inheritance as putting the cart before the horse, since the right to bequeath property was only a particular aspect of the property system itself. Secondly, Marx held that the state is not the independent source of all social evil but merely the instrument by which existing privileges are maintained. On this point the disagreement was not essential, for Marx, like Bakunin, considered that existing political institutions would have to be destroyed, while Bakunin agreed that the state had arisen historically as an instrument of private property, though he also held that in the course of time it had become an independent force and a necessary bulwark of the class system. The dispute therefore came down to whether the socialist revolution could do away with all forms of statehood at the outset. Marx believed that the state of the future would not be concerned with 'governing people' but with 'administering things', i.e. organizing production. To Bakunin, this amounted to extreme etatism: there could be no centralized economic administration without political centralization and therefore slavery. Thirdly, Marx's strategic plan included political activity within existing systems parliamentary and other, and permitted temporary alliances with the democratic bourgeoisie when its interests happened to coincide with those of the proletariat; whereas for Bakunin the only kind of political activity that revolutionaries should under take was to destroy all forms of statehood. Fourthly, Bakunin's idea of completely free economic activity carried on by small autonomous communes appeared to Marx no better than a Proudhonist Utopia, and open to the same objections: on the one hand the natural tendency is for production to be centralized, and, on the other, an economy composed of independent units would be bound to reproduce the system of competition and capital accumulation.

Marx's ideas on all these questions altered and matured over a long period. It was not until after the Paris Commune that he came round to the view, which was to be central in Lenin's version of Marxism, that the existing state machine must be destroyed. Bakunin's Swiss follower Guillaume welcomed this as signifying Marx's conversion to anarchism; but he was mistaken, as Marx remained convinced of the necessity of a central economic administration, though he believed the future state would have no political functions. It is true, however, that Marx did not explain clearly on what basis social life would be organized when the state had been abolished and the whole economy centralized. Bakunin himself had only the crudest ideas of political economy, believing simply that once people were free of the state their natural solidarity and bent for co-operation would come into play and conflicts of interest would be impossible. He envisaged democracy on the lines of the Swiss cantons and villages in which the whole adult population assembled from time to time to decide matters of common concern; but his writings give no indication of how this could be applied on the scale of a province, a country, or the whole world, assuming that representative democracy was done away with.

In these disputes Marx's strength lay in the field of economic criticism and in his conviction that a system of independent productive units would mean reviving all the harmful aspects of a commodity economy. Bakunin, on the other hand, had a strong point in his criticism of the overt or implied etatism in Marx's programme. He raised the very real question to which Marx gives no anwer: how can a centralized economic power be imagined without political coercion? And, if the future society is still divided into rulers and ruled, how can it fail to recreate the system of power privilege, which has a natural tendency to perpetuate itself? These objections were to recur frequently in criticisms of Marx by anarchists and syndicalists. It is clear enough that Marx did not himself imagine socialism as a despotic system in which the political apparatus would maintain its privileges on the basis of a monopoly of the means of production; but he did not answer Bakunin on this point, and the latter deserves credit for being the first, as it were, to infer Leninism from Marxism.

Bakunin believed naïvely that men, left to themselves, would behave as they ought and would live in harmony, since evil did not come from human beings but from the state and private property; he did not explain how man, being good by nature, had come to create such an evil system. Marx for his part thought the question of natural goodness irrelevant and naïve. He was concerned with the Promethean expansion of the human race in its growing mastery over nature, and he thought personal development had no meaning except in relation to the development of the species. He was far from being an advocate of despotism, but he failed to answer the charge that it was implicit in his system.

The First International was destroyed by internal conflicts on the one hand and, on the other, by the Franco-German War and the Paris Commune. The Commune was not the child of the International, still less of the Marxists. Most of its leaders were Blanquists, while those members of the International who joined it were mainly Proudhonists. Marx saw from the beginning that it was doomed to failure, but after the defeat and massacre of the Communards he composed a pamphlet entitled *The Civil War in France* in which, besides paying tribute to their heroism, he analysed the significance of their spontaneous movement from the viewpoint of the future of communism. The Paris Commune, being in a sense the first proletarian regime in history, had, by a natural process as it were, exemplified some of the basic principles of the future socialist society: the replacement of the standing army by an armed citizenry, the transformation of the police into a popular organ, the electivity of all magistrates and officials, a maximum wage, free education, the disestablishment and expropriation of the Church. Nevertheless, Marx did not regard the Commune as specifically either socialist or proletarian; Engels in 1891 referred to it as a dictatorship of the proletariat, but Marx never did so. (Its name, of course, is simply French for the Paris municipality, and has no ideological significance.) In February 1881, in a letter to F. Domela Nieuwenhuis, Marx said expressly that the majority of the Commune was not socialist and that its only right and possible course had been to compromise with Versailles in the interest of the French people as a whole.

The defeat of the Commune gave encouragement to reaction throughout Europe and accentuated the dissensions which broke up the First International. Workers' organizations in France and Germany were subjected to harassment, and the International lost the effective support of the British trade unions, who had joined it for tactical rather than ideological reasons and were chiefly concerned to establish a legal position for themselves within the existing order. At its London Conference in September 1871 the International endorsed Marx's call for combined political and economic action by the working class and for independent workers' parties in all countries; the congress at The Hague in September 1872 showed that Marx's followers were a majority in the General Council. But the International was by now fatally weakened by dissensions and persecution, and was incapable of directing workers' organizations in conditions that differed widely from one country to another. On a proposal by Engels the General Council transferred itself to New York, where the organization lingered on for a few years before it was formally dissolved in 1876. A rival International formed by Bakunin's followers fell to pieces a year later; however, throughout the 1870s anarchism was stronger than Marxism, not only in Spain and Italy but also in France.

Apart from the conflict of influences in the International it may be said that from the 1860s Marxism was the most important of the rival socialist ideologies, in the sense that doctrines and programmes throughout the world defined their position by reference to it. Marxism presented the most consistent and elaborate body of doctrine, and this was due in part to the publication at Leipzig in 1867 of the first volume of Marx's *Capital*. This volume reverted, *inter alia*, to the problems discussed in the *Critique of Political Economy* (1859), and revealed the sources of exploitation by analysing the basic phenomena of the capitalist economy: commodities, exchange- and use-value, surplus value, capital, wages, and accumulation. The fundamental thesis of *Capital* is that exploitation derives from the sale of labour–power by hired workers. Labour is a commodity of a special kind in that the value of its product is much greater than the cost of reproducing it, i.e. of the worker's subsistence; and the exploitation that this involves can only be done away with by abolishing wage-labour.

Marx intended to finish the second and third volumes of his work in a short time. The second was to analyse the circulation of capital and the market, while the third was to deal with the sharing of profit among different groups of exploiters, the origin of the average rate of profit, the tendency of the rate of profit to fall, and the transformation of surplus profit into ground rent. Parts of these volumes were written before the first was published, but although Marx continued working on them till 1878 they were not completed at the time of his death. The manuscripts, arranged and edited by Engels, were published in 1885 and 1894, while *Theories of Surplus Value* was published by Kautsky as the fourth volume of *Capital* in 1905–10.

After the break-up of the International, and as the hope of an early European revolution once more receded, Marx concentrated on theoretical work to the extent permitted by frequent illness, visits to health resorts, money troubles, and domestic misfortune. He read extensively, but in his last years was almost incapable of writing; however, he continued to follow closely the development of European socialism. In 1875 the two German workers' parties, the Lassallists and the Eisenach group, united to form the Socialist Workers' party. Their programme elicited a devastating attack by Marx in the form of a letter to the Eisenach leaders: this *Critique of the Gotha Programme*, first published by Engels in 1891, repeated Marx's objections to Lassallian socialism and contained more trenchant formulations than are found elsewhere in his works on such matters as the state, internationalism, and the nature of a proletarian authority. The document had little effect on the final version of the programme, but it became one of the principal texts invoked by the revolutionary wing of the Second International against reformism and revisionism: its use of the phrase 'dictatorship of the proletariat' made it especially valuable to Lenin and his followers. In 1880 Marx helped Jules Guesde to draft the programme of the French workers' party; in 1881–2 he wrote some letters on the prospects of revolution in Russia, which were afterwards much debated and disputed over by Russian Marxists.

Marx died in London on 14 March 1883. Some of his papers were published posthumously by Engels; after the latter's death in 1895 a vast amount of material remained in the hands of Bernstein and Bebel, who did not do much to make it available. Mehring republished some articles of the 1840s which were difficult of access, and also published the manuscript of Marx's doctoral dissertation, though without the preliminary notes. Bernstein published portions of *The German Ideology*. The first edition of Marx's correspondence, by Mehring and Bernstein, was inaccurate and incomplete. Kautsky, as already mentioned, published *Theories of Surplus Value* and (in 1903) the Introduction to the *Grundrisse*. A great deal of work was done in collecting scattered manuscripts and letters, and publishing them in scholarly form, by David Ryazanov, who created the Marx–Engels Institute in Moscow and was its director until 1930. He also founded the great critical edition of the works of Marx and Engels (*M.E.G.A.*), which, although never completed, made available several texts that were previously unknown, including the whole of *The German Ideology*, the Paris Manuscripts of 1844, and Engels's *Dialectic of Nature*.

Engels survived Marx by twelve years. During the long period of their friendship and collaboration he was content to remain in Marx's shadow, regarding the latter as the founder of scientific socialism and modestly underrating his own contribution. None the less, subsequent generations of Marxists made more use, in expounding and advocating their doctrine, of Engels's writings than of Marx's, always excepting the first volume of *Capital*. Engels was a man of great breadth of knowledge and intellectual capacity. Besides history, politics, and philosophy, which occupied the bulk of his time, he wrote numerous articles on military problems and the technical aspects of current operations of war, and also followed developments in natural science from the viewpoint of his own philosophical reflections. As a writer he is much more digestible than Marx; he endeavoured more than once to set out the main ideas of scientific socialism in a generally accessible form, and his works were widely read by socialists everywhere.

His first important work after 1848 was *Der deutsche Bauernkrieg* (*The German Peasant's War*, 1850), on the subject of the sixteenth-century rising under Thomas Münzer. Based on the history by Wilhelm Zimmermann published in the 1840s, this work attempted to interpret the most important popular rising in German history in terms of the class struggle and to suggest analogies between it and the revolutionary situation of 1848–9. Engels's views on the events of those years, in which he himself took part, were summed up in a series of articles published in 1851–2 over Marx's signature in the *New York Daily Tribune*, entitled *Revolution and Counter-Revolution in Germany*; these were first published in book form (still attributed to Marx) in 1896.

Among Engels's best-known works is *Herrn Eugen Dührings Umwälzung der Wissenschaft*, known as the *Anti-Dühring* (1878). Dühring (1833–1921), who was blind, was dismissed from his lectureship at the University of Berlin for his violent attacks on academic philosophy. His writings were popular among the German social democrats, and for a time he was regarded as one of the party's chief theorists. Engels, who considered Dühring a dangerous influence, attacked his views in a sharply polemical work in which he gave a clear exposition of dialectical materialism as the basis of Marxian economics, and of scientific socialism as opposed to the utopian tradition. The *Anti-Dühring* became a kind of Marxist handbook after Dühring himself had been quite forgotten (though Nazi propagandists were to revive his memory on account of his anti-Semitic views).

After Marx's death Engels, who had moved to London in 1870, devoted much of his energy to editing the remaining portions of *Capital*, but also wrote philosophical works of his own. In 1886 he published in *Die Neue Zeit* an article on 'Ludwig Feuerbach and the End [*Ausgang*] of Classic German Philosophy', in which he related scientific socialism to the German intellectual tradition; this, too, is one of the most popular expositions of Marxism. It was republished in book form in 1888 together with Marx's *Theses on Feuerbach*, which had not previously appeared.

Another classic work by Engels is *The Origin of the Family, Private Property and the State* (1884). This made considerable use of the work of Lewis H. Morgan, who for the first time had systematically analysed primitive society on the basis of direct observation of North American Indians, and in *Ancient Society* (1877) outlined a theory of the stages of human development from savagery to civilization. Using these and other works, Engels endeavoured to present the origins of the basic institutions of civilized life.

In the early seventies Engels conceived the plan of a critique of vulgar materialism which would apply the dialectical method to scientific observation. He wrote some chapters and notes for this work between 1875 and 1882, but did not succeed in completing it. All this material, finished and unfinished, was first published in 1925 in Moscow under the title *Dialectic of Nature*.

The works mentioned here constitute that part of Engels's literary output—a small proportion of the whole—which has become widely read on account of its systematic character and the permanence of its themes. Along with *Capital*, these works are the basic source from which three or four generations of readers have imbibed their knowledge of scientific socialism and its philosophic background.

Engels died in London on 5 August 1895. Unlike Marx, he is not buried there; he was cremated by his own wish, and his ashes, in an urn, were cast into the sea off Beachy Head.

CHAPTER XII

Capitalism as a Dehumanized World. The Nature of Exploitation

1. The controversy as to the relation of Capital *to Marx's early writings*

MARX's exposition of the functioning and prospects of the capitalist economy cannot be studied in isolation from his anthropological ideas and his philosophy of history. His theory is a general one embracing the whole of human activity in its various interdependent spheres. The behaviour of human beings in all ages—whether active or passive, whether intellectual, aesthetic, or engaged in labour—must be understood integrally or not at all. *Capital* is the culmination of a series of works in which Marx applied his basic theory of dehumanization to the phenomena of economic production and exchange. His successive 'critiques'—the Paris Manuscripts of 1844, *The Poverty of Philosophy* (1847), *Wage Labour and Capital* (1849), the *Grundrisse* (1857–8), the *Critique of Political Economy* (1859), and finally *Capital* itself (1867)— are more and more elaborate versions of the same thought, which may be expressed as follows. We live in an age in which the dehumanization of man, that is to say the alienation between him and his own works, is growing to a climax which must end in a revolutionary upheaval; this will originate from the particular interest of the class which has suffered the most from dehumanization, but its effect will be to restore humanity to all mankind.

There is no doubt that Marx's terminology and the mode of his expositon underwent changes between 1844 and 1867, and there has been much discussion as to how far these correspond to changes in his ideas. In particular, it has been suggested that the theory of a 'return to species-essence', which is prominent in the texts of 1843–4, and which implies a normative, anthropological view, was abandoned by the later Marx in favour of a structural description.

Some commentators, such as Landshut and Meyer, Popitz and Fromm, consider that the early writings express a richer, more universal philosophical theory and that the later ones, by comparison, are intellectually more restricted. Many others, such as Sydney Hook, Daniel Bell, and Lewis Feuer, maintain that there was a break in the development of Marx's ideas and that *Capital* differs from the Paris Manuscripts not only in scope but also in substance; this view is denied by such critics as Calvez, Tucker, McLellan, Fetscher, and Avineri. A distinct but closely related question is whether, despite the frequent sharpness of Marx's attacks on Hegel, his ideas were in fact derived from Hegelian sources, and whether in this respect too there was a breach in his intellectual development. Croce, Löwith, and Hook maintain that he parted company with Hegelianism after 1844, while Lukács, Fetscher, Tucker, and Avineri hold that he was inspired by it more or less consciously to the very end. These views are equally compatible with a sympathetic or unsympathetic approach to any particular 'phase' of Marx's thinking, or to the whole of it. Other critics again, such as Jordan, believe that Marx's relationship to Hegel went through different stages: that a short period of fascination was followed by radical criticism and the almost complete abandonment of Hegelianism, but that he subsequently reverted to a middle view.

The literature of this controversy already amounts to a considerable library, and we cannot study the arguments in detail here. It may, however, be briefly stated why the present author agrees with those who hold that there is no discontinuity in Marx's thought, and that it was from first to last inspired by basically Hegelian philosophy.

It must be made clear that the question is not whether Marx did or did not change during his forty years as a writer, since he obviously did change in many respects. Nor is it whether the whole substance of *Capital* can be found in the Paris Manuscripts by anyone who chooses to look hard enough—for Marxism without the theory of value and surplus value is clearly not the same as Marxism with this theory elaborated. The question is whether the aspects of his early thinking which Marx subsequently abandoned are important enough to justify the idea of an intellectual break, and whether the theory of value and its consequences are a basic innovation, either contrary to Marx's philosophy of the early forties or in no way anticipated therein. To this question we would reply as follows.

The fundamental novelty of *Capital* consists in two points which entail a wholly different view of capitalist society from that of the classical economists with their labour theory of value. The first of these is the argument that what the worker sells is not his labour but labour-power, and that labour has two aspects, the abstract and the concrete. But this view is itself the final version of Marx's theory of dehumanization, first sketched in 1843–4. Exploitation consists in the worker selling his labour-power and thus divesting himself of his own essence: the labour process and its results become alien and hostile, a deprivation of humanity instead of a fulfilment. In the second place, having discovered the dual nature of labour as expressed in the opposition between exchange-value and use-value, Marx is able to define capitalism as a system in which the sole object of production is to increase exchange-value without limit; the whole of human activity is subordinated to a non-human purpose, the creation of something that man as such cannot assimilate, for only use-value can be assimilated. The whole community is thus enslaved to its own products, abstractions which present themselves to it as an external, alien power. The deformation of consciousness and the alienation of the political superstructure are consequences of the basic alienation of labour—which, however, is not a 'mistake' on history's part but a necessary precondition of the future society of free beings in control of the vital process of their own lives.

In this way *Capital* may be regarded as a logical continuation of Marx's earliest views; and this continuity is attested by his reference, in the Afterword to the second edition of Volume I (1873), to his criticisms of Hegel 'almost thirty years ago', i.e. no doubt to the Manuscripts themselves.

Admittedly, such expressions as 'man's recovery of his own species-essence' and 'the reconciliation of essence and existence' do not appear in Marx's writings after 1844. This, as we have seen, is best explained by his controversy with the German 'true socialists', who regarded not only socialism itself but the movement towards it as all humanity's concern, putting their faith in action by all social classes and not only the proletariat with its special interests. Marx, however, once he had come to the conclusion that socialism would be achieved not by humanitarian sentiment but by the paroxysm of the class struggle and if necessary by revolutionary force, from then on avoided any expressions which might suggest the idea of class solidarity or imply that the world could be transformed by ideals and emotions which transcended class enmity. Nevertheless, his original intention remained the same. He still believed that socialism was the concern of humanity as a whole and would do away with classes and privilege; and, though he was naturally most moved by the oppression of the working class, he analysed the process of dehumanization and reification from the point of view of the capitalist as well.

The idea of man's recovery of his own self is in fact comprised in that of alienation, which Marx continued to employ: for alienation is nothing but a process in which man deprives himself of what he truly is, of his own humanity. To speak in these terms implies, of course, that we know what man 'truly' is as opposed to what he empirically is: what the content of human nature is, conceived of not as a set of features which may be empirically ascertained but as a set of requirements that must be fulfilled in order to make human beings genuinely human. Without some such standard, vague though it may be, 'alienation' has no meaning. Accordingly, whenever Marx uses this term he presupposes, expressly or otherwise, a non-historical or prehistorical norm of humanity. This, however, is not a collection of permanent, unchanging qualities belonging to some arbitrary ideal, but a conception of the conditions of development enabling men to display their creative powers to the full, untrammelled by material needs. The fulfilment of humanity is not, in Marx's view, a matter of attaining some final, imagined perfection, but of freeing man for ever from conditions that hamper his growth and make him the slave of his own works. The idea of freedom from alienation, and thus of alienation itself, requires a preliminary value-judgement and an idea of what 'humanity' means.

The term 'alienation' still occurs frequently in the *Grundrisse* (1857–8) but is less common in Marx's writing thereafter, and is seldom used in *Capital*. This, however, is a change of language and not of substance; for the process whereby man's labour and its products become alien to him is described in *Capital* in terms which clearly show that Marx has in mind the same phenomenon as that described in the Manuscripts.

It is important to note, in Marx's early criticism of Hegel, that he did not at any time identify alienation with externalization, i.e. the labour process whereby human strength and skill are converted into new products. It would clearly be absurd to speak of abolishing alienation in this sense, since in all imaginable circumstances men will have to expend energy to produce the things they need. Hegel, as we have seen, did identify alienation with externalization, and he could therefore only conceive man's final reconciliation with the world by way of abolishing the objectivity of the object. To Marx, however, the fact that people 'objectivize' their powers does not mean that they become the poorer by whatever they produce; on the contrary, labour in itself is an affirmation and not a denial of humanity, being the chief form of the unending process of man's self-creation. It is only in a society ruled by private property and the division of labour that productive activity is a source of misery and dehumanization, and labour destroys the workman instead of enriching him. When alienated labour is done away with, people will continue to externalize and 'objectivize' their powers, but they will be able to assimilate the work of their hands as an expression of their collective ability.

Again, there appears to be no contradiction between the young Marx's praise of the self-affirmation that a productive worker enjoys or may enjoy, and the argument in the third volume of *Capital* that future progress would consist in the gradual reduction of necessary work, i.e. that involved in satisfying elementary physical needs. The time thus saved was not to be spent in idleness but in free creative activity, the earnest, absorbing toil which, for Marx, was typified by that of the artist. Man would continue to assert his humanity in the form of labour, but would spend less and less time producing food, clothing, and furniture and more and more on the products of art and science.

There is also good ground for saying that Marx continued to hold the view he expressed in 1844 that man is acquainted with nature not as it is in itself but through the medium of a socially created system of needs. In one of his last works, a commentary (written in 1880) on Adolph Wagner's handbook of political economy, he argues that man regards the external world as a means of satisfying his needs and not a mere object of contemplation, and that the features he perceives in it and embodies in language, in other words all his conceptual categories, are related to his practical requirements. It is clear from this that Marx never accepted the view that the world in itself is simply reflected in human minds and that the images found there are then transformed into abstract concepts.

It can be argued, on the other hand, that the Romantic idea of man once more achieving unity with nature does not appear in Marx's writings after 1844, and it may seem from the *Grundrisse* that he shifted to a utilitarian or similar viewpoint. In one of his many descriptions (such as we find both in *The Communist Manifesto* and in *Capital*) of the tremendous part played by capitalism in advancing civilization, he says that capital for the first time made it possible for men to 'assimilate' nature in a universal way, i.e. to treat it as an object of use and not of idolatry. But here too it is difficult to speak of a real change of view. Marx himself did not share the idolatrous view of nature that he commended capitalism for destroying, or regard the world in its primal, untamed state as deserving of human worship. He believed that man perceives and organizes the world in accordance with his needs, and that as humanity progresses so nature becomes more humanized, more obedient, and less incalculable. The expression of his view may have changed, but not the view itself.

As we have already said, the publication of the *Grundrisse* went a long way to refute those who held that there was a major discontinuity in the evolution of Marx's ideas. It was clear, in particular, that his theory of value and of money combined harmoniously with the concept of alienation. No doubt two separate traditions are here synthesized: those of Hegel and of the classic British economists, whom Marx began to study while still in Paris. It was in fact one of his greatest achievements to express the theory of alienation, derived from Bauer, Feuerbach, and Hess, in conceptual categories which he took over, though with substantial modification, from Ricardo.

2. *The classical economic tradition and the theory of value*

The theory of value, which is the core of *Capital*, has a history which may be traced back to Aristotle's time. It was of interest for both theoretical and practical reasons. The theoretical question is: since goods are exchanged for one another at a particular rate, they must have some property that makes them quantitatively comparable, despite all their differences of quality; what, then, is this common feature which reduces the multiplicity of things to a single measure? The practical question, which was much debated in the Middle Ages, is that of a 'just price'. Although expressed in normative terms—how should the just price of a given article be determined?—this is really the same question as how to define the conditions of 'equivalent exchange', in which the purchaser pays the price to which the seller is 'really' entitled. This was directly related to another question frequently posed by medieval theologians, moralists, and political writers: was it lawful to lend money at interest, and, if so, on what ground? Clearly, the question of a 'just price' and of interest could only be answered by determining what constituted the 'real' value of a commodity, and how it was to be measured.

The idea that the value of an article is to be measured by the amount of labour that went into its production was advanced by various thinkers before the eighteenth century. Marx, who had made a thorough study of the history of the problem, took as the starting-point of his own theory two classic works which he regarded as the bases of economic science: Adam Smith's *Inquiry into the Nature and Causes of the Wealth of Nations* (1776) and Ricardo's *Principles of Political Economy and Taxation* (1817).

Smith's main work is devoted *inter alia* to the question of how national wealth increases and how it can be measured objectively, regardless of price fluctuations. He assumed that increasing wealth was desirable, and sought to prove that state intervention in production and trade could only impede its growth. He distinguished between productive and non-productive labour, including in the former not only agricultural labour (as did the physiocrats) but all occupations which involved the processing of material objects for useful purposes—i.e. excluding services, administration, political and intellectual activities, etc.

—and which produced 'surpluses' that could be used for further production. The question of how to measure the value of a product depended, in Smith's view, on how the national product was calculated. He distinguished the use-value of an object, i.e. its power to satisfy a human need, from its exchange-value, which was the proper subject-matter of economics; for it was clear that some objects, such as air, were of great use but were not objects of exchange, while others, though of very small use, fetched enormous prices.

However, Smith argued, exchange-value is not the same as the actual price of a commodity; on the contrary, it is necessary to find out in what conditions prices correspond to 'real' value, and what causes them to vary from it. The real or 'natural' value of an article is measured by the quantity of work that has gone into producing it; such, at least, was the case in primitive societies, where goods were exchanged on the basis of labour-time, for example the time spent in hunting game. But in modern societies other factors come into play besides labour, namely capital and land; so that the value or 'natural price' of a product includes remuneration for the worker, a return on the capital used, and an element for rent. The distribution of profits between capitalists, landowners, and workers is thus in accordance with nature. The increase of wealth is in the general interest of all classes engaged in production: Smith did not believe that wages were bound to gravitate to the bare subsistence level, as Malthus and, at least for a time, Marx were subsequently to contend. It is also in the interest of all for market prices to be as close as possible to 'natural' ones, and the market itself ensures automatically that they will tend to this level despite fluctuations; artificial regulation of the market by administrative action is more likely to disturb the process than to assist it. The market also provides a common measure for unequal forms of human labour, which must be rewarded not simply on a time basis but according to the complexity of the task and the skill put into it.

Smith did not indicate any way in which 'natural' prices and national revenue could be calculated independently of market prices. Nevertheless, his work was the first attempt to arrive at a complete system of categories applicable to the analysis of economic activity, on the premiss that that activity obeys laws of its own, independent of human volition, and is regulated by the 'invisible hand' of the market. *The Wealth of Nations* is one of the most important documents in the history of liberalism, though Smith came to modify in some respects his belief in the automatically favourable effects of competition and the market; he did not, moreover, draw a clear line as yet between economic and moral aspects.

Ricardo put somewhat different questions from those of Smith, but for a time at least used the same instruments of analysis. He was less interested in calculating the national income than in discovering the basis of its distribution among different classes. He believed that in theory the value of commodities could be expressed in terms of labour units (machines too being treated as the sum of labour involved in making them), but he recognized that such a calculation was impracticable in respect of large-scale economic processes. He also perceived a contradiction between the dependence of prices on labour and the tendency for rates of profit to even out as regards different branches of production; for it is clear that the amount of capital per labour unit varies in different sectors of industry, so that there cannot be a uniform rate of profit if prices are proportionate to labour input. In the last resort, the labour theory of value was not so important to Ricardo as it subsequently was to Marx.

Ricardo saw, much more clearly than Smith, the conflict of interest between capitalists and wage-earners. He recognized that technical progress might lead to a fall in employment and so reduce the workers' total income. He was also inclined to share Malthus's view that wages tend to fall to the bare subsistence level, as otherwise the workers will breed more children, the supply of labour will increase, and wages will drop once more.

Marx regarded the works of the classic British economists as a model of unprejudiced analysis, endeavouring without sentiment to discover the actual mechanisms of social life. He understood, indeed, that their doctrine was grounded in economic liberalism and the belief that it was 'natural' for owners of land and capital to be rewarded for their share in production. But what interested him in Smith and Ricardo was their description of the interrelation between the various elements in the production process: investment, population growth, wages, food costs, foreign trade, etc. The classic economists believed, like Hegel, that one could not understand much of human society by observing people's intentions in their individual relationships; the laws that governed its working were not intended by anyone, but it was they and not men's thoughts which determined human behaviour.

Marx, however, used the theory of value in a different way from any of the economists who preceded him. Instead of applying it to the question of how the national product is to be estimated or how it is distributed, he used the theory chiefly to investigate the nature of exploitation in a society based on private property.

Thus, apart from the two points already mentioned (the twofold character of labour, and the contention that what the wage-earner sells is not labour but labour-power), the theory of value was transformed by Marx in two other essential respects. In the first place, unlike Ricardo, he held that labour is not only the measure of value but its only source. Secondly, he maintained that the phenomenon of exchange-value is not a natural and inseparable part of society or civilization, but is a historical and transitory form of the organization of production and exchange. Such are the four main respects in which Marx altered the traditional theory of value.

Marx spent many years amending, correcting, and completing his economic doctrine. As Ernst Mandel has shown, his first notes of 1844–5 indicate that he then regarded Ricardo's theory of value as erroneous because it failed to explain the maladjustment of supply and demand, and hence economic crises, and also as morally suspect because it implied that the natural value of human labour was defined by the subsistence level.

Marx arrived at his own formulation of the theory of value by various stages; we shall not, however, pursue them here, but will describe the theory as it appears in its final form in *Capital*.

3. *The double form of value and the double character of labour*

At the outset of *Capital* Marx observes that every useful object may be looked at from a qualitative or a quantitative point of view: we may either consider the properties that make it useful as bread, cloth, furniture, etc., or simply the amount of work, of whatever kind, that has gone into it. In this way human products have a double value, or rather two incommensurate kinds of value: use-value, the characteristics that enable them to satisfy human needs, and the value which derives from the amount of labour-time that has gone to make them. When articles are compared with each other in the process of exchange, their value takes the form of exchange-value. Particular useful objects are thus endowed with abstract exchange-value, the crystallization of labour-time irrespective of the difference between one form of labour and another. It is only labour as such that constitutes exchange-value. Objects that are useful but are not man-made (natural resources, water-power, virgin soil, and forests) have no value, even though they have a price—a point which Marx explains later in the context of surplus value.

As exchange-values, things are comparable quantitatively in terms of the amount of labour-time embodied in them; they can thus form the object of an exchange in which they are reduced to the homogeneous aspect of labour-time. However, this does not mean the time actually employed in making them: it could not be the case that one loaf of bread was worth twice as much as another because the baker was less skilful or had less good equipment and therefore took twice as long to make it. What we are concerned with is not actual labour but socially necessary labour-time, defined as the average amount of time necessary to produce a given article at a particular historical stage of human ability and technical progress. This necessary labour-time is the quantitative standard of the relative values of things, permitting them to be bought and sold at a determinate rate. Goods embodying the same amount of work in this sense have the same value, however different their uses and physical qualities.

It is clear that the possession of use-value is a necessary, though not a sufficient, condition of the possession of exchange-value: no product can be exchanged, and thus become a commodity, unless it satisfies some need and is good for something. To put it another way: a thing does not become an exchange-value without assuming the character of a commodity, and it does not become a commodity without entering into the process of exchange. People have been making useful things since the dawn of history, but until there is a system of exchange based on homogeneous labour-time there are no commodities and no exchange-value. Exchange-value is not an intrinsic quality of objects, but derives from their involvement in the social process of circulation and exchange. Products are converted into values by being exchanged for one another.

The general form of value results from the joint action of the whole world of commodities, and from that alone. A commodity can acquire a general expression of its value only by all other commodities, simultaneously with it, expressing their values in the same equivalent; and every new commodity must follow suit. It thus becomes evident that, since the existence of commodities as values is purely social, this social existence can be expressed only by the totality of their social relations, and consequently that the form of their value must be a socially recognized form. (*Capital*, I, Ch. I, 3C, 1)

The commodity form of objects is thus the effect of a particular kind of social nexus, viz. the situation in which people engaged in an exchange confront each other as private owners,

whose will resides in those objects and who behave in such a way that each does not appropriate the other's commodity and part with his own except by means of an act done by mutual consent ... All commodities are non-use-values for their owners, and use-values for their non-owners. Consequently, they must all change hands. But this change of hands is what constitutes their exchange, and the latter puts them in relation with each other as values, and realizes them as values. Hence commodities must be realized as values before they can be realized as use-values. (*Capital*, I, Ch. II)

The quality of things that we call value, which is unknown to nature and conferred on them by the conditions of human society, is the basis, in Marx's theory, of the twofold character of human labour. On the one hand, labour is a concrete activity of a specific kind, issuing in a specific product; on the other hand, it is labour in general, the simple expenditure of human labour-power. This abstract, homogeneous labour is the true creator of exchange-value, while differentiated labour creates use-value. In considering the production of commodities, i.e. production for exchange, we abstract our attention from the difference between a baker's work and that of a spinner or a woodcutter, treating them as identical from the point of view of the exertion of labour-power for a time that can be exactly measured. In this way the most complex forms of labour are reduced to labour *tout court*, or labour-time. It is thus that disparate products can be compared and exchanged, and that a change in productivity affects the total amount of use-value created, but not that of exchange-value. When technology improves, the same amount of effort produces more goods, but the value of each article falls correspondingly, so that the total sum of values remains the same. At whatever stage of technical development, society produces the same quantity of values in the same amount of working time.

Since all products of labour manifest their value only in exchange, i.e. in comparison with one another, it follows that any of them is equally suitable as a standard with which to measure the others. The appearance of a universal standard of value in the form of money was possible owing to the prior existence in things of the abstract quality created in the process of exchange. The fact that in the course of time precious metals acquired a privileged position as standards of value was due to the physical properties of uniformity, divisibility, resistance to corrosion, etc., which made them more suitable than other things previously used as money, for example cattle. In itself gold is no different from any other commodity as an exchange-value, and its value derives not from any magic properties but from its being the product of abstract human labour; it had first to be a commodity like any other, before it was promoted to the role of a universal standard. Yet in money—considered as a standard of value, a means of payment, exchange, and accumulation—exchange-value somehow becomes autonomous and its origin in labour is lost to view. The fact that the products of labour can be appropriated in the form of money creates the illusion that money or gold is an intrinsic, original source of wealth. Quoting, in *Capital*, the tirade against gold by Shakespeare's Timon which he had used in the 1844 Manuscripts, Marx observes: 'Just as every qualitative difference between commodities is extinguished in money, so money, like the radical leveller that it is, does away with all distinctions. But money itself is a commodity, an external object, capable of becoming the private property of any individual. Thus social power becomes the private power of private persons.' (Ibid., Ch. III, 3a.)

When considering exchange-value in itself we make the fictitious assumption that goods are exchanged according to their value. The creation of money, however, introduces the factor of price, i.e. the amount of currency for which other articles are exchanged. When value is converted into price, goods express their quantitative relationship to one another in the form of a quantitative relationship to money. It thus becomes possible for value and price to diverge, i.e. for goods to be exchanged at a rate higher or lower than their value expressed in money terms.

The price-form, however, is not only compatible with the possibility of a quantitative incongruity between magnitude of value and price, i.e. between the former and its expression in money, but it may also conceal a qualitative inconsistency, so much so that, although money is nothing but the value-form of commodities, price ceases altogether to express value. Objects that in themselves are no commodities, such as conscience, honour etc., may be offered for sale by their holders and thus acquire, through their price, the form of commodities. Hence an object may have a price without having value . . . On the other hand, the imaginary price-form may sometimes conceal either a direct or an indirect real value-relation; for instance, the price of uncultivated land, which is without value because no human labour has been incorporated in it. (*Capital*, I, Ch. III, 1)

The money-form thus makes possible, and actually brings about, an incongruity between value and the price which is supposed to express it. As Marx states in Volume III of *Capital*, the sum total of the prices of the whole social product must be equal to the sum of its values; but, in a commodity economy, this equation not only permits but actually presupposes inequality in particular cases, i.e. prices that tend to equal values but constantly fluctuate above or below them. The contrast between values and prices expresses the basic contradiction of capitalist production and exchange. This inequality, however, is not the explanation of profit: selling an article above its value is not the true origin of profit, only the origin of one form of it. The phenomenon of profit must be explained on the assumption that all commodities are sold at their true value. This sounds paradoxical, but, as Marx observes in *Wages, Price and Profit*, 'it is also a paradox that the earth moves round the sun and that water consists of two highly inflammable gases. Scientific truth is always paradoxical if judged by everyday experience, which catches only the delusive appearances of things'.

4. Commodity fetishism. Labour-power as a commodity

Before investigating the source of profit, however, we may note the effect of the money form on human thought-processes. Neither the exchange of commodities nor the existence of money is a sufficient condition of capitalist production, which requires in addition the free sale of labour-power and a production system aimed essentially at increasing exchange-value. But the commodity and money-form assumed by objects is the root of the particular illusion which Marx calls commodity fetishism, and which accounts for a large part of the false consciousness of human beings in regard to their own social existence.

The essence of commodity fetishism is that, in measuring the output of energy by labour-time, we introduce into the products of labour the measure which originally relates to the life-process itself. Thus the mutual relations of human beings as exchangers of goods take on the form of relations between objects, as though the latter had mysterious qualities which of themselves made them valuable, or as though value were a natural, physical property of things.

The relation of the producers to the sum total of their own labour is presented to them as a social relation, existing not between themselves but between the products of their labour. This is the reason why the products of labour become commodities, social things whose qualities are at once perceptible and imperceptible by the senses ... The existence of things *qua* commodities, and the value-relation between the products of labour which stamps them as commodities, have absolutely no connection with their physical properties and with the material relations arising therefrom. Here it is a particular social relation between men that assumes, in their eyes, the imaginary form of a relation between things. To find an analogy we must have recourse to the mist-enveloped regions of the religious world, where the productions of the human brain appear as independent beings endowed with life, and entering into relation both with one another and with the human race. (*Capital*, I, Ch. I, 4)

This process whereby social relations masquerade as things or relations between things is the cause of human failure to understand the society in which we live. In exchanging goods for money men involuntarily accept the position that their own qualities, abilities, and efforts do not belong to them but somehow inhere in the objects they have created. In this way they are the victims of the deformation of consciousness called alienation, and more particularly of reification, which confers objective reality on social relationships. Marx no longer uses the term 'alienation', but the description of the phenomenon is the same as in his earlier works, and so is the analogy with religion which he owes to Feuerbach.

Commodity fetishism, then, is the inability of human beings to see their own products for what they are, and their unwitting consent to be enslaved by human power instead of wielding it. Fetishism contains in embryo all other forms of alienation—the autonomy of political institutions which turn into instruments of oppression, the autonomy of creations of the human brain in the shape of religious fantasies: in short, the whole sum of man's enslavement to his own works. All social progress—scientific development and the organization of labour, improved administration and the multiplication of useful products—turn against man and are transformed into quasi-natural forces. Every genuine advance only serves to increase man's subjugation, as though to confirm Hegel's doctrine of the contradictions of progress.

However, the deluded consciousness which mistakes social relations for things finds particular expression in a phenomenon typical of the capitalist mode of production, namely the reification of labour-power—a situation in which human persons, real subjects, appear in the context of labour as commodities bought and sold on the market according to the rules dictated by the law of value.

As we have seen, the socialists had argued from Ricardo's labour theory of value that exploitation consisted in labour being sold at too low a price, and that the cause of social injustice was this non-equivalent exchange between the wage-earner and the capitalist. What must be done, therefore, was to reorganize production and exchange on a basis of equivalence, so that labour was sold at its true value.

However useful this reasoning might be for the purpose of agitation among the workers, Marx regarded it as quite erroneous. Exploitation, in his view, did not consist in the worker selling his labour below its value. To explain the phenomena of profit and exploitation it was necessary to start from the principle of equivalent exchange in the circulation of commodities as well as in the sale of the particular commodity known as labour-power. For—and this is the corner-stone of Marx's analysis of capitalism in its mature form—wage-labour is based on the sale of labour-power, not the sale of labour. Labour creates values, but does not itself possess value. To elucidate this, Marx propounds the question of the origin of capitalist profit. How is it that the owner of the means of production can get more exchange value out of them than he puts into the whole production process? How is it that a man with money can, simply because it is his, multiply it by lending at interest? How is it that a landowner is entitled to rent without any expenditure of labour on his part? It might appear to the simple-minded that capital is an autonomous source of value with a mysterious power of self-multiplication: a view which supports the theory of three independent sources of value, namely land, capital, and labour. Theories of this kind are used to justify the capitalist system and to suggest that capitalists, landowners, and workers have a common class interest as co-producers. They are, however, based on a confusion of thought, as in Condillac's theory that value is increased by the exchange process itself. It is true that the excess of a commodity's value over the cost of producing it is only realized in circulation, in the act of exchange, and this has given rise to the delusion that it originates in the act of exchange. But value, being exclusively the effect of the work of production, cannot be increased by merely commercial operations. Some socialists have argued that a merchant who buys cheap and sells dear is, in effect, a swindler, and that all his profit would at once disappear in conditions of equivalent exchange. But the fact is that profit can exist even when the exchange is strictly equivalent: it does not arise from circulation, though it only manifests itself when goods are exchanged. A man with money can multiply it thanks to the fact that there is on the market a particular commodity whose use-value is a source of value, and which creates exchange-value as its use-value is realized, i.e. in the process of consumption. This commodity is labour-power or capacity for labour, 'the aggregate of those mental and physical capabilities existing in a human being, which he exercises whenever he produces a use-value of any description' (*Capital*, I, Ch. IV, 3; English edn., Ch. VI). Wage-labour is the sale of labour-power for a fixed time. For this exchange to take place there must be a class of wage-earners who are free in a double sense: legally free to dispose of their labour-power and sell it to whomever they like, and also free from ownership of the means of production, i.e. possessing nothing but their labour-power and consequently obliged to sell it. This situation, in which the free wage-earner sells his labour-power to the owner of the instruments of production, is the characteristic feature of capitalism. It is a system which had a beginning in history and will have an end, but meanwhile it has revolutionized the whole historical process.

The value of labour-power is determined in the same way as that of any other commodity, by the amount of labour-time necessary to reproduce it. The reproduction of labour-power consists in maintaining the labourer in a condition in which he is able to work and to rear a fresh generation of non-property-owning producers. In other words, the value of labour-power is the value of the products necessary to keep the labourer and his children alive and able-bodied. Consequently, the sale of labour-power is an equivalent exchange when the wage-earner receives, in return, an amount equal to the cost of his subsistence. This amount is not determined solely by the physiological minimum but also by needs that vary historically; yet the physiological minimum constitutes the lower limit of wages. Thus the utopian socialists are wrong to argue that exploitation arises because the worker sells his labour for less than its value. As long as his wage enables him to remain alive and fit, he has not sold his labour for less than its value; the exchange is an equivalent one.

But this does not mean there is no such thing as exploitation. On the contrary, it is much more prevalent than the utopians thought, but it is due, not to a non-equivalent exchange between the seller and buyer of labour-power, but to the fact that at a certain technological level the application of labour-power can create exchange-values far greater than the values of the products necessary to maintain it. Or, to put if differently, the working day may be much longer than would be necessary to produce the commodities that keep the workman in an active state. The use-value of labour-power consists in the fact that it creates an exchange-value greater than its own. As in any purchase, the seller of labour-power parts with its use-value, which he makes over to the capitalist, in return for its exchange-value. The owner of the means of production pays the value of a day's work and acquires the right to use the worker's labour-power for anything up to twenty-four hours. The excess of the value so created over the cost of the worker's maintenance is 'surplus value', and this is acquired by the capitalist even in conditions of equivalent exchange. If half the worker's day corresponds to the value of the products necessary to reproduce his labour-power, the other half is unrequited labour—i.e. the consumption of labour-power (for the labour *is* the consumption) which creates the surplus value acquired by the owner of the means of production. This explains how exploitation can be consistent with equivalent exchange, and also why there is bound to be a class struggle against exploitation—a struggle which cannot be won simply by raising wages, but only by abolishing the whole system of wage-labour.

The capitalist maintains his rights as a purchaser when he tries to make the working day as long as possible, and to make, whenever possible, two working days out of one. On the other hand, the peculiar nature of the commodity sold implies a limit to its consumption by the purchaser, and the labourer maintains his right as seller when he wishes to reduce the working day to one of definite normal duration. There is here, therefore, an antinomy: right against right, both equally bearing the seal of the law of exchanges. Between equal rights, force decides. Hence it is that in the history of capitalist production the determination of what is a working day presents itself as the result of a struggle between collective capital, i.e. the class of capitalists, and collective labour, i.e. the working class. (*Capital*, I, Ch. VIII, 1; English edn., Ch. X, 1)

The system of wage-labour, in which the capitalist buys labour-power for the time during which it is exercised, obscures the division of the working day into the work necessary to reproduce labour-power and the extra, unpaid labour that creates surplus value. To outward appearance the employer pays for the whole of the worker's labour, but in fact he does not; the situation is the reverse of that which obtains under slavery, the slave appears to be working entirely for his master when in fact part of his working day is devoted to producing the values necessary for his own maintenance. In regular conditions of serfdom, on the other hand, the serf's labour for his lord and the work he does for his own benefit are clearly divided in time, and it is clear which part of his labour is unremunerated. The unrequited labour-time of the wage-earner is concealed in the homogeneous process of production, and it is necessary to analyse the situation to discover the source of surplus value. The capitalist expends a certain sum on the worker's wages, and the values created over and above that sum accrue to him as profit, which, however, only becomes actual in the circulation of commodities. The sum total of these surplus values is called 'absolute surplus value'; the ratio between it and the total amount of capital expended by the employer on wages is called 'relative surplus value'.

5. *The alienation of labour and of its product*

The one and only source of value, then, is productive labour, the shaping of material objects that satisfy human needs. All secondary forms of capital—that of merchants, bankers, and landowners—are used in the acquisition of surplus value, but play no part in its production. 'Industrial capital is the only mode of the existence of capital in which the latter's function consists not only in the appropriation of surplus value or surplus product, but likewise its creation' (*Capital*, II, Ch. I, 4). Industrial capital includes the organization of transport. 'The actual transport industry and expressage can be, and in fact are, industrial branches entirely distinct from commerce; and purchasable and saleable commodities may be stored in docks or in other public premises, the cost of storage being charged to the merchant by third persons in so far as he has to advance it ... The express company owner, the railway director and the shipowner are not "merchants"' (Ibid. III, Ch. XVII). Transport and storage, then, are part of production; but no commercial activity in the strict sense, i.e. no act of exchange, can endow commodities with additional value. Only the workman processing or transporting commodities, or of course the peasant labourer, creates new exchange-values and increases the sum total of value at the community's disposal.

We have thus discovered the social nexus on which the whole edifice of capitalist production is based, namely the commodity character of labour-power. The fact that labour-power is a commodity means that man functions as a thing, that his personal qualities and abilities are bought and sold like any other commodity; his brains and muscle, his physical energy and creative powers are reduced to a state in which only their exchange value counts for anything. This reification, the turning of a personality into a thing, is the measure of human degradation under capitalism. In this part of *Capital* Marx returns to the ideas he formulated as far back as 1843, when he saw in the working class the epitome of dehumanization and also the embodied hope of a restored humanity. In Chapter I of *Wage Labour and Capital* (1849) he wrote: 'The exercise of labour power, labour, is the worker's own life-activity, the manifestation of his own life. And this life-activity he sells to another person in order to secure the necessary means of subsistence. Thus his life-activity is for him only a means to enable him to exist. He works in order to live.' So in *Capital*:

The means of production are at once changed into means for the absorption of the labour of others. It is no longer the labourer that employs the means of production, but the means of production that employ the labourer. Instead of being consumed by him as material elements of his productive activity, they consume him as the ferment necessary to their own life-process, and the life-process of capital consists only in its movement as value constantly expanding, constantly multiplying itself.' (*Capital*, I, Ch. IX; English edn., Ch. XI)

In Volumes I and III of *Capital* Marx returns again and again to the theme of the alienation of labour—the vital productive process is nothing to the worker except as a means of maintaining himself—and the alienation of the fruits of labour: the objectification of the worker's energy, creating surplus value for others, is to him only the means of perpetuating his own poverty and dehumanization.

The relationships of capital ... place the labourer in a condition of utter indifference, isolation and alienation vis à vis the means of incorporating his labour ... The labourer looks at the social nature of his labour, at its combination with the labour of others for a common purpose, as he would at an alien power; the condition of effecting this combination is alien property, the squandering of which would be totally indifferent to him if he were not compelled to economize with it. (*Capital*, III, Ch. V, 1)

Capitalist production is in itself indifferent to the particular use-value and distinctive features of any commodity it produces. In every sphere of production it is only concerned with producing surplus value, and appropriating a certain quantity of unpaid labour incorporated in the product of labour. And it is likewise in the nature of the wage-labour subordinated to capital that it is indifferent to the specific character of its labour and must submit to being transformed in accordance with the requirements of capital and to being transferred from one sphere of production to another. (ibid., Ch. X)

Capitalism separates the product of labour from labour itself, the objective conditions of the productive process from human subjectivity. The worker creates values but has no way of realizing them for himself or enriching his life by appropriating them as use-values.

Since ... his own labour has been alienated from himself by the sale of his labour power, has been appropriated by the capitalist and incorporated with capital, it must, in the [production] process, be realized in a product that does not belong to him. As the process of production is also the process by which the capitalist consumes labour power, the labourer's product is incessantly converted, not only into commodities, but into capital, into value that sucks up the value-creating power, into means of subsistence that buy the person of the labourer, into means of production that command the producers. The labourer therefore constantly produces material, objective wealth, but in the form of capital, of an alien power that dominates and exploits him; and the capitalist as constantly produces labour power, but in the form of a subjective source of wealth, separated from the objects in and by which it can alone be realized—an abstract source, existing only in the labourer's person; in short, the capitalist produces the labourer, but as a wage-labourer. This incessant reproduction, this perpetuation of the labourer, is the *sine qua non* of capitalist production. (*Capital*, I, Ch. XXI; English edn., Ch. XXIII)

Consequently, the worker's vital functions are realized outside the production process, and it is only when not at work that he belongs to himself; as a worker, he belongs to the capitalist and functions only as a living reproducer of capital. This corresponds precisely to the picture drawn by Marx in the Paris Manuscripts. Even the worker's individual consumption, though motivated by his private needs, is, from the point of view of the economic process, a part of the activity of reproducing his labour-power, like greasing a wheel or supplying a steam-engine with coal. 'The labourer exists to satisfy the needs of self-expansion of existing values, instead of, on the contrary, material wealth existing to satisfy the needs of the labourer's development. As, in religion, man is governed by the emanation of his own brain, so in capitalist production he is governed by the work of his own hand.' (*Capital*, I, Ch. XXIII, 1; English edn., Ch. XXV, 1.) Since surplus value merely goes to swell the mass of existing capital, labour confers no kind of ownership. The right of property turns into its opposite: for the capitalist it becomes the right to appropriate values created by others, while for the worker it means that his own product does not belong to him. Consequently, the exchange relationship is completely illusory.

In the worker's situation we observe in its most blatant form the enslavement of man by his own works and by technical progress.

Machinery, considered alone, shortens the hours of labour, but in the service of capital it lengthens them; in itself it lightens labour, but when employed by capital it heightens the intensity of labour; in itself it is a victory of man over the forces of nature, but in the hands of capital it makes man the slave of those forces; in itself it increases the producers' wealth, but in the hands of capital it makes them paupers. (*Capital*, I, Ch. XIII, 6; English edn., Ch. XV, 6)

The effect of disjoining human labour from property, and creating a situation in which the worker's personal life is external to his work, is that the social process of production cannot take the form of a community. Co-operation itself is alienated *vis-à-vis* the co-operating producers: it presents itself to them as a form of compulsion, not alleviating their mutual isolation but intensifying it. 'The behaviour of men in the social process of production is purely atomic, and hence their relations to one another in production assume a material character independent of their control and conscious individual action. The chief manifestation of this is that products in general take the form of commodities.' (*Capital*, I, Ch. II.) Here Marx again repeats an idea from the Manuscripts. The alienation of labour is the source of the commodity form of production, not the other way about; by the same token it is the source of capital, i.e. of the value which increases itself by surplus value thanks to the purchase of labour-power.

6. *The alienation of the process of socialization*

The social character of labour under capitalist conditions is thus apparent only; it is a technological process, not a human one, and does nothing to overcome the isolation of producers.

The connection between [the workers'] various labours appears to them, ideally, in the shape of a preconceived plan of the capitalist, and practically in the shape of the authority of the same capitalist, the powerful will of one who subjects their activity to his own aims ... Being independent of one another, the labourers are isolated persons who enter into relations with the capitalist, but not with one another. This co-operation begins only with the labour process, but by that time they have ceased to belong to themselves. Once involved in that process, they are incorporated with capital. As co-operating members of a working organism, they are merely a special mode of the existence of capital. Hence the productive power developed by the labourer when working in co-operation is the productive power of capital. (*Capital*, I, Ch. XI; English edn., Ch. XIII)

Thus the characteristic and essential function of capitalism which consists in the exchange of variable capital (i.e. capital used to pay employees) for the labour-power of human beings is the true cause of producers being turned into things and prevented from forming a human community; for their community takes the form only of enforced co-operation between elements of their personal existence which have already been sold in the shape of labour-power, and are no longer their property. 'It is a result of the division of labour in manufactures that the labourer is confronted with the intellectual potencies [*Potenzen*] of the material process of production as the property of another and as a power to which he is subjugated.' (*Capital*, I, Ch. XII, 5; English edn., Ch. XIV, 5.) Whatever contributes to increasing man's power over the forces of nature likewise contributes, under the special conditions of wage labour, to destroying the producer himself; this applies both to technical progress and to the increased division of labour.

The division of labour in manufacture ... not only increases the social productive power of labour for the benefit of the capitalist instead of the labourer, but it does this by crippling individual labourers. It creates new conditions for the lordship of capital over labour. While, therefore, on the one hand it presents itself historically as a factor of progress and a necessary phase in the economic development of society, on the other hand it is a refined and civilized method of exploitation' (*Capital*, I, Ch. XII, 5; English edn., Ch. XIV, 5)

The mechanical automaton ... is the subject, and the workmen are merely conscious organs, co-ordinated with the unconscious organs of the machine and, together with them, subordinated to the central moving power.... The lightening of the labour, even, becomes a kind of torture, since the machine does not free the labourer from work, but deprives the work of all interest ... It is not the workman that employs the instruments of labour, but the instruments of labour that employ the workman ... By means of its conversion into an automaton, the instrument of labour confronts the labourer during the labour process itself in the guise of capital, of dead labour that enslaves the power of living labour and pumps it dry.' (Ibid., Ch. XIII, 4; English edn., Ch. XV, 4)

The division of labour becomes a fragmentation of man himself, shackled for life to part-activities whose function of creating use-value is of no concern to him, since the subjective purpose of his work is not to produce useful articles but to satisfy his own elementary needs. Indeed, the capitalist system prefers a stupid, mechanized worker who has no human skills beyond ability to perform the task imposed upon him.

But it is not only the worker who is turned into an instrument for the increase of capital; the same thing happens to the personality of the capitalist. Marx says in his Preface that he is concerned with human beings only as personifications of economic categories, embodiments of particular class-relations and class-interests. This is of course simply a methodological principle, excluding psychology from economic analysis and examining, not the motives of actions, but the laws which govern them and which, like those of natural science, do not depend on anyone's intentions. But this approach is only possible because the factual situation is such that the motives of individual capitalists are only manifestations of the tendency of capital to multiply itself, so that the capitalist as such is literally nothing but an embodiment of capital with no subjective or human qualities. 'As capitalist, he is only capital personified. His soul is the soul of capital. But capital has one single life impulse, the tendency to create value and surplus value and to make its constant factor, the means of production, absorb the greatest possible amount of surplus labour. Capital is dead labour, which, vampire-like, only lives by sucking living labour, and lives the more, the more labour it sucks.' (*Capital*, I, Ch. VIII, 1; English edn., Ch. X, 1.) 'Free competition brings out the inherent laws of capitalist production, in the shape of external coercive laws having power over every individual capitalist.' (Ibid. 5.) In the production process the worker and the capitalist are living representatives of variable capital and constant capital respectively, and this causes them to behave in a predetermined fashion. For the same reason, the utopian reformers are mistaken in thinking that the capitalist system can be changed by appealing to the goodwill or human feelings of the exploiters. The capitalist's personal character and intentions play no part in the economic process; he is subject to a force which inexorably shapes his ends, at all events where action on a socially significant scale is concerned. In capitalist production neither the worker nor the capitalist is a human being: their personal qualities have been taken away from them. Thus, when the class-consciousness of the proletariat evolves from awareness of poverty to a revolutionary consciousness and a sense of its historical mission to destroy capitalism, by the same token the worker becomes a human individual once again, throwing off the domination of exchange-value which turned him into a mere object. As for the capitalists, they cannot as a class take up arms against their own dehumanization, since they rejoice in it and in the wealth and power it brings. Thus, although both sides are equally dehumanized, only the wage-earner is spurred by this state of affairs to protestation and social combat.

It can thus be seen that, in Marx's view, not poverty but the loss of human subjectivity is the essential feature of capitalist production. Poverty indeed has been known throughout history, but awareness of poverty and even the revolt against it are not sufficient to restore man's subjectivity and membership of a human community. The socialist movement is not born of poverty, but of the class antagonism which arouses a revolutionary consciousness in the proletariat. The opposition between capitalism and socialism is essentially and originally the opposition between a world in which human beings are degraded into things and a world in which they recover their subjectivity.

7. *The pauperization of the working class*

The law which governs the sale of labour-power does not appear in itself to entail that the workers will remain poor or grow poorer. If they sell their labour-power at its true value—and there is nothing in capitalism to prevent this—it might seem that their standard of living can be maintained or even improved, inasmuch as that value is partly determined by non-physiological needs that vary from one period of history to another. But in fact the workers grow more and more impoverished, owing to the accumulation of capital. Moreover, their impoverishment is not only relative, involving a decreasing proportionate share of socially created values, but is also absolute: the working class either receives a diminishing sum total of values, or at all events is increasingly down-graded in the social scale.

All methods for raising the social productiveness of labour are brought about at the cost of the individual labourer; all means for the development of production transform themselves into means of domination over, and exploitation of, the producers; they mutilate the labourer into a fragment of a man, degrade him to the level of an appendage of a machine, destroy any remnant of attraction in his work and turn it into a hated toil; they estrange from him the intellectual potentialities of the labour process in the same degree as science is incorporated in it as an independent power [*Potenz*]; they distort the conditions under which he works, and subject him during the labour process to a despotism the more hateful for its meanness; they transform his lifetime into working-time, and drag his wife and child beneath the wheels of the Juggernaut of capital. But all methods for the production of surplus value are at the same time methods of accumulation; and every extension of accumulation becomes again a means for the development of those methods. It follows therefore that in proportion as capital accumulates, the lot of the labourer, be his payment high or low, must grow worse. The law that always equilibrates the relative surplus population, or industrial reserve army, to the extent and energy of accumulation, this law finally rivets the labourer to capital more firmly than the wedges of Vulcan did Prometheus to the rock. It establishes an accumulation of misery, corresponding with accumulation of capital. Accumulation of wealth at one pole is therefore at the same time accumulation of misery, agony of toil slavery, ignorance, brutality and moral degradation at the opposite pole, i.e. on the side of the class that produces its own product in the form of capital. (*Capital*, I, Ch. XXIII, 4; English edn., Ch. XXV, 4)

Marx put the point with equal clarity in *Wages, Price and Profit*: 'The general tendency of capitalist production is not to raise but to sink the average standard of wages, or to push the value of labour more or less to its minimum limit.' Hence, while the workers' economic struggle against pauperization may modify the downward trend of wages, and while it is necessary and important in itself, it cannot affect the basic development of capitalism or achieve the liberation of the proletariat.

The doctrine of the impoverishment of the proletariat is one of those that have excited most controversy among twentieth-century Marxists. The different references to the subject in Marx's works are by no means unequivocal. In his earlier writings, such as *Wage Labour and Capital* and the *Manifesto*, he appears to have believed in absolute impoverishment, or at least that wages in a capitalist economy were constantly governed by the principle of the physiological minimum. In the *Grundrisse*, however, he observes that the value of labour-power is partly determined by cultural factors, including the increase of needs to which capitalism itself gives rise: the satisfaction of previously unknown needs becomes part of the minimum living standard. In *Wages, Price and Profit* he also emphasizes that the conception of the minimum standard varies according to the traditions of different countries; and in the same work he introduces the idea of a relative fall in wages, i.e. a fall in workers' incomes compared with those of capitalists. The passage just quoted from *Capital* ('the lot of the labourer, be his payment high or low') is often used to argue that Marx finally abandoned the theory of absolute pauperization. But a distinction must be made between the level of wages and other factors governing the standard of living. The sense of the quoted passage is that whether wages are 'high' or 'low', the worker's position is bound to deteriorate both relatively and absolutely; not necessarily in terms of food and clothing, but by spiritual degradation and increasing subjection to economic tyranny.

The conclusion thus is that (1) Marx abandoned the theory that wages are bound to fall to, or remain at, the bare

subsistence level; (2) he continued to believe in absolute impoverishment as far as the worker's spiritual and social degradation were concerned; and (3) he maintained the doctrine of
relative impoverishment. This doctrine, however, as we may see
from Marx's writings and from later discussions among his
followers, can be defined in at least three ways. It may mean,
firstly, that total wages constitute a diminishing proportion of
the national product; or, secondly, that the average worker's
income decreases constantly in proportion to the average
capitalist's income; or, thirdly, that the worker earns an ever-
decreasing amount relative to his own growing needs. Clearly
these situations are not interdependent, and any one of them
could exist without the other two. It also appears clear that the
first might result from various causes, for example a relative fall
in the working-class population, in which case it would be misleading to speak of impoverishment. In the third situation impoverishment is defined by subjective criteria which cannot be
measured: if, for any reason, consumer aspirations are rapidly
rising, any or all classes of the population may feel 'impoverished' except for a handful of the very rich, who need not
be members of the bourgeoisie in the strict sense.

It is clear, however, that Marx was determined to find in
capitalism a relentless tendency to degrade the worker, and that
he resisted facts which indicated that the worker was getting
better off. Bertram Wolfe has pointed out that in the first
edition of *Capital* various statistics are brought down to
1865 or 1866, but those for the movement of wages stop
at 1850; in the second edition (1873) the statistics are brought
up to date, again with the exception of those on wages, which
had failed to bear out the impoverishment theory. This is a rare
but important case of disingenuousness in Marx's treatment of
factual data.

In the twentieth century, discussion could not blink the
obvious fact that there was no such thing as absolute pauperization in the capitalist economy. The question thus arose whether
this also meant that Marx was at fault in his whole theory of
accumulation and of the functioning of capitalism. Those who
wished to defend his doctrine, and who believed that the theory
of absolute pauperization flowed inevitably from it, were at
pains to show that despite appearances such pauperization did

exist. However, this point of view is seldom met with among Marxists today. Others argue that although the working class, by exerting pressure on the capitalists, has obliged them to lower the rate of profit, this does not mean there has been any change in the nature of capitalist production or the dehumanization that it inevitably brings. As Marx pointed out, wages and working hours are limited in two directions. On the one hand, the worker's elementary physical needs must be met if he is to go on living and if capitalist production is to subsist; on the other, the maximum wage level is determined by the success of the proletariat's struggle at any given time and the amount of pressure it is able to exert on the bourgeoisie. Hence, although Marx's forecast of absolute pauperization has proved wrong, this is not because of any flaw in the doctrine of accumulation and the tendency of capital to increase without limit, but only because Marx underrated the power of the working class to exert pressure within the capitalist framework.

In general, however, it must be borne in mind that material pauperization was not a necessary premiss either of Marx's analysis of the dehumanization caused by wage-labour, or of his prediction of the inescapable ruin of capitalism. That prediction was based on his belief that the internal contradictions of capitalism would destroy the system by bringing about an intensified class struggle, irrespective of whether material poverty increased or not.

8. The nature and historical mission of capitalism

As we saw, the essential characteristic of capitalism in Marx's eyes was its unlimited urge to multiply exchange-value, the insatiable appetite for self-increase by the exploitation of surplus labour. Capital is indifferent to the specific nature of the goods it produces or sells; it is interested in their use-value only in so far as it may serve to increase their exchange-value. Again and again Marx refers in his chief work to the 'wolfish hungering after surplus value' that is the hallmark of capitalism. Societies in which commercial exchange was practised for the purpose of acquiring use-value could not be characterized by this limitless hunger for growth. People who produce commodities to trade against things they want for themselves are, in effect, producing in order to create use-values. But the circulation of money as capital is, on the contrary, an end in itself, for the expansion of value takes place only within this constantly renewed movement. The circulation of capital has therefore no limits. As the conscious representative of this movement, the possessor of money becomes a capitalist ... The expansion [*Verwertung*] of value which is the objective basis of circulation, becomes his subjective aim; and it is only in so far as the appropriation of ever more and more wealth in the abstract becomes the sole motive of his operations that he functions as a capitalist, that is as capital personified and endowed with consciousness and a will. Use-value must therefore never be looked upon as the real aim of the capitalist; neither must the profit on any single transaction. The restless never-ending process of profit-making alone is what he aims at. (*Capital*, I, Ch. IV, 1; English edn., Ch. IV)

It is understandable, therefore, that the capitalist system required as a precondition the generalization of the monetary form of value, which sets no limit to the possibility of accumulation. The capitalist, however, 'fanatically bent on making value expand itself, ruthlessly forces the human race to produce for production's sake; he thus forces the development of the productive powers of society, and creates those material conditions that alone can form the real basis of a higher form of society based on the full and free development of every individual'. (*Capital*, I, Ch. XXII, 4; English edn., Ch. XXIV, 3.) It is not even the case that the capitalist behaves in this way for the sake of his own consumption; on the contrary, as a rule he regards enjoyment as the destruction of value and a form of waste, this kind of ascetic morality being especially common in the first phase of capitalism.

But the same insatiable hunger for exchange-value which degrades and impoverishes the working class is the cause of the amazing technological advance of capitalism.

Production for value and surplus value implies ... the constantly operating tendency to reduce the labour time necessary for the production of a commodity. i.e. its value, below the social average prevailing at the time. The pressure to reduce cost-price to the minimum becomes the strongest lever for raising the social productiveness of labour, which, however, is seen only as a continual increase in the productiveness of capital. (*Capital*, III, Ch. LI)

It is for this reason that former societies could exist for centuries in a state of technological stagnation, reproducing their way of life from one generation to another, whereas capitalism, as the *Communist Manifesto* pointed out, cannot exist without constantly revolutionizing the means of production. Technological progress is vital to it because the expansionist tendencies of capital obliges the entrepreneur to seek higher and higher profits by reducing the labour-time necessary to produce a commodity to a lower level than that which is socially necessary; he then markets his commodity at the current price and in so doing makes a profit higher than the average, i.e. that obtainable in average technological conditions.

When surplus value has to be produced by the conversion of necessary labour into surplus labour, it by no means suffices for capital to take over the labour process in the form in which it has been historically handed down, and then simply to prolong the duration of that process. The technical and social conditions of the process, and consequently the very mode of production, must be revolutionized before the productiveness of labour can be increased. Only thus is it possible to decrease the value of labour power and reduce the portion of the working day necessary for the reproduction of that value. (*Capital*, I, Ch. X; English edn., Ch. XII)

'Modern industry never looks upon and treats the existing form of a production process as final. Its technical basis is therefore revolutionary, while all earlier modes of production were essentially conservative.' (Ibid., Ch. XIII, 9; English edn., Ch. XV, 9.) For this reason 'the capitalist mode of production presents itself historically as a necessary condition of the transformation of the labour process into a social process.' (Ibid., Ch. XI; English edn., Ch. XIII.)

Capitalism, in short, is the necessary historical condition of progress in technology and the organization of labour. The 'wolfish hunger' for surplus value lies at the root of modern industry and modern co-operative methods, although that progress has been attained at the cost of unspeakable suffering, exploitation, poverty, and dehumanization. Fearful as are Marx's descriptions of the victimization of adults and children by the capitalist system, he regards that system not as a historical mistake which could have been avoided if someone, long ago, had devised a better form of social organization, but as a necessary condition of the re-establishment of a true community of mankind. Hence, though he believed the economic struggle of the proletariat to be indispensable, he did not regard it as an end in itself but, above all, as a means of hastening the revolutionary process. The accumulation of capital, by aggravating the workers' poverty, was also bringing closer the day of their liberation. For the hope of destroying capitalism did not lie only in spontaneous action by the working class. The internal contradictions of the system were bringing about a situation in which it could no longer subsist, and this was due to the self-increasing process which was its most vital principle.

9. *The distribution of surplus value*

In the first volume of *Capital* Marx analyses capitalist production in isolation from the process of circulation and the distribution of profit. He distinguishes between the rate of profit and the rate of surplus value, the former being the ratio of the surplus value obtained in production to the whole of the capital expended —i.e. constant capital (the value of the raw materials, equipment, etc.) plus variable capital (that spent on wages). The defenders of capitalism generally address themselves to the rate of profit, since the capitalist is interested in the ratio between his total investment and the resulting increase of value; exploitation of the worker is a means of maximizing value, not an end in itself. But, according to Marx, the degree of exploitation is not to be measured by the rate of profit but by the rate of surplus value, i.e. the ratio of surplus value to variable capital only: for it is this which shows how much of the value produced by the worker accrues to him and how much he forfeits to the capitalist by the sale of his labour-power. If, for example, the value he creates in a working day is double the price of his labour-power, i.e. the amount of variable capital expended, then the rate of surplus value, or the degree of exploitation, is one hundred per cent. Variable capital alone creates surplus value; the condition of its doing so, however, is the existence of constant capital, 'dead labour' in the form of equipment and the materials of production. There is no linear relationship between the rate of profit and the rate of surplus value; one may rise while the other falls, or vice versa.

The realization of surplus value depends, in reality, on circulation as well as production: the capitalist must sell his product in order to enjoy the excess of value over the production cost. But this complicates the issue in many ways, for commodities do not automatically find a purchaser and there is no guarantee that production, which is not planned on a social scale, will coincide with social demand. As Marx shows in the second volume of *Capital*, the circulation of commodities affects the rate of profit: it takes place over a period of time, during which larger or smaller portions of capital are inactive. Thus the surplus value created by the capital used in production is diminished to the extent of this inactivity, expressed, for example, in raw material stocks or unsold goods. The more rapid the capitalist's turnover, the greater the surplus value and the rate of profit. The market is a race to turn goods into money, in conditions where demand and supply are never exactly matched and consequently prices are never the same as values.

Indeed, capitalist production could not exist if commodities were sold at their true value. The rate of profit varies in different branches of production: different amounts of capital are necessary to hire the same number of workers and thus produce a given quantity of surplus value. Depending on variations in the 'organic composition' of capital (the ratio of the variable to the constant element), and on the different time that it takes for capital to circulate in different spheres of production, there would be vast differences in the rate of profit, i.e. the ratio of the increase of surplus value to the whole of the capital invested. Capital of course flows to where the rate of profit is highest. If there is too much capital in a particular branch of production compared with the market's absorptive power, the product will remain unsold; circulation will be impeded or slowed down, thus reducing the rate of profit and diverting capital to other branches where its value-producing power will be greater. This constant movement of capital creates the 'average rate of profit' applying to all branches of industry despite differences in the organic composition of capital. Competition evens out the rate of profit, but in so doing causes the price of commodities to diverge considerably from their value.

But the entrepreneur does not enjoy the whole of the profit accruing from production. Part of it is taken by the merchant, who does not help to produce surplus value but enables the producer to realize his profit. In this way the average rate of profit is affected by commercial capital. Again, the lending of money at interest does not mean that capital increases by some innate power. Interest is a share in the surplus value created by industrial capital, and reflects the fact that the circulation period affects the rate of profit. By borrowing money the capitalist is able to put into production an additional amount of value; he shares the resulting profit with his creditor, and thus the average rate of profit determines the rate of interest.

A further share in the distribution of profit (the absolute value of which equals the absolute quantity of surplus value produced) is assigned to the landowner. Marx regards agriculture, for the purpose of his argument, as a purely capitalist form of production, a kind of industry in which the entrepreneur invests means of production and employs a work force in the same way as the factory owner. The farmer divides his profit with the owner of the land, who receives part of the surplus value in the form of rent: it makes no difference here whether the land is arable or used for building etc. Thus rent also constitutes a share of the surplus value created by wage-labourers, and land is no more an independent source of increased value than is capital. The landowner is privileged inasmuch as the supply of land is limited and he can therefore demand a share in the profits of industrial capital. Rent is thus a by-product of the capitalist economy; and this also explains the fact that land can have a price, though it has no value. The price of land is anticipated rent, arising from the landowner's power to demand a share in the profits of capital though he has played no part in creating them; just as the price of a slave in antiquity was an anticipation of the surplus value to be got out of his labour.

CHAPTER XIII

The Contradictions of Capital and Their Abolition. The Unity of Analysis and Action

1. The falling rate of profit and the inevitable collapse of capitalism

In its striving after unlimited growth (Marx's argument continues), capitalism involves itself in an inextricable contradiction. As technology progresses and the amount of constant capital increases, less and less work is necessary to produce the same volume of goods; the ratio of variable to constant capital decreases, and so does the average rate of profit. This law of the diminishing rate of profit is a universal feature of capitalist production. On the one hand, capital increases only thanks to the growth of surplus value, and its chief concern is to maximize this value in proportion to the resources used; on the other hand, it is obliged by competition and improved technology to create conditions which constantly lower the rate of profit. It strives to prevent this effect by increasing exploitation, lengthening the workers' day and paying them less than their labour-power is worth. Another factor which helps to maintain profits is that increased productivity, while it tends on the one hand to lower profits, also creates a 'reserve army of labour'—a state of relative overpopulation which forces workmen to compete with one another and so depresses the wage level. The profit rate is also assisted by foreign trade in so far as it helps to lower the price of some ingredients of constant capital or reduce the cost of subsistence. Nevertheless, in spite of all these factors the profit rate tends increasingly to decline. The effect of this is to aggravate exploitation and encourage the concentration of capital, as small capitalists find it harder to make ends meet and are swallowed up by big ones. The fall in the profit rate also leads to overproduction, excess capital, relative overpopulation, and economic crises. The employers' alarm in this situation springs from the feeling

'that capitalist production encounters, in the development of its productive forces, a barrier which has nothing to do with the production of wealth as such. This peculiar barrier testifies to the limitations and the merely historical, transitory character of the capitalist mode of production; it shows that it is not an absolute method of producing wealth, but rather, at a certain stage, prevents it increasing any further' (*Capital*, III, Ch. XV, 1).

The law of the falling rate of profit is, according to Marx, one of the internal contradictions of capitalism which are bound to lead to its downfall; but he never argued, as has been alleged, that the fall in the rate of profit would in itself make capitalism an economic impossibility. A falling profit rate is quite compatible with an increasing total volume of profit, and it is hard to see how it could be the direct cause of the system breaking down. The principal factor working against a fall in the profit rate is a decline in the value of the components of constant capital, owing to the same technical progress which reduces the relative importance of wages in production costs—this being a basic aspect of Marx's analysis. In view of the difficulty of quantifying the factors working in either direction, there is no firm ground for asserting that those tending to produce a fall in the rate are stronger; and the alleged 'law' appears to be no more than an expression of Marx's hope that capitalism would be destroyed by its own inconsistencies. Only empirical observation, and not deduction from the nature of the profit rate, can tell us whether it does tend permanently to decline; and such observation is not found to confirm Marx's theory.

Marx often repeats (for example, in *Capital*, I, Ch. XXI (English edn., Ch. XXIII); III, Ch. LI) that the capitalist production process reproduces the social conditions that estrange the worker from his own labour and its product, and perpetuates itself by depriving producers of a share in the values they create. This does not mean, however, that the process can go on *ad infinitum*. The fall in the rate of profit and increasing accumulation create artificial overpopulation; at the same time the fall in the rate slows down accumulation and provides an incentive to reactivate it by every possible means, with the result that capital repeats the very processes that it desires to prevent. The upshot is a paradoxical situation in which there is both an excess of capital available for production and an excess of working-class population. Consumption cannot keep pace with the increase of production that springs from the insatiable greed for surplus value, since this greed itself prevents a corresponding increase in the purchasing power of the masses. The sum total of wealth produced is by no means too great for real needs, but it is chronically more than the market can absorb. The falling rate of profit is a constant obstacle to the development of the productive force of labour. As capital accumulates, so it is more and more concentrated as a result of small producers being driven out of business. Capital overcomes its contradictions by means of periodic crises of overproduction which ruin the mass of small owners and wreak havoc among the working class, after which the balance of the market is restored for a time. These crises, due to the anarchic character of production and the fact that its sole purpose is to increase exchange-value, are an essential feature of the capitalist economy.

It is not the case, as spokesmen of the working class have often maintained, that crises could be averted by raising wages and enabling the market to absorb more goods, and that therefore it is in the employers' interest to pay higher wages; this is refuted, as Marx argues in the second volume of *Capital*, by the fact that crises regularly occur after a period of relative prosperity and rising wages, which would prevent them if the argument were correct. The fact is that the greed for expansion is such that the market cannot possibly go on absorbing the products of capitalism—especially as the bulk of these, in value terms, consists of means of production, which do not become easier to sell because of a rise in wages. Economic crises involve the squandering of the community's wealth on a vast scale, demonstrating that capitalism cannot cope with its own contradictions. They are the expression of a conflict between the technological level and social conditions of technical progress, between the forces of production and the system within which they work. The capitalist, controlling the means of production and concerned only to increase surplus value to the utmost, is no longer, as he was in the first stage of accumulation, an organizer who plays a necessary part in efficient production; as often as not, he now leaves it to others to run his enterprise for him. Property and management are increasingly disjoined. As production becomes more and more social in character, the private appropriation of the fruits of labour is seen to be more and more anachronistic.

Thus grows the power of capital—the alienation, personified by the capitalist, of the conditions of social production from the real producers. Capital comes increasingly to the fore as a social power, whose agent is the capitalist and which no longer stands in any possible relation to what the labour of a single individual can create. It is an alienated, independent force which stands opposed to society as an object and as the means whereby the capitalist wields his power. The contradiction between the general social force into which capital develops, on the one hand, and the power of individual capitalists over the social conditions of production, on the other, becomes ever more irreconcilable; yet it contains the germ of a resolution of the situation, in the shape of a transformation of the conditions of production into general, common, social conditions. (*Capital*, III, Ch. XV, 4)

Capital seeks frantically for new markets and endeavours to expand into non-capitalist areas, but the more its productive capacity increases, the more obvious is its conflict with the narrow limits of consumption. Marx holds that capitalism is doomed from the purely economic point of view, independently of the class struggle, since the contradiction, inherent in its production system, between use-value and exchange-value is bound to cause ever-recurring crises. As Engels put it,

We have had many of these revulsions, happily overcome hitherto by the opening of new markets (China in 1842), or the better exploiting of old ones, by reducing the cost of production (as by free trade in corn). But there is a limit to this, too. There are no new markets to be opened now; and there is only one means left to reduce wages, namely, radical financial reform and reduction of the taxes by *repudiation of the national debt*. And if the free-trading mill-lords have not the courage to go the length of that, or if this temporary expedient be once exploded, too, why they will die of repletion. It is evident that, with no chance of further extending markets, under a system which is obliged to extend production every day, there is an end to mill-lord ascendancy. And *what next?* 'Universal ruin and chaos,' say the free-traders. *Social revolution and proletarian ascendancy, say we.* (*Democratic Review*, London, Mar. 1850)

The issue here raised was to be much debated by Rosa Luxem-

burg and her critics: was capitalism bound to collapse as soon
as it could no longer expand into non-capitalist markets? If so,
there was a well-defined limit beyond which capitalism could
not endure, whether or not one held (as Marx, Engels, and Rosa
Luxemburg did) that it would not simply destroy itself like an
erupting volcano but would have to be overthrown by the revolutionary working class. While Engels appears to answer the
question in the affirmative, it does not seem to be a necessary
deduction from Marx's views that capitalism can no longer exist
when there are no further non-capitalist markets left to conquer.
All that follows is that capitalism must be destroyed by its own
inconsistencies, especially the conflict between private ownership
and the development of instruments of production and technical
co-operation; that it is becoming a brake on technological
progress, which it did so much to foster in the past, and that
this fact must be its downfall.

The proletarian revolution, Marx believed, would spring from
the same antagonism, *mutatis mutandis*, as had the bourgeois
revolutions. At a certain stage bourgeois technology had become
irreconcilable with the social conditions of feudalism—the
restrictive guild system, local and hereditary privileges, and
checks on the free employment of labour. In the same way, as
technology progressed, the bourgeoisie itself had created a
situation which was bound to ruin it as a class, doing away with
capitalist ownership and, finally, all class differences.

Along with the constantly diminishing number of the magnates of
capital, who usurp and monopolize all the advantages of this process
of transformation, grows the mass of misery, oppression, slavery, degradation, exploitation; but with this too grows the revolt of the
working class, a class ever increasing in numbers and disciplined,
united, organized by the very mechanism of capitalist production
itself. The monopoly of capital becomes a fetter upon the mode of
production, which has sprung up and flourished along with it and under
it. Centralization of the means of production and socialization of labour
at last reach a point where they become incompatible with their
capitalist integument. This integument is burst asunder. The knell of
capitalist private property sounds. The expropriators are expropriated.
(*Capital*, I, Ch. XXIV, 7; English edn., Ch. XXXII)

2. *The economic and political struggle of the proletariat*

It is clear from this that economic analysis alone brought Marx to the conclusion that capitalism was beyond reform, and that despite all political and economic struggles the working class would remain in bondage as long as the capitalist production system continued. As Marx and Engels wrote in 1850 in an appeal from the Central Committee to the Communist League: 'For us the issue cannot be the alteration of private property but only its abolition, not the smoothing over of class antagonisms but the abolition of classes, not the improvement of existing society but the foundation of a new one.' In articles on 'The Housing Question' published in the Leipzig *Volksstaat* in 1872–3, Engels wrote: 'As long as the capitalist system of production exists, it will be absurd to attempt to solve the housing question or any social question affecting the workers. The solution is to destroy the capitalist system of production.' It might seem that as no social question could be solved under capitalist conditions and as the blind onrush of the system was leading it to its doom, Marx and Engels were in effect, as their reformist critics held, adopting a position of 'the worse, the better', i.e. welcoming exploitation and poverty because they brought the revolution closer. This touches on a crucial problem of Marxist theory, the relationship between the 'objective', quasi-natural laws of economics on the one hand and free human initiative on the other. If capitalism is to transform itself into socialism by a spontaneous explosion independent of human will, there is no need to do anything but wait until its contradictions reach their height and the system chokes itself by its own expansion. In actual fact, however, capitalism can only be abolished when the class-consciousness of the proletariat is sufficiently developed. Perhaps Marx's clearest statement of this view is in an article on Russian policy towards Turkey, published in the *New York Daily Tribune* on 14 July 1853:

There exists a class of philanthropists, and even of socialists, who consider strikes as very mischievous to the interests of the 'working-man [*sic*] himself', and whose great aim consists in finding out a method of securing permanent average wages. Besides the fact of the industrial cyclus, with its various phases, putting every such average wages [*sic*] out of the question I am, on the very contrary, convinced that the alternative rise and fall of wages, and the continual conflicts between masters and men arising therefrom, are, in the present organization of industry, the indispensable means of holding up the spirit of the labouring classes, of combining them into one great association against the encroachments of the ruling class, and of preventing them from becoming apathetic, thoughtless, more or less well-fed instruments of production. In a state of society founded upon the antagonism of classes, if we want to prevent Slavery in fact as well as in name, we must accept war. In order to rightly appreciate the value of strikes and combinations, we must not allow ourselves to be blinded by the apparent insignificance of their economical results, but hold, above all things, in view their moral and political consequences. Without the great alternative phases of dullness, prosperity, over-excitement, crisis and distress, which modern industry traverses in periodically recurring cycles, with the up and down of wages resulting from them, as with the constant warfare between masters and men closely corresponding with those variations in wages and profits, the working-classes of Great Britain, and of all Europe, would be a heart-broken, a weak-minded, a worn-out unresisting mass, whose self-emancipation would prove as impossible as that of the slaves of Ancient Greece and Rome.

Marx's position is thus clear: the disarray of capitalist production affords the opportunity for the working class to organize itself in a movement of protest and become conscious of its own revolutionary future. The laws of capitalism that operate against the workers may be weakened in their effects, but cannot be neutralized as long as the system lasts. Hence the economic struggle cannot be expected to yield triumphant results. Its main purpose is to foster the political consciousness of the proletariat; for, as Marx writes in *Wages, Price and Profit*, 'In its merely economic action, capital is the stronger side.' The economic struggle is above all a preparation for the decisive political struggle, not an end in itself. At the same time, the political movement is not an end in itself either, but a means of economic liberation, as was emphasized in the Rules of the First International (1871): 'The economic emancipation of the working class is the great end to which every political movement ought to be subordinate as a means.' Thus, while Marx held that 'though temporary defeat may await the working classes, great social and economical laws are in operation which must eventually insure their triumph' ('The English Middle Class',

New York Daily Tribune, 1 Aug. 1854), he did not draw the conclusion that the workers could sit back and await final success as a gift from History. On the contrary, political consciousness prepared by the economic struggle was an indispensable condition of success. 'Economic laws' in themselves were sufficient to ensure the possibility of victory, but political initiative had its place as an autonomous factor in the historical process. We find here, in a more specific form, a theme present in Marx's writings since the earliest period. In the class-consciousness of the proletariat, historical necessity coincides with freedom of action; the opposition between human will and the 'objective' course of events ceases to exist, the dilemma of utopianism and fatalism is resolved. The working class, and it alone, enjoys the privilege that its hopes and dreams are not condemned to beat against the wall of inexorable destiny; its will and initiative are themselves part of the necessary course of history. This means in practical terms that the economic struggle is a means to political action (the main point on which the reformists dissented from Marxism), while political action is a means to economic emancipation after the revolution; for, under socialism, there will in any case be no separate sphere of political life.

It is absurd, therefore, to say that from the Marxist point of view the working class should welcome crises, unemployment, and falling wages as steps leading to the destruction of capitalism. On the contrary, it must combat the effects of crises, while realizing that it is impossible for capitalism, by reforming itself, to obviate the enslavement of the proletariat. The workers' task is not to invite economic disasters but to use them, when they occur, for revolutionary purposes. In the same way the expropriation of small owners, including peasant ones, is an inevitable law of capitalist accumulation ('The smallholder, like any other survival of an outdated production system, is doomed to extinction and to become the proletariat of the future'— Engels, 'The peasant problem in France and Germany', *Die Neue Zeit*, Nov. 1894); but it does not follow that socialists should do their best to ruin the peasantry, only that they should take advantage of the inevitable process to increase their own political strength. In short, in the political struggle and in the economic struggle which is an instrument of it, the proletariat must defend its own interests from a strictly class point of view; but by doing so it becomes the champion of humanity as a whole, since the revolution that it brings about leads to the socialization and thus the liberation of mankind. In exactly the same way, the bourgeois revolutions inspired by the interest of a single class furthered the cause of all humanity. As Marx wrote in 'The Bourgeoisie and the Counter-revolution' (*Neue Rheinische Zeitung*, 11 Dec. 1848), 'The revolutions of 1648 and 1789 were not English and French revolutions, but European ones. They were not just the victory of a particular social class over the old political order, but the proclamation of a political order for a new European society.' The service they rendered to humanity did not consist, however, in liberating society by allowing capital to develop freely; what has happened is that the immense progress of technology and political organization has prepared the way for a socialist revolution, which can only take place in the conditions created by capitalism.

Capitalism creates the preconditions of the new society not only by revolutionizing technology and evolving new forms of co-operation: as we read in Volume III of *Capital*, joint-stock companies in which property and management are separate, and likewise co-operative factories, are to be regarded as 'transitional forms' or instances of the abandonment of the capitalist mode of production within the system itself. In this sense socialism is not simply the negation of capitalism but also a continuation of it and of the socializing process based on the technological development of the present age.

3. The nature of socialism, and its two phases

Capitalism, then, creates the necessary preconditions of socialism. Its historical mission was to bring about a tremendous development of technology due to the unbridled urge to increase exchange value to the maximum. By constantly transferring masses of workers from one occupation to another, capitalism calls for a certain versatility in the working class and thus creates conditions for an upheaval in which the division of labour will be abolished: cf. *Capital*, I, Ch. XIII (XV in English edn.), 9. But, as Engels wrote, 'it is only at a certain stage, and in modern conditions a very advanced one, of the development of social conditions that production can be increased to a point at which the abolition of class differences can constitute a real progress and can be achieved permanently without causing a standstill or a regression in the community's method of production' ('Social Conditions in Russia', *Volksstaat*, 1875). Socialism reaps the harvest of capitalism, and without the latter it could only be an empty dream. The new society will arise out of the catastrophe towards which capitalism is swiftly yet unconsciously tending.

'The working class has conqured nature; now it must conquer man' (Marx in the *People's Paper*, 18 Mar. 1854). This is a concise expression of the Marxian idea of socialism. To 'conquer man', as Marx indicated on many occasions, is to create conditions in which men are in full control of their own labour process and its physical and spiritual product, so that the results of their actions cannot in any circumstances turn against themselves. Man ruling over himself and no longer subjected to material forces of his own creation—man identified with the social process, overcoming the opposition between blind necessity and his own free behaviour—such will be the effect of the socialist revolution. Socialism, as we saw, does not consist essentially in abolishing material poverty or the luxurious consumption of the bourgeoisie, but in abolishing human alienation by doing away with the division of labour. If the bourgeois standard of living were equated with that of the workers, this in itself would not bring about any significant change. It is not a question simply of redistributing the same income produced in the same old way. Nor, as Marx emphasized against the Lassalleanists, is it a matter of the worker receiving for his own benefit the whole of the value that he creates, for that is an impossibility. There are many occupations that create no value, yet are socially necessary and must be preserved in the socialist system. So, as we read in the *Critique of the Gotha Programme*, there can be no question of demanding 'the undiminished proceeds of labour'. Considerable sums must always be deducted from the social product for the renewal of consumed values, the expansion of production, insurance against emergency, administrative costs, collective consumption (schools and hospitals), and the care of those unfit to work. The basic difference between the capitalist and socialist modes of production is that in the latter the system of wage labour, i.e. the sale of labour-power, is abolished and the whole of material production is devoted to use-value. In other words, the scale and character of production in all its branches will be governed purely by social needs and not by the desire to accumulate the maximum exchange-value; and this, of course, requires the social planning of production.

By abolishing the capitalist form of production, the length of the working day could be reduced to the necessary labour time. But, even in that case, the latter would extend its limits. On the one hand, because the notion of 'means of subsistence' would considerably expand and the labourer would lay claim to an altogether different standard of life. On the other hand, because a part of what is now surplus labour would then count as necessary labour, viz. the labour of forming a fund for reserve and accumulation. (*Capital*, I, Ch. XV [English edn., Ch. XVII], IV)

The distinction between necessary and surplus labour would in fact lose its meaning in socialist conditions: not all labour would be directly remunerated in the form of wages, but all of it would accrue to society in the collective satisfaction of various needs.

But the liberation of humanity does not consist solely in the satisfaction of material needs, however much their scope may be extended, but rather in achieving a full and many-sided life for all. This is why Marx was so concerned to abolish the division of labour, which crippled human beings physically and spiritually and condemned individuals to a stultifying one-sidedness. The prime task of socialism is to liberate all the powers latent in every human being, and develop his personal abilities to the utmost in the social context. This being so, in what sense are we to understand the claim that socialism is the 'final' state of man? As Engels wrote in *Ludwig Feuerbach*, I: 'Just as knowledge is unable to reach a perfected termination in a perfect, ideal condition of humanity, so is history unable to do so.' Socialism is not 'final' in the sense of a stagnant society providing for the satisfaction of a fixed total of needs and therefore containing no incentive to development. But it is 'final', according to Marx, in that it would ensure that society was in full control of the conditions of its own existence, so that there was no occasion for any further transformation; there would no longer be any difference between rulers and ruled, and no limitation on human creativity. Socialism does not mean that human development and creativity cease to exist, but that there are no longer any social restrictions upon them. However, the development of creative forces does not mean simply, or even chiefly, the increase of material wealth. The well-known passage in Volume III of *Capital* is very significant here:

The realm of freedom actually begins only where labour which is determined by necessity and mundane considerations ceases; thus in the very nature of things it lies beyond the sphere of actual material production. Just as the savage must wrestle with Nature to satisfy his wants, to maintain and reproduce life, so must civilized man, and he must do so in all forms of society and under all possible modes of production. With his development this realm of physical necessity expands as a result of his wants; but, at the same time, the forces of production which satisfy these wants also increase. Freedom in this field can only consist in socialized man, the associated producers, rationally regulating their interchange [*Stoffwechsel*] with Nature, bringing it under their common control, instead of being ruled by it as by a blind force; and achieving this with the least expenditure of energy and under conditions most favourable to, and worthy of, their human nature. But it nevertheless still remains a realm of necessity. Beyond it begins that development of human energy which is an end in itself, the true realm of freedom, which, however, can blossom forth only with this realm of necessity as its basis. The shortening of the working day is its basic prerequisite. (*Capital*, III, Ch. XLVIII, III)

We thus have a schema of the values that Marx associated with the socialist transformation. Socialism as a mode of organization consists in removing the obstacles which prevent human beings from developing their creative abilities to the full. This free expansion in all spheres is the true purpose of humanity. The production of physical requirements belongs to the 'realm of necessity', and the time spent therein is the measure of man's dependence on nature. This of course cannot be completely overcome, but its effect can be minimized and, more important, it is possible to eliminate the forms of compulsion connected specifically with social existence, i.e. so to order things that social life will be a fulfilment of individuality and not a curb on it. This identification of personal and collective existence will not be a matter of compulsion—for then it would be contrary to its own premises—but will spring from the consciousness of each and every individual, who will regard his own life as value-creating in respect of others. There will no longer be a gap between social and private life—not because the individual is absorbed into a single, grey, uniform collectivity, but because social life will no longer create forms that are alienated from the individual: it will cease to arouse antagonisms and will present itself to every man as his own personal creation. Social relations will become transparent to all instead of wrapped in the mystification of religious forms.

The religious reflex of the real world can only then finally vanish, when the practical relations of workaday life offer to man none but perfectly intelligible and rational relations with his fellow-man and Nature. The life-process of society, i.e. the process of material production, does not strip off its mystical veil until it takes the form of production by freely associated men, and is consciously regulated by them in accordance with a settled plan. This, however, demands for society a certain material groundwork, or set of conditions of existence, which in their turn are the spontaneous product of a long and painful process of development. (*Capital*, I, Ch. I, 4)

The socialist movement thus leads to a revolution without precedent in history—the greatest transformation of all and, in the sense explained above, the final one. Socialism is *novissimus*, the end of history as it has hitherto been known and the beginning of the adventure of mankind. It makes a radical break with the past and has no need of any existing tradition to justify it or bring it to self-awareness. 'The social revolution of the nineteenth century can only create its poetry from the future, not from the past. It cannot begin its own work until it has sloughed off all superstitious regard for the past. Earlier revolutions have needed world-historical reminiscences to deaden their awareness of their own content. In order to arrive at its own content the revolution of the nineteenth century must let the dead bury their dead.' (*The Eighteenth Brumaire of Louis Bonaparte*, I.)

From 1848 onwards Marx went through alternate phases of expecting an early European revolution and reconciling himself to a longer wait. Every new period of disturbances, war, or economic depression increased his hopes. Shortly after 1848 he gave up the optimistic conviction that the death-knell of capitalism had already sounded; instead, he told the advocates of 'direct action' that the workers had fifteen, twenty, or fifty years of hard struggle to face before they would be ready for power. Again and again he was encouraged by political or economic crises to hope that in one place or another, in

Germany, Spain, Poland, or Russia, a revolutionary spark might touch off a fire that would sweep across Europe. In accordance with his theory he expected most from the more developed countries, but he also hoped at times that even backward Russia might witness the breaking of the storm that would herald the worldwide transformation. Among his followers many sterile disputes arose as to the conditions which, according to the doctrine, were most likely to presage a world revolution of the proletariat. Marx himself did not formally indicate what these were, and his scattered remarks on the subject over many years do not form a consistent whole. It is evident that there was a conflict in his mind between revolutionary impatience and the theory that capitalism must first attain its 'economic maturity' —which, he apparently thought, had not taken place in any European country except Britain—and one or other of these viewpoints prevailed according to the turn of events. He never indicated, however, by what signs economic maturity was to be recognized. In 1871–2, moreover, he took the view that in advanced countries such as Britain, the U.S.A., and Holland the transition to socialism might be achieved by peaceful propaganda, without violence or rebellion.

All in all, Marx came in time to believe that there could be no immediate transition to a socialist system as he imagined it. In the *Critique of the Gotha Programme* he observed that there would have to be an intermediate period between the revolution and the final realization of socialist hopes. In the first stage, human rights would be proportionate to labour. 'This equal right is an unequal right for unequal labour. It recognizes no class differences, because everyone is only a worker like everyone else; but it tacitly recognizes unequal individual endowment and thus productive capacity as natural privileges. It is therefore a right of inequality in its content, like every right.' The transitional period would bear the mark of the society out of which it had grown. Economically it would be based on the principle 'To each according to his labour'; politically it would be a dictatorship of the proletariat, a system in which a particular class exercised authority and used force for the purpose of abolishing class distinctions. Only in the higher phase of communist society, when men would no longer be enslaved by the division of labour and when the difference between physical and intellectual work would be done away with—when the development of productive forces ensured a sufficiency for all, and labour was the most vital requirement of a man's being—only then could the slogan become a reality, 'From each according to his ability, to each according to his needs.'

Although Marx did not leave any detailed description of the organization of the future society, its basic principle is clear: socialism stands for complete humanization, restoring man's control over his own powers and his own creative energy. All its specific features can be derived from this principle: the gearing of production to use-value, the abolition of the division of labour in so far as it impedes the acquisition of a diversity of skills (but not, of course, in the sense of reverting from industry to craftsmanship), the dismantling of the state apparatus as distinct from the administration of production, the abolition of all social sources of inequality (equality, as Engels wrote, means doing away with class differences but not with individual ones) and of all social conditions that in any way restrict human creativity. It is significant that, according to Marx, the abolition of capitalism 'does not re-establish private property for the producer, but gives him individual property based on the acquisitions of the capitalist era: i.e. on co-operation among free workers and their possession in common of the earth's resources and the means of production, themselves produced by labour'. (*Capital*, I, Ch. XXIV, 7; English edn., Ch. XXXII.) 'Individual property' stands in contrast to capitalist property: the latter is non-individual in the sense that its transformation and growth are not controlled by particular human beings and that it develops its own laws in the anonymous force of capital, subjugating even the capitalist himself. Socialism, by contrast, is the return to a situation in which only individual human subjects truly exist and are not governed by any impersonal social force; property is individual, and society is no more than the assemblage of the individuals who own it. The notion that Marx regarded socialism as a system for depressing individuals into a Comtean universal being deprived of all subjectivity is one of the absurdest aberrations to which the study of his work has given rise. What can be said with truth is that in Marx's view personality is not a mere matter of self-experience on the lines of the *cogito ergo sum*, since there is no such thing as pure self-knowledge apart from consciousness of the social life in which the individual has his being. The contrary supposition could only arise in conditions in which intellectual work has been so completely severed from productive work that the links between them were forgotten. Every individual was a social being; man realized himself in the community, but this did not mean that the latter derived its creative forces from any other source than that of personal, subjective existence.

4. *The dialectic of* Capital: *the whole and the part, the concrete and the abstract*

These views do not in any way conflict with Marx's over-all analysis of capitalism. Throughout history material forces have dominated human beings, and in considering capitalist society each separate element must be related to the whole and each phenomenon treated as a phase in a developing process. In *Capital* Marx more than once recalls this global aspect of his method of inquiry. No economic act, however trivial, such as the buying and selling that occurs millions of times a day, is intelligible except in the context of the entire capitalist system.

Every individual capital forms but an autonomous fraction, endowed with individual life, as it were, of the aggregate social capital, just as every individual capitalist is but an individual element of the capitalist class. The movement of the social capital consists of the totality of the movements of its fractional parts, the turnovers of individual fragments of capital. Just as the metamorphosis of the individual commodity is a link in the series of metamorphoses of the commodity-world—the circulation of commodities—so the metamorphosis of the individual capital, its turnover, is a link in the circulation of social capital. (*Capital*, II, Ch. XVIII, I)

Accordingly, the existence of an average rate of profit means that each capitalist makes a profit proportionate to his share of the aggregate social capital and not to the organic composition of capital in his particular branch of production. The whole functioning of the capitalist economy is geared to creating the maximum exchange-value in conditions of the interdependence of every link in the process of production and the circulation of capital; the economy has become a single process and can only be understood as such.

But the dialectical rule that a phenomenon can only be

understood in relation to the whole does not mean that the starting-point of analysis must be an empirical 'whole' untouched by theory, a mere confused jumble of perceptions. On the contrary, such a 'whole' is incapable of being an object of cognition. It is the function of analysis to reproduce the concrete on the basis of abstractions, viz. of the simplest social categories, which take shape in the first instance as isolated phenomena and are only afterwards enriched by the perception of their mutual relations. The argument is summarized in a passage in the Introduction to the *Grundrisse*:

It would seem to be correct to begin with the real and concrete, the real precondition, and thus, in economics, to begin with e.g. the population, which is the foundation and subject of the entire social act of production. However, on closer examination this proves false. The population is an abstraction if I leave out, for example, the classes of which it is composed. These are in turn an empty phrase if I am not familiar with the elements on which they are based, e.g. wage labour, capital etc ... Thus, if I were to begin with the population, this would give a chaotic picture of the whole; I would then, by means of further definition, move analytically towards ever more simple concepts, from the imagined concrete towards ever thinner abstractions until I had arrived at the simplest formulations ... The economists of the seventeenth century always begin with the living whole, with population, nation, state, several states etc.; but they generally conclude by discovering through analysis a small number of determinant, abstract, general relations such as division of labour, value, money etc. Once these individual factors were more or less firmly established and abstracted, there began the economic systems, which ascended from simple relations, such as labour, division of labour and exchange value, to the level of the state, exchange between nations and the world market. This is obviously the scientifically correct method. The concrete is concrete because it is the concentration of many determinations, hence unity of the diverse. It appears in the process of thinking, therefore, as a process of concentration, as a result and not a point of departure even though it is the point of departure in reality and hence also the point of departure for observation and conception.... Abstract determinations lead towards a reproduction of the concrete by way of thought. In this way Hegel fell into the illusion of conceiving the real as the product of thought concentrating itself, probing its own depths, and unfolding itself out of itself; whereas the method of rising from the abstract to the concrete is only the way in which thought appropriates the concrete, reproduces it as the concrete in the mind. But it is by no means in this way that the concrete itself comes into being.

Thus in Marx's view the order of exposition of social phenomena is the reverse of the order of factual observation. The former begins with the simplest and most abstract qualitites of social life, for example value, and from these it reconstructs concrete phenomena in the form in which they are assimilated by the mind and subjected to theory. The 'whole' thus reproduced is not the chaotic mass of direct perception but a conceptually linked system. To achieve this result we make use, as in any other science, of the method of ideal situations which assume for the sake of argument certain simple relationships undisturbed by any outside factor, so that their complexity may afterwards be analysed.

In this way Marx attempts to transfer to political economy the basic method of modern science which originated in Galileo's perception that mechanics cannot be an account of actual experience (as was believed by the empiricists of the sixteenth and seventeenth centuries, including Gassendi), but must presuppose ideal situations that never occur in actual experimental conditions: namely situations involving limiting values, like the investigation of the course of a projectile discharged in a vacuum so that there is no air resistance, or the movement of a pendulum on the supposition that there is no friction at the point of suspension, etc. This method is universally acknowledged although the conditions it assumes are imaginary ones: there is no such thing in nature as a vacuum, a perfectly elastic body, an organism affected by only a single stimulus at a time, and so on, but these must be assumed in order to measure the deviations from the norm that take place in empirical circumstances. In the same way Marx begins by considering the creation of value in a notional society consisting purely of a bourgeoisie and a proletariat, and examines the process of the creation of surplus value abstracting from circulation and the variations it causes; then he considers circulation in isolation from supply and demand, and so forth.

In reality, supply and demand never coincide; ... but political economy assumes that they do, so that we may study phenomena in their fundamental relations, in the form corresponding to their conception, that is independently of the appearances caused by the movement of supply and demand. The other reason for proceeding thus is to find out and to some extent record [*fixieren*] the actual tendencies of these movements. Since the deviations are of an opposite nature, and since they continually succeed one another, they balance out through their mutual contradiction. (*Capital*, III, Ch. X)

There is, however, an essential difference between the use of this method in physics and in political economy. In the case of Galileo's pendulum the limiting conditions were such that deviations could be observed in experimental conditions. But nothing of the sort is possible with complex social phenomena, where there are no instruments to measure the deviation of reality from the ideal model. Hence Marx's exposition in *Capital* has given rise to the question: is he describing a real society or a purely theoretical one (apart, of course, from the historical passages which clearly relate to particular, non-recurring situations)? From some of his remarks it might be inferred that he was not analysing capitalism as it actually was, but only a schema with no real existence. In that case the analysis would be, so to speak, in the air, since we do not know how to compare the model with historical reality or in what way the two are related. But it cannot actually have been Marx's intention to describe an 'ideal' capitalist society (in the sense of a theoretical one, not of course in the normative sense) irrespective of whether it explained the workings of the real one or, above all, how it was likely to develop. What theoretical or practical use could there be, for instance, in saying that in the capitalist 'model' the profit rate must fall or the classes become polarized, if, owing to interference of one sort or another, this does not happen in a real capitalist economy? The model is only of value if it enables us to say: 'Capitalism under such and such conditions would undergo such and such changes, but as the conditions are affected in certain ways, the changes will take place somewhat differently, as follows . . .' But this is precisely what we cannot say; for, if capitalism in real life undergoes different changes, in some respects at least, from theoretical capitalism, then, even if we can explain the differences *ex post facto*, the analysis of the model has been no good to us. At all events, it is very doubtful whether Marx regarded the diminishing profit rate or the polarization of classes as merely tendencies of 'ideal' capitalism which, according to circumstances, might or might not occur in practice. He certainly believed that the rate of profit was bound to fall in real-life capitalism and that the middle classes were historically bound to die out. Attempts to interpret *Capital* as relating only to 'ideal' capitalism sometimes serve as a means of resisting the empirical evidence that refutes Marx's predictions, which are thus represented merely as statements of what would happen in a non-existent ideal form of capitalism. But such interpretations protect Marxism against the destructive results of experience only by depriving it of its value as an instrument of real-life social analysis.

The laws of physics serve to explain observational data by postulating unreal limiting values. The ideal conditions expounded by Marx, on the other hand, are intended to display the 'essence' of reality underlying 'appearances': this may be seen from the passage quoted above and from other statements, including his remark that there would be no need for science if the essence of things always coincided with appearances. But, it may well be asked, what is the status of an 'essence' that may be contradicted by phenomena, and how do we make sure that we have discovered it when, *ex hypothesi*, we cannot do so by empirical observation? The fact that, for example, the existence of atoms and genes was accepted before it was confirmed by direct observation is not a sufficient answer. Atoms and genes bore a clear logical relation to empirical data and served to explain factual observations; they were not the result of mere abstract deduction. In the case of discoveries that purport to explain the 'essence' of things it is important to inquire whether their status is like that of atoms in the time of Ernst Mach (who questioned their existence) or genes in the time of T. H. Morgan, or, on the contrary, like that of 'phlogiston' in the seventeenth and eighteenth centuries—a mere verbal pseudo-explanation which there can be no question of confirming empirically.

It is certain, however, that Marx's holistic approach to social phenomena, relating all categories to a single system, permeates every stage of his analysis. He emphasizes again and again that the qualities he is concerned with have no 'natural being' discernible by perception, but only a 'social being', and that value in particular is not a physical attribute but a social relation which takes on the form of a quality of things. 'In the analysis of economic forms, neither microscopes nor chemical reagents are of use. The force of abstraction must replace both. But in bourgeois society the commodity-form of the product of labour—or the value-form of the commodity—is the economic cell-form' (*Capital*, I, Preface to the first German edn.). 'The value of commodities is the very opposite of the crude materiality of their substance: not an atom of matter enters into its composition. Turn and examine a single commodity by itself as we will, yet in so far as it constitutes an object of value it is impossible to grasp it. If, however, we bear in mind that the value of commodities has a purely social reality, and that they acquire this reality only in so far as they are expressions or embodiments of one identical social substance, viz. human labour, it follows as a matter of course that value can only manifest itself in the social relation of commodity to commodity.' (Ibid., Ch. I, 3.) Value is not something that inheres in a commodity independently of its circulation; it is not accessible to perception, being the crystallization of abstract labour-time—a fact that appears in the relation between commodities on the market, compared as objects of exchange. 'In a sort of way, it is with man as with commodities. Since he comes into the world neither with a looking-glass in his hand nor aṣ a Fichtean philosopher to whom "I am I" is sufficient, man first sees himself reflected in other men. Peter only recognizes himself as a man by first relating himself to Paul as being of like kind.' (Ibid., Ch. I, 3A, 2a.) When a coat expresses the value of a quantity of cloth it does not denote an innate property of the two things but their value, which is purely social in character. (Ibid., 2b.) 'The form of wood, for instance, is altered when we make a table out of it; yet, for all that, the table continues to be that common, everyday thing, wood. But, so soon as it takes the form of a commodity, it is changed into something material yet transcendent [*ein sinnliches übersinnliches Ding*].' (Ibid., Ch. I, 4.)

These arguments contain, it will be seen, an anti-naturalistic premiss according to which social life creates new qualities that are irreducible to those of nature and inaccessible to direct perception, yet are real and determine historical processes. They are not, strictly speaking, new attributes of natural objects, or they are such only in conditions of commodity fetishism; they are inter-human relations which create laws of their own. Such relations cannot be explained *à la* Feuerbach as continuations or specific forms of those existing in pre-human nature. They form complexes which obey their own laws, and they confer on the human beings concerned qualities which cannot be discovered in the non-human world. In this sense the human individual cannot be understood, either by himself or by theoretical analysis, as a mere natural being, or in any way except as a participant in the social process. Thus it is the case, as Marx wrote in 1843, that 'For man, the root is man himself.' Objects, moreover, when involved in human relationships, become different from what they are 'in themselves'. 'A negro is a negro; he only becomes a slave in certain relationships. A cotton-spinning jenny is a machine for spinning cotton; it becomes capital only in certain relationships. Torn from these relationships it is no more capital than gold in itself is money, or sugar the price of sugar.' (*Wage-Labour and Capital*, III.)

We can thus understand more precisely Marx's idea of a return to humanity as a result of socialist revolution. Under socialism, when all useful labour is subordinated to use-value, a cotton-spinning machine is indeed a machine for spinning cotton, an instrument used by human beings to provide themselves with clothing. It is also the crystallization of a certain amount of human labour-time, but it does not constitute exchange-value, at least in the more advanced phase of socialist society, because products in general will not be exchanged by value but distributed in accordance with real need. Hence what happens to the machine, as to any other product, does not depend on its relation to other objects in terms of value. Things which in a commodity economy are humanized in appearance, i.e. assume qualities that are in fact human relations, lose this appearance under socialism and are humanized in reality: they are acquired by people as objects of use and become true individual property. Man continues to be a 'political animal', or a city-dwelling one (Marx refers expressly to Aristotle's phrase); he realizes his creative possibilities as social values, but under socialism abstractions cease to dominate human beings. In this sense socialism is a return to the concrete. The process of inversion whereby objectified labour increasingly extends its power over living labour, so that human activity is not merely a matter of objectification but primarily of alienation, is, as Marx explains in the *Grundrisse*, a process inherent in society itself and not only in the imagination of workers and capitalists. This inversion is indeed a historical necessity without which productive forces could not have developed as they have done, but it is by no means an absolute necessity of all production.

With the suspension of the immediate character of living labour, as merely individual, or as general merely internally or merely externally, with the positing of the activities of individuals as immediately general or social activity, the objective moments [*Momente*] of production are stripped of this form of alienation; they are thereby posited as property, as the organic social body within which the individuals reproduce themselves as individuals, but as social individuals. (*Grundrisse*, III, 3)

5. *The dialectic of* Capital: *consciousness and the historical process*

The dialectic method of *Capital*, however, does not consist merely in regarding every part of capitalist reality as a component of a whole that functions according to its own laws. It is a no less important feature, and indeed the principal one in Marx's view, that every existing form is considered as a stage in a continuing process, i.e. phenomena are observed in terms of historical evolution. Marx never gave a separate exposition of his dialectic— like Hegel's, it cannot be described in isolation from its subject— but from time to time he indicates its general character in the course of a specific argument. One of the most frequently quoted passages is in the Afterword to the second German edition of *Capital*, where he says: 'My dialectic method is not only different from the Hegelian but is its direct opposite. To Hegel the thought-process, which he actually transforms into an independent subject under the name of the Idea, is the demiurge of the real world, and the real world is only the external, phenomenal form of the Idea. With me, on the contrary, the ideal is nothing else than the material world reflected in the human mind and translated into forms of thought.' In the same Afterword he quotes with approval an account of his method given by a Russian reviewer of *Capital* in 1872, who observed that 'Marx treats the social movement as a process of natural history governed by laws independent of human will, consciousness and intentions', and that in his system each historical period has its own laws, which give way in due course to those of the next. However, Marx says, his dialectic 'in its comprehension and affirmative recognition of the existing state of things, at the same time comprehends the negation of that state and its inevitable breaking up; it regards every historically developed social form as in fluid movement, and therefore takes into account its transient nature no less than its momentary existence; it lets nothing impose upon it, and is in its essence critical and revolutionary.'

However, the doctrine of the transitoriness of social phenomena is not in itself a sufficient basis for analysis. The whole of history must in addition be interpreted in relation to its highest forms; in particular, former systems can only be understood in terms of their outcome in bourgeois society.

Bourgeois society is the most developed and the most complex historic organization of production. The categories which express its relations, the comprehension of its structure, thereby also afford insight into the structure and the relations of production of all the vanished social formations out of whose ruins and elements it built itself up ... Human anatomy is the key to the anatomy of the ape. The intimations of higher development among the subordinate animal species can be understood only when the higher development is already known. In the same way the bourgeois economy supplies the key to the ancient, etc.; but not after the manner of those economists who blur all historical differences and see bourgeois relations in all forms of society. One can understand tribute, tithe etc. if one is acquainted with ground rent; but one must not identify them. (*Grundrisse*, Introduction)

Not only are bygone social forms intelligible only in relation to present ones, but present-day society can only be understood in the light of the future, i.e. the form that will take its place after its inevitable collapse. In this important respect Marx's thought differs from Hegel's, which was essentially confined to interpreting the past. The idea of extending the dialectic into the future and interpreting the present in terms of its own dissolution was adopted by Marx from the Young Hegelians.

From time to time Marx, in *Capital*, invokes Hegelian formulas. For example, having argued that in given social conditions an accumulation of value can only be described as 'capital' if it is large enough to employ wage-labour, he cites this as an instance of the Hegelian transformation of a quantitative change into a qualitative one: value, beyond a certain quantitative level, acquires the power to command living labour and to create surplus value. Again, having described capitalist property as the negation of individual private property based on labour, he refers to socialism as the 'negation of a negation'—i.e. the return to individual property, based, however, on joint ownership of the means of production instead of on private ownership.

To Marx as to Hegel, however, the dialectic is not a collection of rules independent of one another and of the subject-matter to which they are applied. If it were simply a method applicable to any subject and capable of being expounded in isolation, there would be no reason for Marx to say that his own dialectic was contrary to Hegel's because of the latter's idealism; for its laws could be formulated in the same way whether history was interpreted in an idealist or a materialist fashion. But, in Marx's view, the relationship of consciousness to the historical process is part of the very content of the dialectic. Whereas for Hegel the dialectic was a history of the diffraction of ideas in the course of which the mind comes to understand being as its own creation, for Marx it is a history of the material conditions of life in which mental and institutional forms are vested with an apparent autonomy before returning, as they are bound to do, into union with their substructure. The dialectic as a means of understanding the world is secondary to the actual dialectic of the world itself, inasmuch as the theory of the dialectical movement of social reality is aware of its own dependence on the historical process that gave it birth. Marx repeats several times that a theory reflecting working-class interests can only spring from observation of the changing situation of the workers. The theory is in fact the self-conscious superstructure of that situation; it knows itself to be merely a reflection of the real historical process, a product of social praxis and not an independent contemplation of it. The Marxian dialectic ends in the 'unity of subject and object', but in a different sense from Hegel's. It restores to man his true function as a conscious historical subject, by abolishing the situation in which the results of his free, conscious initiative are turned against himself. The subject will be in full command of the process whereby he objectifies himself in production and creative work; this objectification will not degenerate into alienation; real human individuals will possess the work of their own hands and will no longer be subjected to an independent objectified power. The course of history will be completely governed by the conscious human will; the latter will know itself for what it is, namely consciousness of the life-process. The historical process and the free development of consciousness will be one and the same.

Marx's dialectic is a description of the historical evolution leading to this unity of consciousness and social Being. As with Hegel, it is the description of a movement in which contradictions arise and are overcome, giving place to fresh contradictions. The advance through contradictions is essential to the dialectical interpretation of the world. But these are not logical contradictions or a different term for social conflict: the latter has been with us throughout history, but no one built a dialectical system of interpretation on it. Class antagonism in conscious political forms is an effect of the contradictions underlying an unconscious, 'objective' process. In Hegel's theory, concepts as they developed revealed internal contradictions, the resolution of which gave rise to higher forms of consciousness. In Marx's view, contradictions 'occur' in the historical process independently of whether they are translated into consciousness or conceptual forms; they consist in the fact that a phenomenon gives rise to situations contrary to its own nature and basic tendency. The most important feature of the dialectic of the internal contradictions of capitalism is Marx's analysis of the falling profit rate and of economic crises, in which he shows that the urge to maximize the rate of profit defeats its own object by increasing the amount of constant capital and so causing the profit rate to fall steadily. The same urge to increase surplus value in absolute terms leads to crises and the collapse of capital, despite the 'inborn tendency' of capital itself (a different matter from the intentions of capitalists, which are secondary here). Thus capital, which originally displayed a single, undifferentiated tendency, gives rise to phenomena which work in the opposite direction to itself, and the contradiction finally reaches a point, in spite of all efforts, at which capital can no longer exist. This is analogous to the Hegelian disjunction of concepts, but it is a pattern that history develops of its own accord, independently of anyone's consciousness. Consciousness, indeed, has figured in the process up to now only as a complex of delusion and mystification. The return to the unity of subject and object does not mean, as with Hegel, depriving the world of its objective character and of objectivity altogether; man will still objectivize his powers by means of labour and will still be confronted by independent Nature. What it does mean is depriving social phenomena of their thing-like character, their independence of real, individual human subjects. The dialectic that explains this process is the consciousness of the working class raised to the level of intellectual understanding.

Having come this far, we can define the Marxian dialectic in its entirety as follows. The dialectic is the consciousness of the working class, which, aware of its own condition and its opposition to bourgeois society, perceives the entire functioning of that society, and the whole of past history, as a recurrent process of the emergence and resolution of contradictions. The dialectical consciousness, by a process of abstraction, strips social phenomena of their contingent character and apprehends their basic structure; it relates every component of the historical process to the whole, and in this way understands itself likewise. In its final stage it reflects the intensified contradictions which—including itself *qua* dialectical consciousness—will be swept away in a revolutionary explosion; this event will terminate the prehistory of the human race and restore the unity of society as subject and object of history, or, to put it another way, the unity of the consciousness of history with history itself.

It may be seen from this formulation that the dialectic is not a method, like those of mathematics, that can be applied to any subject-matter in any conditions. It exists as a method only in so far as it is conscious of its functional relation to the class-situation that it reflects, and in so far as it not only understands history but at the same time anticipates it by the revolutionary abolition of existing contradictions. The dialectic cannot exist outside the practical struggle for a future society whose ideal image it contains within itself.

It can be seen why Marx need not, and indeed cannot in terms of his own method, provide socialism with an ethical basis, i.e. present it simply as a collection of desirable values. This is not because he thought of socialism merely as a 'historical necessity' and was not interested in whether it was good or bad, nor because he took the absurd view that it was people's duty to follow the course of history wherever it led. The reason why ethical justification is irrelevant is that in Marx's theory the understanding of bourgeois society comes into being as a practical act, or rather it is the reflection in consciousness of revolutionary action and cannot appear independently of it. It is foreign to Marx's conception to divide his theory into separate elements of fact, obligation, and method—to determine first what the world is like, then what it should be like to satisfy certain norms, and finally by what means it can be transformed. The capitalist world presents itself to the proletariat in the latter's act of understanding which springs from the practical act of destroying it. The workers' movement came into being before the theory which reflects its real though at first unconscious tendencies: when the theory takes shape, it does so as the self-knowledge of the movement. Those who adopt the theory do not thereby come into possession of a set of values in the form of an external imperative; they become aware of the aim they were in fact pursuing, though they had no clear theoretical understanding of it. There is no room here for the process of fixing an aim and then considering how to attain it, as with a technical problem where the objective is arbitrarily given and the rational solution is devised afterwards, or again as in the moralizing socialism of the utopians. In Marx's theory, awareness of the aim takes the form of an act in which the participants in the historical process acquire theoretical insight into the means they have already begun to employ. Since men do in fact strive to liberate themselves from oppression and exploitation, and afterwards become aware of their action as part of the 'objective' movement of history, they have no need of a separate imperative telling them that they should strive for liberation in general or that freedom from oppression is a good thing. It is only in action that man becomes aware of himself—although he may be deceived, and indeed notoriously has been, as to the true content of that self-knowledge. The movement to free mankind from slavery recognizes itself at once for what it is and identifies its own position as a fighting movement; it could not put to itself the question 'What is the fight for?' without first ceasing to fight, and if it did that it would cease to exist. The dichotomy between fact and value, observation and appreciation has no place here. It belongs to those whose ideals and dreams go far beyond reality and are not anchored in history—the Epicureans who see a great gulf fixed between themselves and the real world. But, in the case of the working class, the understanding of the historical world and its practical transformation are a single undifferentiated act: there is not and cannot be a separate perception of what is and what ought to be. Understanding history and participating in it are one and the same, and require no separate justification. The dialectic is a rule of observation, but it is also the self-knowledge of the historical process; it cannot escape this role by setting itself up as an instrument for the mere observation of history, still less of the natural world in general.

6. Comments on Marx's theory of value and exploitation

Marx's theory of value has been much criticized from several points of view, especially that of its unsuitability for empirical analysis. This objection was voiced by Conrad Schmidt and after him by Böhm-Bawerk (to whom we shall return), Sombart, Struve, Bernstein, and Pareto, and in recent years by Joan Robinson and Raymond Aron. Some lines of argument recur frequently in the various criticisms. We cannot go into all the details here, but will mention the main points.

To begin with, it has been observed that value in Marx's sense is unmeasurable, i.e. it is impossible to state the value of any commodity in units of necessary labour-time. This is so for two independent reasons. The first is that the value of any product, on Marx's theory, includes the value of the tools and materials used to make it, those used to make the tools and materials in question, and so on *ad infinitum*. It is true that, according to Marx, instruments do not create fresh value but only transfer to the product part of the value crystallized in them; but, if we have to calculate the value of the product in units of labour-time, we should have to reduce the value of the tools to such units also, which is clearly impossible. The second reason is that different kinds of work cannot be reduced to a common measure. Human labour involves varying degrees of skill, and on Marx's showing we should have to add to the quantity of labour expended in making the product the amount of labour which went into the worker's training; but this too is impossible. The usual Marxist defence that the labour market automatically reduces complex and simple labour to a common measure is of no avail here, since it means that value cannot be calculated independently of price, which is exactly the point of the objection. In any case the price of labour-power (assuming, with Marx, that it is labour-power and not labour which is exchanged in a commodity economy) depends, like other values, on numerous factors and especially the laws of supply and demand; so there is no reason to suppose that wage differences between skilled and unskilled labour correspond to the amount of labour-time required to produce a skilled workman.

If value cannot be calculated independently of price, there is no way of verifying the statement that the actual prices of commodities fluctuate around their real value. Marx of course knew that prices are determined in practice by various factors, including labour productivity, supply and demand, and the average rate of profit. If he disregarded these in the first volume of *Capital*, it was for methodological reasons and not because he thought value and price were the same thing; thus he cannot be reproached with inconsistency as between Volume I and Volume III, which deals *inter alia* with the average rate of profit. But the point is that it is impossible to measure quantitatively the respective effect of the various factors on market prices. If Adam Smith thought that primitive men exchanged products in accordance with the time it had taken to make them, or if Engels strove to maintain that this still happened in the late Middle Ages, the Marxian theory of value is still in no better case. If we accept these historical statements we can only assert that while they are true for primitive economies, in a developed commodity economy labour-time is one of the factors determining price, but not the only one. Yet Marx, while aware of the other factors, maintained that real value is determined only by socially necessary labour-time. In other words, he was not answering the question 'What determines prices?' but the question 'What is value?' We have, in that case, to investigate the meaning of the latter question and whether it is possible to give a reasoned answer to it.

A second difficulty that is often raised is how we can imagine a proof of the assertion that the 'real' value of a commodity (what the Middle Ages called a 'just price', and the classical economists a 'natural price') is determined by labour-time. What does Marx mean, in fact, when he speaks of the 'law of value'? A natural law is generally a statement that certain phenomena occur in certain circumstances; but it is not clear that Marx's definition of value can be expressed as a law. The most general statement that might deserve this name, though it cannot have a quantitative character, would be that variations in the productivity of labour generally affect prices. But this is not the same as Marx's theory, which holds not that labour-time affects prices but that it is the only constituent of value. This is not a law, but an arbitrary definition which cannot be proved and is of no use for the empirical description of economic phenomena. As there is no transition from value to price, so there is no transition from the theory of value to the description of any actual economic process.

Many Marxists, such as Lukács, have maintained, for example, that the ruin of small firms by large ones is a confirmation of the law of value or even proves that Marx's 'abstract labour' is a genuine economic phenomenon. This, however, is a misuse of words. If small firms fail to compete with large ones because of lower productivity, this can be explained by the notion of production costs without bringing in the law of value. If labour-consuming techniques are replaced, in many cases at least, by others that are less so, this can be explained by the analysis of prices, which, unlike values, are an empirical phenomenon. To state that the 'law of value' operates in such cases does not make the process easier to understand—especially when we do not know what is meant by the 'law of value' if it is something different from a definition of value, which is certainly not a law.

For this reason economists of an empirical turn consider Marx's theory of value to be useless, as it cannot be applied to the empirical description of phenomena. Their point is not that Marx gave the wrong answer to the question 'What is real value?', but that this question has no meaning in economic science if it refers to anything but the factors governing prices. On this ground Marx's theory has been criticized as 'metaphysical' in the pejorative sense given to this term by the positivists: i.e. it claims to reveal the 'essence' hidden beneath the surface phenomena, but provides no way of empirically confirming or refuting what it says. The objection that Marx was, in this sense, hunting after the 'substance' of value has been denied by Marxists who point out that he defined it as a social relation having no existence except in the exchange of commodities. But this is not a good answer to the objectors, even if they use the word 'substance' improperly. Marx, it is true, expressly rejects the idea that exchange-value is immanently present in a commodity, independently of the social process of exchange to which it is subjected. But if we distinguish value from exchange-value we can say that any commodity 'represents' or is the embodiment or vehicle (or any similar metaphor) of the total sum of labour that has been put into it, while exchange-value is the manifestation of value as between goods on the market. Exchange-value thus depends on there being a commodity economy (and in this sense, according to Marx, it is a transient historical phenomenon) and also on the existence of value itself, which is 'crystallized labour time'. The existence of value does not depend on the system of production and exchange; men have always expended labour on making various objects, and value, in consequence, is an immanent quality of things, manifested in certain social conditions as exchange-value. But if Marx's 'law' is meant to signify anything more than two logically independent empirical statements—that most useful objects are the fruit of labour, and that labour-time is one element in price; if it is supposed to mean that there is a 'real', unmeasurable value independent of price, then this is no better than a 'latent property' of the type condemned by science since the seventeenth century. Yet there can be no doubt that Marx did mean more than the above two statements, and intended to throw light on the true nature of value and exchange-value. The assertion that true value is crystallized labour-time is on a par with the statement that opium puts people to sleep because it is soporific. We are told of a hidden quality that manifests itself empirically (opium puts people to sleep, goods are exchanged), but the information does not explain the empirical phenomena or enable us to predict them better than we can without it.

There is another formula that might seem to give content to the law of value, viz. Marx's statement that the sum of prices equals the sum of values. This too, however, is not supported by any argument, and its meaning is not clear. If objects are sold which possess no value—for example, land, the price of which is anticipated rent—this must mean that the equality of prices and values is not actual at any particular moment, but only over a period of time that is not and cannot be determined. In this sense the statement has no definable meaning, and in any case it is not clear how it could be verified, since value cannot be quantitatively expressed.

As an interpretation of economic phenomena Marx's theory of value does not meet the normal requirements of a scientific hypothesis, especially that of falsifiability. It may, however, be defended on a different basis, as a piece of philosophic anthropology (or, as Jaurès put it, social metaphysics)—a continuation of the theory of alienation and an attempt to express a feature of social life which is important to the philosophy of history: namely that when human skill and effort are transformed into commodities they become abstract vehicles of currency and are subject to the impersonal laws of the market over which producers have no control. The theory of value, then, is not an explanation of how the capitalist economy works, but a critique of the dehumanization of the object, and therefore of the subject, in a system wherein 'everything is for sale'. On this view the theory is part of the Romantic attack on a society enslaved by the money-power.

It should be observed that those analyses of Marx's which can be checked empirically with some degree of rigour, such as the falling rate of profit or the schemata of reproduction in Volume II of *Capital*, do not (whatever Marx himself believed) depend logically on the theory of value, which can be ignored in appraising them.

As already mentioned, Marx's theory of value includes the statement, peculiar to himself, that labour is not only the measure of value but its only source. Logically the two parts of this proposition are separable: labour might be the measure but not the only source, or vice versa.

The statement that human labour is the only source of value, and the connected distinction between productive and non-productive labour, are not supported by argument either. It is not clear why, when a farmer uses a horse to plough his land, he himself creates new values but the horse does no more than transfer part of its own value to the product. The motive for this arbitrary assertion appears to lie in the conclusion, so important to Marx, that capital does not create value. Marx knew, and indeed emphasized in the *Grundrisse*, that capital as an organizing force greatly increases the productivity of labour; yet he maintained, following Ricardo, that it contributes only to use-value and not to exchange-value. But if so, capital is in fact a source of real wealth, i.e. the increase of usable objects—although the sum total of the values of that wealth will be the same whatever its quantity, if they represent the same number of labour hours (reduced to 'simple labour'). Thus the increase of social wealth has nothing to do with the increase of values. We can imagine a society in which all production was perfectly automatized, so that the society produced no values in Marx's sense, though it produced great quantities of wealth or use-value. There is no logical, physical, or economic reason why such a society should not be based on capitalist ownership, even though it employed no 'living labour' or productive labourers at all.

Thus Marx's ridicule of the idea that money has a magic power of self-multiplication because it can be lent at interest is over-facile. The proposition that capital does not increase values follows logically from Marx's definition of value, and must be assented to if we accept that definition; but there are no sufficient logical or empirical grounds for accepting it. The fact that capital increases use-values by organizing labour is not contrary to Marx's premisses. But, for that very reason, the growth and distribution of social wealth are unrelated to the theory that labour is the only source of value; for the increase of exchange-values, as distinct from the question of prices and the multiplication of commodities, is in itself of no interest to society. What is of interest is the quantity of goods produced, the manner of their sale and distribution and the question of exploitation. But the theory that the workman is the only creator of value throws no light on these matters; it merely serves to arouse indignation at the fact that the 'only real producer' gets so small a share of the result of his work, while the capitalist, who contributes nothing to value, rakes in profits on the strength of being a property-owner. Apart from this moral interpretation it is not clear how the theory is supposed to throw light on the mechanism of the capitalist economy; and, it should be repeated, Marx did not agree with the Ricardian socialists who deduced from the theory of value that the workman was entitled to the equivalent of what his labour produced.

The distinction between productive and unproductive labour

appears in Marx in two forms. In one sense, as we read in the *Grundrisse*, productive labour is labour that helps to create capital; and in this sense the distinction applies only to capitalist production. In another sense productive labour is labour which creates values of any kind, irrespective of social conditions. The distinction has been much debated by Marxists, as the dividing line between the two kinds of labour is very hard to draw. In general we gather from Marx that productive labour is physical labour applied to material objects; but from occasional remarks it appears that he was prepared to count as producers those who did not directly work on the material themselves but enabled others to do so—for example, engineers or designers in factories. In this case, however, the distinction is highly obscure and has given rise in the socialist countries to practical as well as theoretical dilemmas. It could be disputed, for instance, whether a doctor's work was 'productive' or not: from the economic point of view it means restoring or reproducing labour-power and is thus productive, but the same can be said of begetting children, which throws a doubtful light on the argument. Again, a teacher's work may help to produce important industrial skills, so presumably it too creates values. The practical aspect of the question is that in societies that try with greater or less (usually less) success to apply criteria derived from Marx's theories, labour regarded as productive is more highly respected and better paid; so, as long as teachers and medical staff were officially non-productive, the wretched level of their salaries could be theoretically justified. Another consequence of the theory was that the whole services sector was reckoned to be non-productive and was therefore totally neglected in planning.

At the present time, the distinction is more and more anachronistic and its purpose is not very clear. The proportion of those whose work consists in the direct physical processing of material objects grows less as technology improves, and the increase in total wealth depends less and less on the number of such workers.

It is also not clear on what Marx bases his view that what the worker sells is not labour but labour-power. Even if we agree with him that labour, while it is the source of value, has no value of its own, it does not follow that it cannot be sold: according to Marx many objects and activities are sold though they have no value as he defines it. What he probably meant to emphasize was that when the capitalist buys labour-power, according to the laws of capitalist economy he is the owner of the labourer's person for the stipulated time and is entitled to make him work up to and beyond his physical capacity and endurance. But the capitalist's right to exploit the labourer and prolong the working day is not a built-in feature of capitalism as such, but only belongs to an early phase of it. How far it exists in practice depends on legislation and the amount of pressure the working class is able to exert; in the capitalist world at present there is no country where the employer can be said to have such a right. Even if he believes himself entitled to all he can squeeze out of the labourer, legal or other reasons prevent the claim being made good, and it is not clear how Marx's assertion helps towards an understanding of present-day capitalism. Nor is his theory necessary to explain the workers' struggle for shorter hours and fairer wages.

The distinctions and concepts most clearly linked with Marx's theory of value are an ideological expression of his belief that capitalism cannot be reformed and that it tends inexorably to depress wages to the minimum value of labour-power and to work labourers to the physical maximum. (Any rise in wages is due to the increase of needs, to which no limit is set, so that whatever the wage-level it can be maintained that the worker is selling his labour-power at its market value.) However, at the present time when resistance to exploitation has not only been successful but has radically transformed social life, the theory of value and its corollaries merely obscure the picture, as Marxists feel obliged to maintain the validity of laws that bear no relation to the facts. This does not mean, of course, that the capitalist is not out to make the highest profit he can; but this is a common-sense principle which has nothing to do with any particular theory of value.

As for exploitation, it can be defined consistently with Marx's intentions without any logical need to invoke his theory. He explains it as a matter of unpaid labour, i.e. the surplus value appropriated by the capitalist after deducting the cost of materials, wages, and the replacement of constant capital. Yet Marx himself ridiculed the utopians and Lassalle for holding that the worker should receive in the form of wages the whole equivalent of the values produced by him, as this would be impossible in any society. The abolition of exploitation meant, in his view, not that workers should receive the equivalent of what they produced, but that the surplus value they do not receive in wages should accrue to society in the form of fresh investments, emergency reserves, payment for necessary 'unproductive' services, administration, etc., plus provision for those unable to work. But under capitalism surplus values over and above what is consumed by the bourgeoisie do revert to society in all these forms. The moral aspect of exploitation comes to the fore when there is a blatant contrast between bourgeois luxury and the poverty of the workers. But Marx did not contend, like the leaders of earlier people's movements, that it would help to solve social problems if the goods consumed by the bourgeoisie were divided among the whole population. Bourgeois consumption in the face of workers' poverty is a moral issue, not an economic one; the distribution, once and for all, of rich men's wealth among the poor would not solve anything or bring about any real change. Such a measure only made sense as regards landed property, which could be divided among the peasantry and has been so divided in several countries. If the homes, furniture, clothing, and valuables of the bourgeosie were distributed among the poor it would only be an isolated act of revenge, not a solution of social problems—yet this is the only sharing that would result from the socialization of property. For this reason Marx avoided encouraging the facile but false idea that to abolish exploitation meant simply to strip the rich of their movable possessions: this was contrary to his own theory and only served to foster the envious and predatory mentality of peasant movements and the lumpenproletariat.

Exploitation, in fact, does not signify either that the worker receives less than the equivalent of his product, or that incomes in general are unequal—since there is no known way of making them perfectly equal in an advanced industrial society—or even that the bourgeoisie pay for their luxuries out of unearned income. Exploitation consists in the fact that society has no control over the use made of surplus product, and that its distribution is in the hands of those who have an exclusive power of decision as to the use of means of production. It is thus a question of degree, and one can speak of limiting exploitation not merely by increasing wages but by giving society more control over investment and the division of the national income. Bourgeois luxury is not itself exploitation but is a consequence of it: those who control the means of production, and therefore the distribution of surplus product, naturally seize a large share of the cake.

Although this account of exploitation appears to be in accordance with Marx's own views it is hard to reconcile with orthodox Marxism, since it implies that the nationalization of means of production does not necessarily prevent exploitation and, in certain circumstances that have actually occurred, may increase it to a considerable degree. For, if exploitation can be limited in so far as society controls the distribution of surplus product, it must be greater to the extent that the machinery for such control is weakened. If, instead of private ownership, the power to control the means of production and distribution is confined to a small ruling group uncontrolled by any measure of representative democracy, there will be not less exploitation but a great deal more. The important thing is not the material privileges that the rulers keep for themselves, just as it is not important what clothes the bourgeoisie wear or how much caviar they eat; what matters is that the mass of society is excluded from decisions as to the use of means of production and the distribution of income. Exploitation, in short, depends on whether there is or is not effective machinery to enable the workers to share in decisions concerning the product of their labour, and hence it is a question of political freedom and representative institutions. From this point of view socialist communities at the present day are examples not of the abolition of exploitation but of exploitation in an extreme degree, since by cancelling the legal right of ownership they have destroyed the machinery which gave society control over the product of its own labour. In capitalist communities, by contrast—at all events the more advanced ones—this machinery makes it possible to limit exploitation by progressive taxation, partial control of investment and prices, welfare institutions, increasing the social consumption fund, etc., even while private ownership of the means of production continues and exploitation has not been abolished.

CHAPTER XIV

The Motive Forces of the Historical Process

1. Productive forces, relations of production, superstructure

In his description in *Capital* Marx referred to the causal connection between the advance of technology and the unlimited expansionism of capital. At the same time he argued that this tendency could only arise and become universal in certain technological conditions, and not at any period of history without distinction. The functioning and the expansionist tendency of capitalism was a special case of the more general system of relations that had governed social life in all its forms, past and present. Marx's description of that system goes by the name of historical materialism or the materialist interpretation of history. It was first clearly set out in *The German Ideology*, but the best-known general formulation is in his Preface to *A Contribution to the Critique of Political Economy* (1859); the doctrine is also stated in different versions in the popular writings of Engels. Here is Marx's classic exposition:

In the social production which men carry on they enter into definite relations that are indispensable and independent of their will; these relations of production correspond to a particular stage of development of their material forces of production. The sum total of these relations of production constitutes the economic structure of society—the real foundation, on which rises a legal and political superstructure and to which correspond particular forms of social consciousness. The mode of production in material life determines the social, political and intellectual life processes in general. It is not the consciousness of men that determines their being, but, on the contrary, their social being that determines their consciousness. At a certain stage of their development the material forces of production in society come into conflict with the existing relations of production, or—what is but a legal expression for the same thing—with the property relations within which they have been at work before. From forms of development of the forces of production, these relations turn into their fetters. Then begins an epoch of social revolution. With the change of the economic foundation the entire immense superstructure is more or less rapidly transformed. In considering such transformations a distinction should always be made between the material transformation of the economic conditions of production, which can be determined with the precision of natural science, and the legal, political, religious, aesthetic or philosophic—in short, ideological forms in which men become conscious of this conflict and fight it out. Just as our opinion of an individual is not based on what he thinks of himself, so we cannot judge of such a period of transformation by its own consciousness; on the contrary, this consciousness must be explained rather from the contradictions of material life, from the existing conflict between the social forces of production and the relations of production. No social order ever disappears before all the productive forces for which there is room in it have been developed; and new higher relations of production never appear before the material conditions of their existence have matured in the womb of the old society itself. Therefore mankind always sets itself only such tasks as it can solve; since, looking at the matter more closely, we shall always find that the task arises only when the material conditions necessary for its solution already exist or are at least in the process of formation. In broad outlines we can designate the Asiatic, the ancient, the feudal and the modern bourgeois modes of production as so many epochs in the progress of the economic formation of society. The bourgeois relations of production are the last antagonistic form of the social process of production—not in the sense of individual antagonism, but of one arising from the social conditions of life of the individuals; at the same time the productive forces developing in the womb of bourgeois society create the material conditions for the solution of that antagonism. The present formation constitutes, therefore, the closing chapter of the prehistoric stage of human society.

In the history of human thought there are few texts that have aroused such controversy, disagreement and conflicts of interpretation as this one. We cannot retrace the whole intricate debate here, but will note some of the main points.

In *Socialism, Utopian and Scientific* (Introduction to English edn., 1892) Engels defines historical materialism as 'that view of the course of history which seeks the ultimate cause and the great moving power of all important historic events in the economic development of society, in the changes in the modes of production and exchange, in the consequent division of society into distinct classes, and in the struggle of these classes against one another'.

Historical materialism is thus an answer to the question: what circumstances have had the greatest effect in changing human civilization?—this word being understood in a broad sense covering all social forms of communication, from categories of thought to the social organization of labour and political institutions.

The starting-point of human history from the materialist point of view is the struggle with nature, the sum total of the means employed by man to compel nature to serve his needs, which grow as they are satisfied. Man is distinguished from other animals by the fact that he makes tools: the brute creation may use tools in a primitive way, but only such as they find in nature itself. Once equipment is perfected to the extent that an individual can produce more goods than he consumes himself, there is a possibility of conflict as to the sharing of the excess product and of a situation in which some people appropriate the fruits of others' labour—that is to say, a class society. The various forms that this appropriation may take determine the forms of political life and of consciousness, i.e. the way in which people apprehend their own social existence.

We thus have the following schema. The ultimate motive force of historical change is technology, productive forces, the whole of the equipment available to society plus acquired technical ability plus the technical division of labour. The level of productive forces determines the basic structure of the relations of production, i.e. the foundation of social life. (Marx does not regard technology itself as part of the 'base', since he speaks of a conflict between productive forces and the relations of production.) The relations of production comprise, above all, property relations: i.e. the legally guaranteed power to dispose of raw materials and the instruments of production and, in due course, of the products of labour. They also include the social division of labour, wherein people are differentiated not by the kind of production they are engaged in, or the particular phase of a production process, but by whether they take part in material production at all or perform other functions such as management, political administration or intellectual work. The separation of physical from intellectual work was one of the greatest revolutions in history. It was able to occur because of the social inequality which permitted some men to appropriate the work of others without taking part in the process of production. The volume of leisure thus created made possible intellectual work, and thus the whole spiritual culture of mankind— the arts, philosophy, and science—is rooted in social inequality. Another component of the 'base', or the relations of production, is the way in which products are distributed and exchanged between producers.

The relations of production further determine the whole range of phenomena to which Marx gives the name of superstructure. This includes all political institutions, especially the state, all organized religion, political associations, laws and customs, and finally human consciousness expressed in ideas about the world, religious beliefs, forms of artistic creation, and the doctrines of law, politics, philosophy, and morality. The principal tenet of historical materialism is that a particular technological level calls for particular relations of production and causes them to come about historically in the course of time. They in turn bring about a particular kind of superstructure, consisting of different aspects which are antagonistic to each other: for the relations of production based on appropriation of the fruits of others' labour divide society into classes with opposing interests, and the class struggle expresses itself in the superstructure as a conflict between political forces and opinions. The superstructure is the sum total of the weapons employed by the classes fighting one another for a maximum share in the product of surplus labour.

2. Social being and consciousness

The objections most frequently raised against historical materialism in the nineteenth century were: (1) it denies the significance of conscious human action in history, which is absurd; (2) it declares that men act only from motives of material interest, which is also contrary to all evidence; (3) it reduces history to the 'economic factor' and treats all other factors such as religion, thought, feeling, etc. either as unimportant or as determined by economics to the exclusion of human freedom.

Some formulations of the doctrine by Marx and Engels might indeed seem open to these objections. The critics were answered partly by Engels and partly by later Marxists, but not in such a way as to remove all ambiguity. However, the objections lose much of their force if we recall what questions historical materialism does and does not set out to answer.

In the first place, it is not and does not claim to be a key to the interpretation of any particular historical event. All it does is to define the relations between some, but by no means all, features of social life. In a review of Marx's *Critique* in 1859 Engels wrote: 'History often proceeds by jumps and zigzags, and, if it were followed in this way, not only would much material of minor importance have to be included but there would be much interruption of the chain of thought ... The logical method of treatment was therefore the only appropriate one. This, however, is essentially no different from the historical method, only divested of its historical form and disturbing fortuities.' In other words, Marx's account of the dependence of the superstructure on the relations of production applies to great historical eras and fundamental changes in society. It is not claimed that the level of technology determines every detail of the social division of labour, and thus in turn every detail of political and intellectual life. Marx and Engels thought in broad historical categories and in terms of the basic factors governing the change from one system to another. They believed that the class structure of a given society was bound sooner or later to manifest itself in basic institutional forms, but the course of events which brought this about would depend on a multitude of chance circumstances. As Marx wrote in a letter to Kugelmann (17 April 1871), 'World history ... would be of a very mystical nature if "accidents" played no part in it. These accidents fall naturally into the general course of development and are compensated by other accidents. But acceleration and delay are very dependent upon such "accidents", including the "accident" of the character of those who at first stand at the head of the movement.' Engels too, in some well-known letters, warned against exaggerated formulations of so-called historical determinism. 'While the material mode of existence is the *primum agens*, this does not preclude the ideological spheres from reacting upon it in their turn, though with a secondary effect.' (Letter to Conrad Schmidt, 5 Aug. 1890.)

The determing element in history is, in the last resort, the production and reproduction of real life. More than this neither Marx nor I have ever asserted. If therefore somebody twists this into the statement that the economic element is the only determining one, he transforms it into a meaningless, abstract and absurd phrase. The economic situation is the basis, but the various elements of the superstructure—political forms of the class struggle and its consequences, constitutions established by the victorious class after a successful battle, etc.; forms of law, and even the reflexes of all these actual struggles in the brains of the combatants: political, legal, philosophical theories, religious ideas and their further development into systems of dogma—all these exercise their influence upon the course of the historical struggles and in many cases preponderate in determining their form. There is an interaction of all these elements in which, amid all the endless host of accidents ... the economic movement finally asserts itself as necessary. (Letter to Joseph Bloch, 21 Sept. 1890)

In the same way, great individuals who appear to shape the course of history actually come upon the scene because society needs them. Alexander, Cromwell, and Napoleon are instruments of the historical process; they may affect it by their accidental personal traits, but they are unconscious agents of a great impersonal force which they did not create. The effectiveness of their action is determined by the situation in which it takes place.

If, then, we can speak of historical determinism, it is only in the context of major institutional features. The technological level of the tenth century being what it was, there could not have been at that time a Declaration of the Rights of Man or a *Code Napoléon*. As we know, there can in fact be widely differing political systems in societies where the technological level is much the same. Nevertheless, if we consider the essential features of these societies and not the accidental details of personal character, tradition, and circumstance, it will appear from the point of view of historical materialism that in all decisive respects they resemble one another or show a tendency to do so.

As to the reflex action of the superstructure on the mode of production, here, too we must remember the qualification 'in the last resort'. The state may, for instance, act in such a way as either to help or to hinder the social changes required by the level of productive forces. The effectiveness of its action will vary according to 'accidental' circumstances, but in the fullness of time the economic factor will prevail. If we consider history in panoramic form it appears as a tumult of chaotic events, amid which the analyst is able to perceive certain dominant trends, including the basic interrelations of which Marx spoke. It will be seen, for instance, that legal forms approximate steadily to the situation in which they best serve the interests of the ruling classes, and that these interests are constituted according to the mode of production, exchange, and ownership which obtains in the society in question; it will also be seen that philosophies and religious beliefs and observances vary according to social needs and changes in political institutions.

As to the part played by conscious intentions in the historical process, the view of Marx and Engels appears to be as follows. All human acts are governed by specific intentions— personal feelings or private interests, religious ideals or concern for the public welfare. But the result of all these multifarious acts does not reflect the intentions of any one person; it is subject to a kind of statistical regularity, which can be traced in the evolution of large social units but does not tell us what happens to their components as individuals. Historical material- ism does not state that personal motives are necessarily perverse or selfish, or that they are all of a kind; it is not concerned with such motives at all, and does not attempt to predict individual behaviour. It is only concerned with mass phenomena which are not consciously willed by anyone but which obey social laws that are as regular and impersonal as the laws of physical nature. Human beings and their relations are, nevertheless, the sole reality of the historical process, which ultimately consists of the conscious behaviour of individuals. The sum total of their acts forms a pattern of diachronic historical laws, describing the transition from one social system to the next, and also functional laws showing the interrelation of such features as technology, forms of property, class barriers, state institutions, and ideology. 'Men make their own history, but they do not make it just as they please; they do not make it under circumstances chosen by themselves, but under circumstances directly found, given and transmitted from the past.' (*The Eighteenth Brumaire of Louis Bonaparte*, I.)

Strictly speaking, it is wrong to represent materialism as distinguishing various 'factors' in history and then 'reducing' them to a single one or claiming that all the others depend on it. The misleadingness of this approach was pointed out by Plekhanov among others. So-called historical 'factors' are not substantive entities but abstractions. The historical process is one, and all important events are made up of the most varied influences and phenomena: mental attitudes, traditions, interests and ideals. According to historical materialism, on the stage of world history men's opinions, customs and institutions are predominantly affected by the prevailing system of production, exchange, and distribution. This, of course, is an extremely general statement and scarcely does more than signify its opposition to the type of theory which regards institutions and social organization as ultimately the product of opinions or the Spirit of History working towards its goal. Nor does the statement indicate in what way men's 'social being determines their consciousness': many interpretations of this saying are possible, even apart from the rejected notion that men are consciously motivated by nothing but material interest. In particular it is not clear whether the 'determination' is teleological or merely causative. If we say that forms of consciousness such as religious and philosophical doctrines 'reflect' or 'express' the interests of the community or class in which they arise, this may either mean that they serve the interests of that community, i.e. it derives advantage from believing them, or simply that they are what they are because of the community's situation. Marx and Engels explained, for instance, that the ideals of political freedom served the interests of the bourgeoisie because they included the idea of free trade and freedom to buy and sell wage-labour. In this sense it can be said that the idea of freedom was a device to support bourgeois expansionism. But when Engels says that the Calvinistic theory of predestination was a religious expression of the fact that commercial success or bankruptcy does not depend on the businessman's intentions but on economic forces, then, whether we agree with his statement or not, we must regard it as asserting a merely causal connection: for the idea of absolute dependence on an external power (viz. the market in the 'mystified' shape of Providence) does not seem to further the businessman's interest, but rather to set the seal on his impotence. As a rule, however, when the founders of historical materialism interpret the phenomena of the superstructure, they do so in order to show that the ideas, trends, or institutions are not only caused by the interests of the class in question but actually serve those interests, i.e. they are functionally adapted to that class's needs. The analogy, moreover, is with a physical organism rather than with a human purpose. The ideas conduce to the advantage of those who hold them even though, or rather because, the fact that they do so is not perceived or is misunderstood. Part of their function is in fact one of mystification, transforming interests into ideals and concrete facts into abstractions, so that those who make use of them do not understand what they are doing and why.

At this point, of course, the interpretative possibilities of historical materialism begin to show certain limitations. In explaining, for example, the history of religion it accounts not so much for the genesis of a particular idea as for the fact of its becoming widespread. It cannot tell us why a certain conception of Deity and salvation occurred to a Jew living on the confines of the Roman Empire in the time of Augustus and Tiberius, but it purports to explain the social process whereby Christianity spread throughout the Empire and finally prevailed over it. The theory cannot interpret every dogmatic dispute that has arisen among the innumerable Christian sects, but it explains the main tendencies of those sects in terms of the social classes to which their adherents belonged. It cannot account for the appearance and nature of a particular artistic talent, but it can interpret the principal trends in the history of art in the light of the 'world-view' that each represents and the origins of that view in class ideology. The limitations on the use of the theory are important, for it would be a delusion to suppose that the division of society into classes could ever provide an explanation of all its differentiations without exception. Even political struggles and controversies are full of details which cannot be explained by the class conflict, although the method of historical materialism can be applied to fundamental disputes or periods when society is more than usually polarized in class terms.

What, then, in the last resort is the determining influence of the base on the superstructure, and what is the 'relative independence' possessed by the various forms of superstructure according to Engels and most theoretical Marxists? The influence in question relates only to certain features of the superstructure, but they are important ones. For instance, the possessing class in any political system will do its best to order the law of inheritance so as to keep estates intact, and it will be able to do this unhindered if it enjoys full political power. However, even when this class's material interest and the law are thus openly linked, its freedom of action may be limited by accidental circumstances such as the traditional laws and customs of the society in question, or religious beliefs which arose in other times but have not lost their effect. Within the superstructure of class societies there are always antagonistic forces at work, so that political and legal institutions are generally the fruit of a compromise among discordant interests. These, moreover, are as a rule distorted by tradition acting as an independent force, which will be all the stronger in so far as the different elements of the superstructure are not embodied in institutions. The force of tradition will be strongest in purely ideological matters, for example philosophical or aesthetic opinions: here the influence of the base on the superstructure will be relatively weaker than, for example, in the case of legal institutions. It must not therefore be inferred from historical materialism that the relations of production unequivocally determine the whole of the superstructure: they only do so in broad lines, excluding some possibilities and encouraging certain tendencies at the expense of others. Some elements of a given superstructure may persist apparently unchanged through various economic formations, though their significance may be different in different circumstances: this is true of religious beliefs and philosophical doctrines. In addition, elements of the superstructure become autonomous because human needs take on an independent form and instrumental values become ends in themselves. As Marx observed, the sum of needs is not constant but grows with the advance of production. 'The need which consumption feels for the object is created by the perception of it. An object of art, in the same way as every other product, creates a public which is sensitive to art and enjoys beauty. Production thus not only creates an object for the subject, but also a subject for the object.' (*Grundrisse*, Introduction.) 'At the dawn of civilization the productiveness acquired by labour is small, but so too are wants, which develop with and by the means of satisfying them.' (*Capital*, I, Ch. XIV; English edn., Ch. XVI.) It is in no way contrary to Marx's ideas or to historical materialism to hold that aesthetic needs, for instance, have come to require satisfaction for their own sake, as opposed to being merely 'apparent' or subordinate to some other, more fundamental needs. However, if some instrumental values have in this way become independent ones alongside elementary physical needs, the biological conditions of existence, it is quite natural that the process of creating them should largely cease to be dependent on relations which are ultimately based on those elementary needs.

The functional character of various elements of the superstructure is not inconsistent, in Marx's view, with the permanence of the creations of human culture. To explain the immortality of Greek art he suggests that humanity, like the individual, returns with pleasure to the imaginations of its childhood, which it knows to be past for ever but for which it still feels affection. From this it would follow that according to Marx cultural activity is not merely accessory to socio-economic development, but contains values independent of its role in subserving a particular order of society.

Nor should it be supposed that 'Social being determines consciousness' is an eternal law of history. The *Critique of Political Economy* describes the dependence of social consciousness on the relations of production as a fact that has always existed in the past, but it does not follow that it must be so for ever. Socialism, as Marx saw it, was vastly to enlarge the sphere of creative activity outside the production process, freeing consciousness from mystification and social life from reified forces. In such conditions, consciousness, i.e. the conscious will and initiative of human beings, would be in control of social processes, so that it would determine social being rather than the other way about. The maxim, in fact, appears to relate to ideological consciousness, i.e. that which is unaware of its own instrumental character. On the other hand, *The German Ideology* assures us that consciousness can never be anything other than conscious life, i.e. the manner in which men experience situations that arise independently of consciousness. It may be, however, that these two views can be reconciled. The rule that social being determines consciousness can be regarded as a particular case of the more general rule that consciousness is identical with conscious life—a particular case applying to the whole of past history, in which the products of human activity have turned into independent forces dominating the historical process. When this domination ceases and social development obeys conscious human decisions, it will no longer be the case that 'social being determines consciousness'; but it will still be the case that consciousness is an expression of 'life', for this principle is one of epistemology and not of the philosophy of history. Consciousness of life is a function of 'pre-conscious' life, not of course in the sense of Schopenhauer or Freud but in the sense that thought and feeling and their expression in science, art, and philosophy are instruments related positively or negatively to man's self-realization in empirical history. In other words, the situation in which social being determines consciousness is one in which consciousness is 'mystified', unaware of its true purpose, acting contrary to man's interest and intensifying his servitude. When consciousness is liberated it becomes a means of strength instead of enslavement, aware of its own participation in the realization of man and of the fact that it is a component of the whole human being. It controls the relations of production instead of being controlled by them. It is still the expression and instrument of life aspiring towards fullness, but it furthers that aspiration instead of impoverishing life, and is a source of creative energy instead of a brake on it. In short, the liberated consciousness is de-mystified and aware of its contribution to the expansion of human opportunities. Consciousness at all times is an instrument of life, but throughout history up to now (prehistory) it has been determined by relations of production that are independent of the human will. This interpretation, at all events, is consistent with Marx's writings, though he does not anywhere expressly adopt it.

3. Historical progress and its contradictions

The whole of progress up to the present time (so the theory continues) has been beset by an internal contradiction: it has increased man's total power over nature while depriving the majority of the fruits of that power, and enslaving all mankind to objectified material forces. Contrary to Hegel's view, history is not the gradual conquest of social freedom but rather its gradual extinction. 'At the same pace that mankind masters nature, man seems to become enslaved to other men or to his own infamy. Even the pure light of science seems unable to shine except against the dark background of ignorance.' (Speech delivered by Marx at the anniversary of the Chartist organ, the *People's Paper*, on 14 Apr. 1856.) Engels wrote in a similar strain in his *Origin of the Family* (Ch. II): 'Monogamy was a great historical advance, but at the same time it inaugurated, along with slavery and private wealth, that epoch, lasting until today, in which the well-being and development of the one group are attained by the misery and repression of the other.' Again: 'Since the exploitation of one class by another is the basis of civilization, its whole development moves in a continuous contradiction. Every advance in production is at the same time a retrogression in the condition of the oppressed class, that is, of the great majority.' (Ibid., Ch. IX.) 'Indeed, it is only by dint of the most extravagant waste of individual development that the development of the human race is at all safeguarded and maintained in the epoch of history immediately preceding the conscious reorganization of society.' (*Capital*, III, Ch. V, II.)

This negative, anti-human side of progress is an inseparable consequence of alienated labour. But for this very reason, even in the cruellest aspects of civilization we can perceive history working towards the final liberation of man. From this point of view, perhaps the most characteristic of Marx's observations are contained in his articles on British rule in India. After describing its devastating effect on the peaceful, stagnant Indian communities he goes on to say:

Sickening as it must be to human feeling to witness those myriads of industrious patriarchal and inoffensive social organizations disorganized and dissolved into their units, thrown into a sea of woes, and their individual members losing at the same time their ancient form of civilization and their hereditary means of subsistence, we must not forget that these idyllic village communities, inoffensive though they may appear, had always been the solid foundation of Oriental despotism; that they restrained the human mind within the smallest possible compass, making it the unresisting tool of superstition, enslaving it beneath traditional rules, depriving it of all grandeur and historical energies ... We must not forget that these little communities were contaminated by distinctions of caste and by slavery, that they subjugated man to external circumstances instead of elevating man to be the sovereign of circumstances, that they transformed a self-developing social state into never-changing natural destiny and thus brought about a brutalizing worship of nature, exhibiting its degradation in the fact that man, the sovereign of nature, fell down on his knees in adoration of Hanuman, the monkey, and Sabbala, the cow ...

The question is, can mankind fulfil its destiny without a fundamental revolution in the social state of Asia? If not, whatever may have been the crimes of England she was the unconscious tool of history in bringing about that revolution. Then, whatever bitterness the spectacle of the crumbling of an ancient world may have for our personal feelings, we have the right, in point of history, to exclaim with Goethe: '*Sollte diese Qual uns quälen,/Da sie unsre Lust vermehrt?*' ['Should we be grieved by this pain that increases our pleasure?'] (*New York Daily Tribune*, 25 June 1853)

This argument is an important clue to the understanding of the Marxian interpretation of history. We find in it the Hegelian doctrine of the historical mission fulfilled unconsciously, despite their crimes and passions, by particular nations or classes. There is also the idea of the historical mission of humanity, the vocation of mankind as a whole. We see, further, that Marx constantly regarded the historical process from the point of view of the future liberation of mankind, which was the sole touchstone of current events: in particular he attached no importance to the economic conquests of the working class under capitalism except in relation to this ultimate purpose. Finally it should be noted that Marx's historical appraisal of human actions in terms of the part they play in bringing about liberation had nothing to do with a moral judgement: the crimes of the British imperialists were not palliated by the fact that they brought the day of revolution nearer. This is also the viewpoint of the whole of *Capital*, in which moral indignation at the cruelty and villainy of exploitation is found side by side with the conviction that this state of affairs was helping on the revolution. Increasing exploitation was bringing about the downfall of capitalism, but it did not follow that the workers who resisted it were acting 'against history'. However, their action was progressive not because it improved their lot and this improvement was good in itself, but because it helped to develop the workers' class-consciousness, which was a precondition of revolution.

Marx and Engels is believed in the rights of a higher civilization over a lower one. The French colonization of Algeria and the U.S. victory over Mexico seemed to them progressive events, and in general they supported the great 'historical' nations against backward peoples or those which for any reason had no chance of independent historical development. (Thus Engels expected Austria-Hungary to swallow up the small Balkan countries; Poland, as a historical nation, should, he thought, be restored and include in its dominion the less developed peoples to the east—Lithuanians, White Russians, and Ukrainians.) The future liberation on which their historical optimism was based was not merely a matter of abolishing poverty and satisfying elementary human needs, but of fulfilling man's destiny and ensuring his dignity and greatness by giving him the maximum control over nature and his own life. We see how, despite Marx's abandonment of the old formulas about restoring man's nature, his faith in humanity and its fulfilment in the course of history lived on and determined his attitude to current events. Capitalism, through all its negative features and manifold inhumanity, had prepared the technological basis enabling man to escape from the compulsion of material needs and develop his intellectual and artistic faculties as ends in themselves.

The surplus labour of the mass has ceased to be the condition for the development of general wealth, just as the non-labour of the few has ceased to be the condition for the development of the general powers of the human brain. With that, production based on exchange value breaks down, and the direct, material production process is stripped of the form of penury and antithesis. [It is] the free development of individualities, and hence not the reduction of necessary labour time so as to posit surplus labour, but rather the general reduction of the necessary labour of society to a minimum, which then corresponds to the artistic, scientific etc. development of the individuals in the time set free, and with the means created, for all of them. (*Grundrisse*, III, 2, Notebook VII)

Thus the martyrdom of history would not be in vain, and future generations would enjoy the fruits of their predecessors' sufferings.

It should be emphasized that to Marx the concept of 'modes of production' is a basic instrument for the division of history into periods and also for the comprehension of it as a single whole. There is one point, however, which has given trouble to commentators, namely the 'Asiatic mode of production', to which Marx refers in the *Grundrisse* and in certain articles and letters of 1853. The essence of the Asiatic system, found historically in China, India, and some Muslim countries, is that private ownership of land was almost unknown, as geographical and climatic conditions called for an irrigation system that could only be provided by a centralized adminstration. Hence the special autonomous role of the despotic state apparatus, on which the economy largely depended; commerce developed to a very small extent, towns did not exist as centres of trade and industry, and there was scarcely any native bourgeoisie. The traditional village communities lived on through the ages in social and technical stagnation. The gradual dissolution of these communities and of the state despotism was due to European capitalism rather than internal causes.

In Stalin's time orthodox Marxism altogether excluded the 'Asiatic mode of production' from its schema of history, for the following reasons. Firstly, if a large part of humanity had lived for centuries with an economy of a type all its own, there could be no question of a uniform pattern of development for all mankind. The progression from slave-owning through feudalism to capitalism would apply only to one part of the world and not the rest, so that there could be no universally valid Marxist theory of history. Secondly, according to Marx the peculiarities of the Asiatic system were due to geographical factors; but how could the primacy of technology over natural conditions be maintained, if the latter could bring about a different form of social development in a large part of the globe? Thirdly, the Asiatic system was said by Marx to have involved the countries concerned in stagnation from which they were only rescued by the incursion of peoples whose economic development had been on different lines; apparently, then, 'progress' is not a necessary feature of human history but may or may not happen, according to circumstances. Thus the 'Asiatic mode of production' appeared contrary to three of the fundamental principles that orthodox Marxists generally attributed to historical materialism: the primary role of productive forces, the inevitability of progress, and the uniformity of human evolution in society. It might seem that the doctrine applied only to Western Europe and that capitalism itself was an accident—a system that had happened to arise in a particular, not very large part of the world and had subsequently proved strong and expansive enough to impose itself on the whole planet. Marx did not himself draw this influence, although, significantly, he observed at a later stage that the analysis in *Capital* applied only to Western Europe. But the conclusion follows naturally enough from what he says about the Asiatic system. It may appear no more than a detail in his philosophy of history, but if it is accepted it calls for the revision of a number of stereotypes, especially those connected with historical determinism and the idea of progress.

4. *The monistic interpretation of social relationships*

Historical materialism, as we have seen, provides a theoretical account of the main determinants and can be used to predict general lines of development, but not specific occurrences. Like any other philosophy of history, it is not a quantitative theory and cannot inform us of the relative strength of the factors at work in a particular social process. It purports to enable us, however, to discern the fundamental structure of any society by analysing its relations of production and the class divisions based directly thereon. As to the meaning of 'relations of production', this does not appear unequivocally from the writings of Marx and Engels. The latter, in *The Origin of the Family*, refers to 'the immediate production and reproduction of life' as including not only the making of tools and means of subsistence, but also the biological reproduction of the species—a doctrine frequently criticized by later Marxists; and in his letter to Starkenburg of 25 January 1894 Engels includes among 'economic conditions' the entire technique of production and transport and also geography. This is not merely a verbal question of the precise definition of such terms as 'relations of production' or 'economic conditions': the point is whether a single type of circumstance determines the whole superstructure, or several independent types. For instance, is the social aspect of the increase of the species, i.e. family institutions and the demographic situation, completely dependent on the mode of production and distribution, or does it present biological or other features with an independent effect on other social phenomena in the realm of the superstructure? Similarly, how far can geography be regarded as an independent factor in social processes? Marx observes in *Capital* (Vol. I, Ch. XIV; English edn., Ch. XVI) that capitalism arose in the temperate zone because the luxuriance of the tropics did not spur mankind to the efforts which gave rise to technology. It thus appears that, to Marx, some natural circumstances are at least a necessary condition of a particular social development. But in that case the level of technology, which in its primitive form has been achieved by all branches of the human race, cannot be a sufficient condition of changes in the relations of production. What has been said of geography applies equally to demographic phenomena. The message of historical materialism would seem to be that a given technology is a sufficient cause of particular relations of production provided certain other conditions of geography or demography are present. In the same way, such relations of production are a sufficient cause of essential features of the political superstructure if certain other conditions are fulfilled, for example as regards the consciousness and traditions of a people or its present situation. Hence the interpretative value of historical materialism appears only in particular analyses in which various concurrent factors can be discerned, and not in the general premisses which only dictate the direction of investigation.

Finally, historical materialism as a set of guidelines drawing attention to a particular type of interrelation must be distinguished from historical materialism as a theory which traces the basic course of human events from the first community to the classless society. This survey of world history is based on the premiss that if developments are considered on a large enough scale they can be explained by changes and improvements in the production of means wherewith to satisfy material needs, and that above a certain technological level these developments take the form of a struggle between classes with conflicting interests.

5. *The concept of class*

In his letter of 5 March 1852 to Joseph Weydemeyer, Marx declares that it was not he who discovered the existence of classes or the class struggle: what he did was to prove that the existence of classes is bound up with particular phases in the development of production, that the class struggle leads to the dictatorship of the proletariat, and that this dictatorship constitutes the transition to a classless society.

Neither Marx nor Engels ever clearly defined the concept of class, and the last chapter of Volume III of *Capital*, which was to treat of this question, breaks off after three or four paragraphs. In it Marx poses the question 'What makes wage-labourers, capitalists and landlords constitute the three great social classes?' It would seem at first sight, he goes on, that they are characterized by the identity of sources of revenue within each class—wages, profit, and ground-rent respectively. But, he goes on, from this point of view doctors, officials, and many others would constitute separate classes defined in each case by their source of revenue: so this criterion is in any case insufficient.

Kautsky, who took up the argument where Marx left off and tried to reconstruct his thoughts, arrived at the following conclusion (*The Materialist Interpretation of History*, IV, i, 1–6). The concept of class has a polarized character, i.e. a class exists only in opposition to another class. (It would therefore be absurd to speak of a one-class society: a society can only be classless or composed of at least two hostile classes.) A collectivity does not become a class simply because its members' revenue comes from the same source; it must also be in a state of conflict with another class or classes over the distribution of revenue. But this is not sufficient either. Since workers, capitalists, and landowners in fact all derive their revenue from the same source, namely the value produced by the workers' labour, and since the way this value is distributed depends on who owns the means of production, it is this ownership that constitutes the ultimate criterion. Thus we have on the one hand the possessing classes who own the means of production and therefore the surplus value created by the workers' labour, and on the other hand the class of the exploited, who own nothing but their own labour-power and are obliged to sell it. On this criterion we can also distinguish intermediate classes of those who, like small peasants or craftsmen, possess some means of production but do not employ wage-labour; they do not enjoy the results of others' unpaid labour, but create values by employing themselves or their families. These classes have a divided consciousness: the ownership of means of production inclines them towards solidarity with the capitalists, but they are likened to the workers by the fact that they live from their own efforts and not from the surplus value produced by others. Capitalism tends constantly to deprive these middle classes of their small possessions and depress them to the status of the working class, allowing only a small minority to enter the ranks of the exploiters.

Marx approached the question of classes from the standpoint of conditions in Britain, while Kautsky had in view Germany and the rest of central Europe. The criterion as to ownership of the means of production and the employment of wage-labour enables us to distinguish between the exploiters, the exploited, and those in between, but it does not distinguish capitalists from landowners, both of whom, by their ownership of the means of production including land, appropriate unpaid hours of surplus labour. In point of fact the class opposition between these two is different from that between them and the workers: for both the possessing classes are interested in maximizing exploitation and surplus value. Hence at times of crisis they present a common front against the proletariat, although the latter may temporarily ally itself with one of them against the other, for example with the bourgeoisie to secure political freedom in situations where feudal institutions retain their power. The ultimate source of the capitalist's and the landowner's revenue is the same—the surplus value created by the workers; and, according to Marx, this is also the case with financiers, merchants, and lenders of money at interest. The exploiting classes differ, however, in their way of appropriating profit. Only industrial capital does so by exchanging objectified labour for living labour, while the landowner or the usurer subsists on rent while taking no part in the process of exchange.

It would seem to be in accordance with Marx's intentions, therefore, to distinguish between primary and secondary criteria in the class division of society. The primary criterion is the power to control the means of production and therefore to enjoy the values created by others' surplus labour. This criterion places on one side all the exploiting classes, i.e. those which profit by surplus labour, including industrial and commercial capitalists and landowners. On the other side are the sellers of labour-power, i.e. wage-earners, and small peasants, craftsmen, etc. using their own means of production. The first category is divided by a secondary criterion into the direct acquirers of labour-power (industrial capitalists) and those who appropriate surplus value indirectly by the possession of land or capital. Within the second category, wage-earners are divided from the rest by the fact that they do not own any means of production.

The primary criterion in its general form is also applicable to pre-capitalistic class formations, such as serfdom and feudalism; while the secondary criteria are peculiar to the capitalist mode of production.

The definition of class is by no means a purely verbal or methodological question. The need for a definition arises from observation of the facts of the class struggle; it is a question of ascertaining the criteria which in practical terms distinguish the groups whose antagonisms define the basic historical processes.

Another essential feature of a class is that it shows spontaneous solidarity in its opposition to other classes, though this does not prevent its members from being rivals to one another. In Volume III of *Capital* Marx describes the economic basis of capitalist class solidarity: as the rate of profit evens itself out in all spheres of production, and every capitalist shares in profit according to the amount of his capital,

in each particular sphere of production the individual capitalist, and capitalists as a whole, are involved in the exploitation of the total working class by the totality of capital and in the degree of that exploitation, not only out of general class sympathy, but also for direct economic reasons ... A capitalist who did not in his line of production employ any variable capital or, therefore, any labour (in reality an exaggerated assumption) would nevertheless be as much interested in the exploitation of the working class by capital, and would derive his profit quite as much from unpaid surplus labour, as a capitalist who (another exaggeration) employed only variable capital and thus invested his entire capital in wages. (*Capital*, III, Ch. X)

The clash of interests between particular capitalists is naturally repressed in situations dominated by antagonism between the exploiters and the exploited as a whole. None the less their individual interests are bound to conflict, and so do those of the workers, for example when there is grave unemployment. But, while the rivalry between capitalists does not in itself harm the interests of capital as a whole, competition between workers does harm the interests of the working class. Hence the latter's class-consciousness is much more important to the realization of its class-interest than that of the exploiters is to theirs.

Finally, an essential feature of Marx's concept of class is that he rejects the utopian–socialist classification according to scale of income or relative share in the whole social product. The utopian division according to wealth is quite alien to Marx's thought. A person's share in the national revenue does not determine his place in the class system but is determined by it. A small craftsman may in some circumstances earn less than a skilled worker, but this does not affect the class they belong to. Luxury consumption is not a determinant of class either, as witness the 'heroic asceticism' of the bourgeoisie in its early period. In the second place, class is not determined by Saint-Simon's distinction between idlers and workers. The capitalist may perform essential functions of management or may hire others to do so for him; which he does may be of importance to the efficiency of his firm, but does not affect his class position. The performance of managerial functions is neither a necessary nor a sufficient condition of belonging to the capitalist class.

An essential condition of the existence of a class is, however, that there should be at least the germ of class-consciousness, an elementary sense of common interest and shared opposition to other classes. A class may indeed exist 'in itself' without being a class 'for itself', i.e. aware of its role in the social process of production and distribution. But before one can speak of class there must be a real community of interest, manifesting itself in practice. If its members are isolated from one another, a class has no more than a potential existence. As Marx wrote in *The Eighteenth Brumaire*, sect. VII:

The small peasants form a vast mass, the members of which live in similar conditions but without entering into manifold relations with one another. Their mode of production isolates them from one another instead of bringing them into mutual intercourse ... In this way the great mass of the French nation is formed by simple addition of homologous magnitudes, much as potatoes in a sack form a sackful of potatoes. In so far as millions of families live under economic conditions that divide their mode of life, interests and culture from those of the other classes, and put them in hostile contrast to the latter, they form a class. In so far as there is merely a loose interconnection among these small peasants, and the identity of their interests begets no unity, no national union and no political organization, they do not form a class. They are consequently incapable of enforcing their class interest in their own name, whether through a parliament or through a Convention. They cannot represent themselves, they must be represented. Their representative must at the same time appear as their master, as an authority over them, as an unlimited governmental power.

On the other hand, the existence of a political class struggle is not, in Marx's view, a necessary condition of the reality of class division. 'In ancient Rome the class struggle took place only within a privileged minority, between the free rich and the free poor, while the great productive mass of the population, the slaves, formed a purely passive pedestal for these combatants' (ibid., Preface to second edn.). None the less, Marx regarded the slaves as a class.

Marx regarded the class division as the essential, but not the only, division in every society in which classes exist. Within each class there are groups whose interests may conflict, for example industrial capital and finance. Among those who draw their revenue from ground-rent there are separate divisions of landowners, mine-owners, and property-owners. The working class is divided by branches of industry and by different degrees of skill and rates of pay. The professions and trades are divided from one another. The intelligentsia, as Marx conceived it, is not itself a class but is divided according to the class whose interest it serves. In short, the divisions of society are infinitely complex. None the less, Marx contended that throughout the history of antagonistic societies—i.e. all except primitive classless communities—class divisions were the chief factor determining social change. The whole sphere of the superstructure—political life, wars and conflicts, constitutional and legal systems, and intellectual and artistic production of every kind—was dominated by the class division and its consequences. Here too, of course, it is only possible to operate with qualitative characteristics, for we cannot measure the relative importance of different forms of social stratification in determining particular aspects of the superstructure.

It would seem to follow that the mere removal of the class division by abolishing private ownership of the means of production would not abolish all sources of social antagonism, but only the most important ones due to different degrees of control over surplus value. Marx believed, however, that the domination of the class system was such that its removal would do away with all other sources of antagonism and bring about a unity of social life in which the freedom of one man would not be limited by that of another.

6. *The origin of class*

As to the origin of class distinctions, a necessary though not a sufficient condition was the achievement of a state of technology in which it was possible to appropriate the fruits of surplus labour. Engels considers this question in *The Origin of the Family* and the *Anti-Dühring*. Dühring had suggested that classes owed their origin to the use of force, and had offered the example of two individuals unequally endowed by nature. Engels opposed this theory, which he thought erroneous and unfounded. Neither property nor exploitation, he said, were the result of violence. Property was based on production exceeding the producer's needs, and exploitation presupposed the inequality of property. As for classes, they had arisen in various ways. In the first place, commodity production led to the inequality of possessions, which were bequeathed from one generation to another and thus made it possible for a hereditary aristocracy to spring up, not by violent means but as a result of custom. Secondly, primitive communities had to entrust their defence to individuals appointed for the purpose, and the offices thus created were the germ of political power. What were at first socially necessary institutions of defence and administration became in time hereditary estates, independent of society and, as it were, above it. Thirdly, the natural division of labour took on a class form when technical progress and economic development made it possible to use slave labour obtained by conquest. Slavery made possible for the first time a real division between industry and agriculture, and hence the whole political system and culture of the ancient world; it was thus a condition of the huge progress of civilization up to the present day. But in all the forms in which class divisions arose, their ultimate origin lay in the division of labour. This was the condition of the whole evolution of mankind, and was therefore the cause of private property, inequality, exploitation, and oppression.

7. *The functions of the state and its abolition*

The division of classes led in time to the creation of a state apparatus. Tracing the development of primitive society on the lines of Morgan's explorations, Engels suggests that the state arose as a result of the breakdown of the democratic organization of the tribe. Several factors were at work in this process: to begin with, the transformation of offices into hereditary estates as already described, and the need to defend fortunes acquired through various contingencies. The state, as an instrument of coercion in defence of class interests, presupposes at least the elements of a class division. The apparatus of authority and the use of force to control slaves are economic in origin. Conquest is one way in which a state may come into being, but in its typical form it arises from class antagonisms within a single community. The state sanctifies acquired wealth and privilege, defending them against the communist tradition of earlier societies and creating conditions in which private fortunes and inequality increase. 'Because the state arose from the need to hold class antagonisms in check, but because it arose, at the same time, amid the conflict of these classes, it is, as a rule, the state of the most powerful, economically dominant class, which, through the medium of the state, becomes also politically dominant, and thus acquires new means of holding down and exploiting the oppressed class' (*Origin of the Family*, IX). In relation to the bourgeois state, this function of defending the privileges of the possessing class is highly conspicuous and vital to its political structure. As Marx and Engels wrote in 1850 in a review of a book by E. de Girardin, 'The bourgeois state is nothing but the mutual insurance of the bourgeoisie against its own individual members and the class of the exploited—an insurance that must become ever more costly and, in appearance, independent of bourgeois society, which finds it increasingly hard to keep the exploited in a state of subservience' (*Neue Rheinische Zeitung, Politisch-ökonomische Revue*, 4, 1850). Hence, although the original, socially necessary functions which, by becoming autonomous, gave rise to political power have still to be performed, it is not they which determine the character of the state. For these functions themselves contain no element of political power, and their autonomization would not have led to the creation of a state apparatus if it were not for the need to defend the privileged classes.

It can also happen in bourgeois society, as Marx observes in connection with Louis Napoleon's *coup d'état*, that the bureaucratic machine asserts its independence of the class it serves. But such situations can also be explained by class interests. The bourgeoisie may give up parliamentary power and entrust the direct exercise of political authority to an autonomized bureaucracy, if this is necessary to maintain its own economic position as a class.

If we define the meaning of the state in this way, two conclusions follow which are of great importance for Marx's doctrine: viz. the disappearance of the state in a classless society, and the necessity of destroying the existing state machine by a revolution.

The first conclusion is evident. Once class division has been abolished, there is no need for the institutions whose function is to maintain it and oppress the exploited classes.

The first act in which the state really comes forward as the representative of society as a whole—the taking possession of the means of production in the name of society—is at the same time its last independent act as a state. The interference of the state power in social relations becomes superfluous in one sphere after another, and then ceases of itself. The government of persons is replaced by the administration of things and the direction of the processes of production. The state is not 'abolished', it withers away. (*Socialism, Utopian and Scientific*, III)

The state is not eternal, but a transient feature of civilization which will disappear with class divisions—as Engels puts it, 'into the museum of antiquities, by the side of the spinning-wheel and the bronze axe' (*Origin of the Family*, IX).

As we see, the abolition of the state does not mean abolishing the administrative functions necessary for the management of production; but these functions will not be an exercise of political power. Putting the matter in this way implies a situation in which all social conflicts have disappeared; and this confirms the interpretation according to which Marx and Engels held that the abolition of class divisions would at the same time abolish all other sources of conflict.

In the second place, the political superstructure as an apparatus of coercion cannot be reformed in such a way as to start serving the interests of the exploited class; it must be destroyed by revolutionary violence. This conclusion, as we have seen, forced itself on Marx at the time of the Paris Commune. The abolition of the bourgeois state is a step towards the abolition of the state in general, but, during the period when the victorious working class is still fighting the exploiters, it must possess its own means of coercion, which for the first time in history will be an instrument of the majority. This is the dictatorship of the proletariat, in which the latter will use force without concealment for the purpose of doing away with class altogether. The transition to a socialist society, even though it is prepared by the development of the capitalist economy, cannot be effected by the economic process alone but only in the realm of the superstructure. The positive prerequisite of socialism in a capitalist economy is a high degree of technology and of co-operation in the productive process; its negative causes are the internal contradictions of capitalism and the class-consciousness of the proletariat. The transition itself is a political and not an economic act; however, according to Marx's aphorism, 'Force is the midwife of every old society pregnant with a new one. It is itself an economic power [*Potenz*]' (*Capital*, I, Ch. XXIV, 6; English edn., Ch. XXXI).

In 1895, a few months before his death, Engels wrote an Introduction to the second edition of Marx's *The Class Struggles in France, 1848–50*, which has been invoked by reformists as a proof that Engels replaced the idea of revolutionary force by that of achieving power for the proletariat by parliamentary means. In this text he states that since the repeal of the Anti-Socialist Law in Germany and in view of the success of social democracy at the polls, 'rebellion in the old style, the street fight with barricades, which up to 1848 gave everywhere the final decision, is to a considerable extent obsolete'. Insurgents were worse off in street-fighting than previously, and in any case a rebellion by a small vanguard could not bring about the transformation of society. This required the conscious, rational participation of the masses, and it was therefore a mistake to sacrifice the most enlightened part of the proletariat in street-fighting: what must be done was to continue the advance by legal means in parliament and in the field of propaganda, accumulating strength for the decisive conflict. 'We, the "revolutionaries", the "rebels", are thriving far better on legal methods than on illegal methods and revolt.'

Engels certainly laid great emphasis on peaceful means of strengthening the workers' movement; and he did not exclude the possibility, in Germany at all events, that power might be achieved by non-violent means. But the change of viewpoint brought about by the electoral success of the German social democrats is not so great as it at first seems. In the first place, Engels confines his hopes to Germany, as Marx had once confined them to Britain, the U.S.A., and Holland. Secondly, he did not think it a foregone conclusion that power would be achieved by parliamentary means: this depended entirely on the attitude of the bourgeoisie, and a violent revolution was still a possibility. Thirdly, while he expected a 'decisive conflict' in the form of a seizure of power by the working class, he believed that this might be a bloodless act owing to the latter's strength, its highly developed consciousness, and its ability to enlist the support of the lower middle classes. He did not reject the idea of a revolution as necessary in principle and inevitable in practice, but he believed that it might be a non-violent one. He did not say expressly that he thought the working class might achieve power simply by obtaining a majority at the polls, and it is hard to be certain whether he envisaged this; but he undoubtedly attached more importance than previously to peaceful instruments of the class struggle. If he did expect that power would be taken over by electoral means, this would signify a radical change in his position; but even in this case we cannot attribute to him the idea of co-operation between classes or the extinction of class conflict.

Whatever the means by which the proletariat was to achieve victory, Marx and Engels always saw the state power as an instrument only: unlike Hegel or Lassalle, they did not regard the state as a value in itself or identify it with society, but saw it as a historical, transient form of social organization. Man's social existence was by no means the same as his political existence; on the contrary, the state as such was the political expression of a situation in which man's powers, embodied in his works, were opposed to him—i.e. social alienation in the highest degree. If the proletariat should need a temporary means of coercion, this would consist in the domination actually exercised by the great bulk of society. But the whole purpose of such domination would be to terminate its own existence and put an end to politics as a separate sphere of life. Marx's theory of the state is thus a repetition and development of what he wrote in philosophical language in 1843 in *The Jewish Question*. Real human individuals, who are the only true 'subjects', will absorb into themselves the species-essence which has hitherto existed in the alienated sphere of political life. The social character of men's individual energies will not express itself as an alienated political creation; men and women will perform their mission in society in a direct manner and not in a realm specially created for the purpose—in short, private and community life will be integrated at the level of each and every human being. Man's species-essence will resolve itself completely into the lives of individuals, and there will be no more distinction between private and public life. The abolition of class divisions is a necessary and sufficient condition of this return to concreteness.

8. *Commentary on historical materialism*

In the foregoing account of the main principles of historical materialism we have been at pains to interpret the doctrine as sympathetically as possible. We have not, for instance, taken in a literal sense certain compressed or aphoristic statements by Marx and Engels which appear to assert dogmatically and without proof that every detail of history is the outcome of the class system, determined in its turn by the technological development of society. When Marx says in *La Misère de la philosophie* that the handmill 'produces' feudal society and the steam-mill capitalism, we are clearly not meant to take this literally. What the handmill and steam-mill produce is flour, and both kinds of mill may coexist in a society which, in its turn, may have predominantly feudal or capitalist features. When Engels said in his funeral oration that Marx's great merit was to have discovered that 'mankind must first of all eat, drink, have shelter and clothing before it can pursue politics, science, art, religion etc.', it is hard to use such expression as a proof of historical materialism or to see why it should be an immortal discovery to repeat the maxim *primum edere, deinde philosophari*. But it would be petty to attack the doctrine on the strength of such formulas in isolation. On the other hand, doubts and objections arise at a more fundamental level. The great majority of Marxist theoreticians have followed Engels in speaking of the 'reciprocal influence' of the base and superstructure, the 'relative independence' of the latter, and the fact that it is determined by economic factors 'in the last resort'. As we have seen, the exact meaning of 'economic factors', 'base', and 'superstructure' is by no means unequivocal, and the statement itself is in any case open to serious controversy. It would seem that to say there is an interaction between the relations of production and the 'superstructure' is to utter a truism which all would accept and which has nothing particularly Marxist about it. Historic events—wars, revolutions, religious changes, the rise and fall of states and empires, artistic trends and scientific discoveries—can be rationally explained by many circumstances, not excluding technology and class conflicts: this is a matter of common sense and would not be denied by a religious believer, a materialist, or any philosopher of history unless he were a fanatical champion of some 'unique factor' or other. That books and plays cannot be understood without knowledge of the historical circumstances and social conflicts of the time was known, long before Marx, to many French and other historians, some of whom were conservative in politics. We must ask then, what exactly is historical materialism? If it means that every detail of the superstructure can be explained as in some way dictated by the demands of the 'base', it is an absurdity with nothing to recommend it to credence; while if, as Engels's remarks suggest, it does not involve absolute determinism in this sense, it is no more than a fact of common knowledge. If interpreted rigidly, it conflicts with the elementary demands of rationality; if loosely, it is a mere truism.

The traditional way out of this unhappy dilemma is, of course, the qualification 'in the last resort'; but Engels never explained precisely what he meant by this. If it only means that the relations of production determine the superstructure indirectly, through other factors, then we may still object that the theory is one of absolute determinism: it makes no difference whether one wheel acts directly on another or whether it does so by means of a conveyor belt. Most probably, however, Engels meant that the determination was not absolute: not all features of a civilization were dictated by the class structure, and not all relations of production by the technological level, but only the chief ones in each case. But then, how do we decide which features are more important and which less? We may choose to regard as important those relationships of which historical materialism tells us, but then we are involved in a tautology or a vicious circle: the base determines those parts of the superstructure that are determined by the base. We can of course say, for instance, that what is important in Verlaine's poetry is not the versification, which is 'contingent' or traditional, but the poet's melancholy, which can be accounted for in terms of the class situation (a stock example of literary history as expounded by the materialist school). But historical materialism cannot tell us why one is important and not the other, except on the ground that it is able to account for the latter—and this is clearly a vicious circle.

Again, if the relations of production determine only some features of the superstructure and not all, the doctrine cannot explain any particular historical phenomenon—for any historical fact is an accumulation of many circumstances—but only certain broad lines of the historical process. This, it appears, was in fact the main intention: not to explain a particular war, revolution, or movement of any kind, but only the fact that one major socio-economic system gave place to another. Everything else—the 'zigzags' and reversals of history, the fact that a process occurred when it did and not a few centuries sooner or later, the particular struggles and efforts that accompanied it—all this would be relegated to the status of contingency, with which the theory need not concern itself. In that case historical materialism could not claim to be an instrument of prognostication. It could state, for instance, in the most general way that capitalism must be replaced by socialism; but when and how this would be, in how many decades or centuries and after what wars and revolutions—as to these 'contingent' aspects it could offer no prediction whatever.

But even if the scope of historical materialism is thus limited, we have not done with the objections to it. The course of history is one and unrepeatable: it does not, therefore, permit the formulation of a rule that, for example, a slave-owning society must everywhere and at all times be superseded by a society based on feudal landownership. If, on the other hand, we say that history consists of many independent processes, since different parts of the world have lived for centuries in more or less complete isolation from one another, this tends rather to refute historical materialism than to confirm it: for Asiatic or Amerindian societies before the European invasions did not in fact evolve in the same manner as our own, and it would be a gratuitous fancy to assert that they would have done so if they had been left alone for long enough.

All the more detailed historical and political analyses by both Marx and Engels show that they were not themselves prisoners of their 'reductionist' formulas but took into account all kinds of factors—demography, geography, national characteristics, and so on. When, for instance, Engels in a letter of 2 December 1893 attributes the absence of a U.S. socialist movement to ethnic considerations, he shows clearly that he did not regard the bourgeois–proletariat conflict as the determinant of all social processes, though he expected it finally to take much the same political form in America as in Europe. If this expectation was not fulfilled, as it has not been for the past eighty years, the fact can of course be ascribed to 'secondary factors', and Marxists can preserve indefinitely their faith in the validity of the doctrine despite accidental interferences. Any failure of their predictions can be explained away by saying that the theory is not a schematic one, that a multiplicity of factors have to be considered, etc. But, if it is easy to dismiss inconvenient facts in this way, it is not because of the theory's profundity but because of its vagueness—a quality which it shares with all universal theories of history that have ever been put forward.

This same vagueness enables the theory to make various unprovable historical assumptions. When Engels says that great men like Alexander, Cromwell, and Napoleon appear when the social situation requires them, this is the merest speculation: for by what signs could such a 'requirement' be recognized, other than by the fact that these men actually did appear? Clearly, a deduction of this kind, based on universal determinism, cannot help us to understand any single phenomenon.

There also exists an even less rigorous interpretation of historical materialism. Marxists have often asserted that, according to the doctrine, relations of production do not bring about the superstructure but 'define' it in the negative sense of limiting the options at society's disposal, without prejudicing its choice between them. If Marx and Engels meant no more than this, the doctrine is again in danger of becoming a truism. All would agree that the legal, political, artistic, and religious forms that we know from history cannot be imagined irrespective of social conditions: to take an example already quoted, the Declaration of the Rights of Man could not have happened among the Aztecs or in the technical and social environment of tenth-century Europe. Yet the fact that some aspects of the superstructure preserve their continuity in spite of profound social changes is relevant to the validity of even this diluted version of historical materialism. Christianity, like Islam, has persisted through many social and economic systems. It has of course changed in many ways, in the interpretation of Scripture, in its organization and liturgy and the development of dogma; it has gone through crises, schisms, and internal conflict. Yet, if the term 'Christianity' can still be used with meaning, it is because it has not changed in every respect and has preserved its essential content despite the vicissitudes of history. Every Marxist, of course, admits that tradition possesses an autonomous force of its own, and there are plenty of passages in Marx to confirm this. But if the objection can be brushed aside in this manner, it merely shows that the doctrine is so imprecise that no historical investigation and no imaginable facts can refute it. Given the variety of factors of all kinds, the 'relative independence of the superstructure', 'reciprocal influence', the role of tradition, secondary causes, and so forth, any fact whatever can be fitted into the schema. As Popper observes, the schema is in this sense irrefutable and constantly self-confirming, but at the same time it has no scientific value as a means of explaining anything in the actual course of history.

Furthermore, it seems highly improbable that any fact or series of facts in the realm of ideology could be explained or understood without reference to other circumstances which are of an ideological or biological nature, or otherwise not covered by Engels's 'in the last resort'. To take a simple example: in the fifteenth century there arose in Catholic Christianity a demand for communion under both kinds, and this was taken up by an important heretical movement (the Utraquists). It is contended, with some truth, that the demand expressed a desire to efface the difference between the clergy and laity, and can thus be regarded as a manifestation of egalitarianism. But the question then arises: 'Why do men desire equality?' To reply 'Because they are unequal' would be a tautological pseudo-explanation. We must therefore take for granted that, at some periods of history, men have regarded equality as a value worth fighting for. If the fight is carried on by men who are starving or otherwise deprived of essentials, we can say that it is explicable on purely biological grounds. But, if it is more than a question of satisfying physical needs, we cannot explain that men fight for equality 'because of economic conditions' without positing the existence of an egalitarian ideology, for otherwise they would have no reason to desire it. Or, to take an even simpler example, already quoted: the possessing classes in any community try to influence legislation so as to minimize death duties, and it is regarded as 'obvious' that they will do this. But in fact their action is due not only to particular relations of production and the existence of private property, but also to concern for their own posterity. This concern, being universal, is considered obvious, but it does not appear in itself to be an economic fact; it can be interpreted biologically or ideologically, but not related to any particular economic system or to the collectivity of systems based on the profit motive.

Both Marxists and their critics have often pointed out that the concept of technical progress as the 'source' of changes in the relations of production is doubtful and misleading. The steam-engine was not created by the stage-coach but by the intellectual labour of its inventors. The improvement of productive forces is obviously the result of mental labour, and to ascribe to it the primacy over the relations of production and, through them, over mental labour is consequently absurd if the words are taken literally. Orthodox Marxists of course reply that technical progress and the intellectual labour that produces it result from the 'requirements' of society, and that the creative mind which devises more perfect instruments is itself an instrument of social situations. But if this were so it would still not mean that the 'primacy' belonged to technical progress; one could speak of a multiplicity of links between intellectual labour and social environment, but this does not entail any specifically Marxist theory of interdependence between the different aspects of social life. In any case, even the idea of society 'requiring' improved technology is of limited application. It is true that modern technical progress is broadly dictated by clear social requirements; but Marx himself points out that there was no incentive to technical progress in pre-capitalist economic forms, because they did not subordinate production to the increase of exchange-value. On what basis, then, do we assume that technical progress 'must' occur and that capitalism was bound to make its appearance? Why could not feudal society have gone on for ever in a condition of technical stagnation? Marxists generally reply: 'Well, capitalism did make its appearance'; but this does not answer the question. If, by saying that it had to appear, they mean simply that it did, this is a misuse of language. If they mean something more, in the shape of 'historical necessity', then the appearance of capitalism does not prove that it was necessary unless, of course, we hold that everything happens because it must—an unproven metaphysical doctrine that anyone is free to hold, but which does not help to explain history in any way whatever.

Considered as a theory explaining all historical change by technical progress and all civilization by the class struggle, Marxism is unsustainable. As a theory of the interdependence of technology, property relationships, and civilization, it is trivial. It would not be trivial if this interdependence could be expressed in quantitative terms, so that the effect of the various forces acting upon social life could be measured. Not only have we no means of doing this, however, but it is impossible to imagine how these forces could be reduced to a single scale. In interpreting past events or predicting the future, we are obliged to fall back on the vague intuitions of common sense.

All this does not mean, however, that Marx's principles of historical investigation are empty or meaningless. On the contrary, he has profoundly affected our understanding of history, and it is hard to deny that without him our researches would be less complete and accurate than they are. It makes an essential difference, for example, whether the history of Christianity is presented as an intellectual struggle about dogmas and interpretations of doctrine, or whether these are regarded as a manifestation of the life of Christian communities subject to all manner of historical contingency and to the social conflicts of successive ages. We may say that although Marx often expressed his ideas in radical and unacceptable formulas, he made a tremendous contribution by altering the whole fashion of historical thought. It is one thing, however, to point out that we cannot understand the history of ideas if we do not consider them as manifestations of the lives of the communities in which they arose, and quite another to say that all the ideas known to history are instruments of class struggle in the Marxian sense. The former statement is universally acknowledged to be true and we therefore think of it as obvious—but it has become so largely as a result of Marx's thought, including his hasty generalizations and extrapolations.

Marx is of course partly to blame, if we may so put it, for the over-simplified and vulgarized notions that can be defended by many quotations from his works. If we believe literally that 'the history of all societies that have existed up to now is the history of class struggles', we can indeed interpret Marxism as contending that every feature of the history of all civilizations in every sphere is an aspect of the class struggle. Whenever Marx himself went into detail, he certainly did not push his hypothesis to such an absurd extreme. He did, however, coin some formulas that gave colour to this simplistic interpretation. It was possible to deduce from these formulas that men were deluded whenever they imagined that they were actuated by anything but the material interests of the class with which they identified, consciously or otherwise; that men never 'really' fought for power or freedom for their own sakes, or for their country as such, but that all these values, aspirations, and ideals were disguises for class-interests. It could likewise be inferred that political bodies did not evolve any interests independent of the classes they 'represented' (despite Marx's observations on bureaucracy), and that if the state appeared to play an autonomous part in social conflicts, it was only (as Marx argued in the case of Louis Bonaparte) the result of a momentary balance of forces in an acute class struggle.

Some contemporary historians and some sociologists, like T. B. Bottomore, suggest that Marxism should be treated not as an all-embracing theory of history but as a method of investigation. Marx himself would not have agreed with this limitation—he regarded his theory as a complete account of world history, past and present—but it is an attempt to rationalize Marxism and strip it of its prophetic and universalist claims. However, the word 'method' also requires qualification.

Historical materialism, conceived loosely enough to be free from the objections we have mentioned, is not a method in the proper sense, i.e. a set of rules that will lead to the same result if applied by anyone to the same material on different occasions. In this sense there is no general method of historical research except, of course, the method of identifying sources. Historical materialism in the sense just defined is too vague and general to be called a method; but it is a valuable heuristic principle, enjoining the student of conflicts and movements of all kinds—political, social, intellectual, religious, and artistic—to relate his observations to material interests, including those derived from the class struggle. A rule of this kind does not mean that everything is 'ultimately' a matter of class-interest; it does not deny the independent role of tradition, ideas, or the struggle for power, the importance of geographical conditions or the biological framework of human existence. It avoids sterile debate on the question of 'determination in the last resort', but it takes seriously Marx's principle that men's spiritual and intellectual life is not self-contained and wholly independent but is also an expression of material interests. If this seems obvious, we repeat, the reason is that Marxism has made it so.

Needless to say, the account we have given considerably limits the validity of Marxism as an instrument for interpreting the past. We now turn to consider it as a means of predicting the future, and here its limitations are perhaps even more serious.

No student can fail to recognize that in Marx's view history as he knew and analysed it derived its meaning not from itself alone but from the future that lay before mankind. We can understand the past only in the light of the new world of human unity to which our society is tending—this is the Young Hegelian point of view, which Marx never abandoned. Marxism, then, cannot be accepted without the vision of the communist future: deprived of that, it is no longer Marxism.

But we should consider on what the prophecy is based. Rosa Luxemburg was the first Marxist to point out that Marx never specified the economic conditions that make the downfall of capitalism inevitable. Even if we accept his view that capitalism will never be able to prevent crises of overproduction, his analysis of such crises and their devastating consequences does not prove that such an uncontrolled system of adjusting production to demand cannot continue indefinitely. Most Marxists have rejected Rosa Luxemburg's theory that capitalism depends for its existence on non-capitalist markets which are ruined by it. But neither poverty, nor uncontrolled production, nor a falling profit rate give ground for believing that capitalism 'must' collapse, still less that the result of its collapse must be a socialist society as described by Marx.

For Marx, it is true, the collapse of capitalism and the communist millennium were 'necessary' in a different sense from that in which capitalism had, as he believed, evolved from feudalism. No one, at the time, had set himself the aim of 'establishing capitalism'. There had been merchants, all of whom wanted to sell dear and to buy cheap. Navigators and discoverers had sailed the seas for adventure or treasures or to enlarge their country's dominions. Later there had been manufacturers out for profit. Every one of these men had been bent on his own interest, but none of them was concerned for 'capitalism', which was the gradual, impersonal outcome of millions of individual efforts and aspirations—an 'objective' process, in which human consciousness was not involved except in a 'mystified' form. But the necessity of socialism, as Marx saw it, was of a different kind. Socialism could only be brought about by men who knew what they were doing; the fulfilment of 'historical necessity' depended on the proletariat being aware of its role in the productive process and its historical mission. In this one privileged case, necessity would take the form of conscious action: the subject and object of historical change were one, and the understanding of society was itself the revolutionary movement of that society.

Although the revolutionary consciousness of the proletariat was identical with the latter's revolutionary movement, it would of necessity arise from the development of capitalist society. The proletariat's historical mission could not be fulfilled unless it were a fully conscious one, in a different way from that of the conquistadors of capitalism; but this consciousness was an inevitable result of the historical process.

Marx was convinced that the proletariat was destined by history to establish a new classless order; but this conviction was not based on any argument. It was not a question of perceiving that the proletariat would go on fighting for its interests against the employers. Awareness of a conflict of interests was not, in Marx's view, the same thing as revolutionary consciousness, which required the conviction that there was a fundamental worldwide opposition between two classes and that this could and must be resolved by a worldwide proletarian revolution. The proletariat was a 'universal' class, not only as the bourgeoisie had been when its aspirations coincided with the needs of 'progress' (whatever this word meant), but also because it restored the universality of the human species; it was appointed to realize the destiny of mankind and terminate 'prehistory' by removing the source of social antagonisms. It was also a universal class in the sense that it would free humanity from ideological mystification, making social relationships transparent to all; it would put an end to the duality that had dominated human affairs from the beginning of time, between the impotent moral consciousness and the uncontrolled, unfathomable course of 'objective' history.

Marx's conviction that the proletariat would evolve a revolutionary consciousness in this sense was not a scientific opinion but an ungrounded prophecy. Having arrived at his theory of the proletariat's historic mission on the basis of philosophical deduction, he later sought empirical evidence for it. The first empirical premiss was his belief that the classes were bound to become increasingly polarized. This at least was capable of verification; it actually proved to be untrue, but even if it had not, it is hard to see how it proved that a worldwide socialist revolution was inevitable. Nor does this conclusion follow from the fact that the working class is the agent of production and is dehumanized to the maximum extent, for in both these respects it is no different from the slaves of antiquity. If it were true that the social degradation of the working class was bound to increase, the prospects for a world socialist revolution, as Marx's critics often pointed out, would not become any the brighter: for how could a class that was kept in a state of ignorance and debility, humiliated, illiterate, and condemned to exhausting labour, find the strength to bring about a universal revolution and restore the lost humanity of mankind? Least of all, as Marx himself maintained, could the proletariat hope for victory because it had justice on its side—at least if past history was to be any guide to the future.

In point of fact Marx did not believe that the proletarian

revolution would be the result of poverty, nor did he ever entertain the idea that an improvement in the workers' condition would affect their 'natural' revolutionary tendency. Later orthodox Marxists did not accept this idea either, though several of them expressed contempt for the 'working-class aristocracy' who, thanks to higher wages and security, came under the ideological influence of the bourgeoisie—which, according to theory, they ought not to have done.

Even if Marx's two premisses that were capable of being tested in practice—that society would approximate more and more to a two-class model, and that the lot of the proletariat could not really improve—had been vindicated by the facts, this would still not have shown that the working class must, by virtue of its position, evolve a revolutionary consciousness; but it would have given ground for thinking that there would be a state of ferment among the proletariat which might lead to the overthrow of the existing property system. Failing the presence of the two factors in question, Marx's prophecy had no firm foundation, which is not to say that it had no social effects. However, the success of political movements that have invoked Marx's doctrine, whether or not they have deformed it in the process, does not prove that the doctrine is true: in the same way, the victory of Christianity in the ancient world, foretold by its own prophets, did not prove the doctrine of the Trinity but, at most, showed that the Christian faith was able to articulate the aspirations of important sections of society. There is no need to demonstrate that Marxism had a powerful effect on the workers' movement, but this does not mean that it is scientifically true. We have no empirical confirmation of Marx's predictions, as there has never been a proletarian revolution of the kind described, brought about by the conditions his theory required— 'contradiction' between productive forces and the relations of production, inability of capitalism to develop technology, etc.

Even assuming, however, that for economic reasons capitalism cannot last indefinitely, it would still not follow that it must be replaced by Marxian socialism. There might instead be a general breakdown of civilization (and the alternative 'socialism or barbarism' suggests that Marx did not always believe in the historic necessity of socialism), or a technologically stagnant form of capitalism, or some other form of society that did not depend on constant technical progress but was not socialist either. Marx's argument that capitalism must collapse because it had lost or would soon lose the capacity for technological improvement involves at least two assumptions: first, that technical progress is bound to continue, and secondly, that the working class is its agent. Both these assumptions are improbable. The first is merely the extrapolation of a historical fact (not a law) that for long periods men have continued to improve the instruments of production; but there is no certainty that they will do this for ever, and there have been times of stagnation and regress. As for the second assumption, in capitalist society the working class is not the exponent of any superior form of technology. The supposition would therefore have to be that socialism will owe its success to a higher degree of labour productivity than is possible under capitalism. This is hardly borne out by the socialist record, nor can it be deduced from anything in capitalism. Altogether, it is difficult to imagine the mechanism of a revolution based on these premisses.

The idea that half a million years of man's life on earth and five thousand years of written history will suddenly culminate in a 'happy ending' is an expression of hope. Those who cherish this hope are not in a better intellectual position than others. Marx's faith in the 'end of prehistory' is not a scientist's theory but the exhortation of a prophet. The social effect of his belief is another question, which we shall examine in due course.

CHAPTER XV

The Dialectic of Nature

1. The scientistic approach

IN the 1860s European intellectual life entered a new phase, as Lothar Meyer, Helmholtz, and Schwann were succeeded by Darwin, Virchow, Herbert Spencer, and T. H. Huxley. The natural sciences appeared to have reached a point at which the unitary conception of the universe was an incontestable fact. The principle of the conservation of energy and the laws governing its transformation were, it appeared, close to providing a complete explanation of the multiplicity of natural phenomena. Studies of the cellular structure of organisms gave promise of the discovery of a single system of laws applying to all basic organic phenomena. The theory of evolution afforded a general historical schema of the development of living creatures, including man with his specifically human attributes. Fechner's studies opened the way to the quantitative measurement of mental phenomena, which had previously been most rebellious to investigation. The day seemed close at hand when the unity of nature, hidden beneath the chaotic wealth of its diversity, would be laid bare to human view. The worship of science was universal; metaphysical speculation seemed condemned to wither away. The methods of the physicists were held to be applicable to all branches of knowledge, including the social sciences.

Engels, who followed the progress of natural science with enthusiasm, shared the hope that a new *mathesis universalis* was about to dawn. A student of Hegel in his youth, he never ceased to admire and respect the great master of dialectics, but he believed that the rational content and value of the latter's speculations would come to light through the development of experimental science, every new stage of which pointed to the dialectical understanding of nature. However, the philosophic interpretation of the new discoveries called for a theoretical examination of the breakdown of former methods, especially the mechanistic viewpoint which had dominated scientific investigation since the seventeenth century but was now an anachronism. From his earliest writings onwards, Engels strove to maintain the strictest possible relation between theoretical concepts and empirical data. This is especially clear in all his works expounding and popularizing the ideas of Marx, who was more concerned with theoretical consistency than with relating his doctrine to the facts of experience.

It is not surprising that Engels was infected by the scientistic enthusiasm of his day, and sought to create an image of the world in which the same basic methods would apply to physical and social science, the latter being a natural prolongation of the former. In his search for unity of method and content, and for concepts linking human history, in Darwinian fashion, with natural history, Engels was close to the positivists of his day. However, he did not propose to find this unity by reducing the whole of knowledge to mechanistic schemata (after the manner of many physicists, such as Gustav Kirchoff), but by discovering dialectical laws appropriate equally to all fields of inquiry. This can be seen in his three most important works written between 1875 and 1886: the *Anti-Dühring, Ludwig Feuerbach*, and *Dialectics of Nature*. The last of these, an unfinished collection of short essays and notes, was begun with a view to controversy with Ludwig Büchner, whose mechanistic materialism was seen by Engels as furnishing the opportunity to formulate a new dialectical materialism; subsequently, however, he went beyond his polemical intention. All three works, unlike those of Marx, deal with questions traditionally regarded as belonging to philosophy, and provide the outline of a doctrinal stereotype which, under the name of dialectical materialism, came to be officially regarded as the 'Marxist ontology and theory of knowledge'. From Plekhanov's time onwards, Marxism was more and more generally defined as a doctrine composed of Engels's philosophical ideas, the economic theory of *Capital*, and the principle of scientific socialism. It has been a matter of dispute for some decades whether these form a consistent whole, and in particular whether Engels's dialectic of nature is in harmony with the philosophical basis of Marx's work.

2. *Materialism and idealism. The twilight of philosophy*

According to Engels—who inherited this view from Leibniz, Fichte, and Feuerbach—the opposition between materialism and idealism is the central question on which philosophy has always turned. In the last analysis it was, in his opinion, a debate concerning the creation of the world. The idealists were those who maintained that spirit (whether a divine creator, or the Hegelian Idea) existed prior to nature, whereas the materialists held the opposite. Berkeleyan subjectivism, according to which being consists in being perceived, falls, of course, on the idealistic side of the division.

Although the history of philosophy is filled with the debate between these two views, they do not occur in identical terms at all periods. There have been times, for example the Christian Middle Ages, when civilization knew nothing of materialism in the strict sense. Yet even in the basic controversies of that time we can detect something akin to materialism in the nominalist view concerning universals, which reveals a certain interest in physical nature and in concreteness. There have also been many doctrines in the history of philosophy which tried to find a compromise or middle way between the two main views, irreconcilable as they are. It is difficult, therefore, to distinguish two main currents expressing the adverse opinions in all their purity and, between them, comprising the whole history of thought. Nevertheless, we always find two conflicting tendencies of which one is closer to the materialist viewpoint or contains more of the elements which usually accompany materialism in its pure form. The fact that idealist or spiritualist tendencies are more frequently met with in philosophy is due, Engels tells us, to the division between physical and intellectual labour, the resulting autonomy of mental pursuits, and the existence of a class of professional ideologists who, in the nature of things, tend to ascribe the primacy to mind rather than matter.

How is the materialistic view to be more closely defined? Since Engels maintains that the essential opposition in philosophy is between nature and spirit, it would seem that both the opposing views express a kind of dualism: so that although the materialists regard mind as genetically secondary to nature, they must also regard it as something separate and different. But Engels does not in fact take this view. He holds that the opposition between nature and spirit is not that of two different substances in a particular genetic relation: consciousness is not a thing in itself, but an attribute of material objects (human bodies) organized in a certain way, or a process which takes place in them. His standpoint is thus a monistic one, rejecting the belief in any form of being that cannot be called material.

But, to know what materialism is, we must first define matter. In some passages Engels appears to take a purely scientistic or phenomenalistic view and to dispense altogether with the category of substance. He says, for instance: 'The materialistic outlook on nature means nothing more than the simple conception of nature just as it is, without alien addition' (*Dialectics of Nature*, 'From the History of Science'); and again: 'Matter as such is a pure creation of thought and an abstraction. We leave out of account the qualitative differences of things when we lump them together as corporeally existing things under the concept 'Matter'. Hence matter as such, as distinct from definite existing pieces of matter, is not anything that exists in the world of sense' (Ibid., 'Forms of Motion of Matter'). From this it would follow that materialism as understood by Engels is not an ontology in the usual sense but an anti-philosophical scientism which sees no need to ask questions about 'substance' and is content with the bare facts of natural science, purged of all speculative additions. From this point of view all philosophy is idealism, an imaginative embellishment of scientific knowledge; and, sure enough, Engels prophesies the decline and extinction of philosophy. 'If we deduce the world schematism not from our minds, but only through our minds from the real world, deducing the basic principles of being from what is, what we need for this purpose is not philosophy but positive knowledge of the world and what goes on in it; and the result of this deduction is not philosophy either, but positive science' (*Anti-Dühring*, I, 3). 'With Hegel philosophy comes to an end: on the one hand because his system sums up its whole development most admirably, and on the other because, even though unconsciously, he showed us the way out of the labyrinth of systems to real positive knowledge of the world' (*Ludwig Feuerbach*, I). 'Modern materialism ... is not a philosophy but a simple conception of the world [*Weltanschauung*] which has to establish its validity and be applied not in a "science of sciences" standing apart, but within the actual sciences. Philosophy is thus "sublated" [*aufgehoben*], that is "both abolished and preserved" —abolished as regards its form, and preserved as regards its real content' (*Anti-Dühring*, I, XIII). 'As soon as each separate science is required to understand clearly its position in the totality of things and of our knowledge of things, there is no longer any need for a special science dealing with this totality. Of all former philosophy, all that now independently survives is the science of thought and its laws, i.e. formal logic and dialectics. Everything else is merged in the positive science of nature and history' (Ibid., Introduction).

Engels thus regards philosophy as either a purely speculative description of the world or an attempt to perceive general connections between phenomena over and above those established by natural science. Philosophy in this sense is to disappear, leaving behind it nothing but a method of ratiocination which has this much in common with 'former philosophy' that it was traditionally considered part of it, though not the most essential part. Engels does not express himself quite unequivocally, but basically his views are in line with the positivism that was widespread in his day: philosophy is a superfluous adjunct to the individual sciences, and there will soon be nothing left of it but the rules of thought, or logic in a broad sense. But there is a different side to this. While, in the passage quoted, Engels speaks of 'dialectics' as meaning simply the laws of thought, he elsewhere uses the term to denote a comprehensive and legitimate system of knowledge of the most general laws of nature, of which our thought processes are a particular exemplification. In this sense, he is a good deal less anti-philosophical than at first appeared. Philosophy, it would seem, is the science of the most general laws of nature; its conclusions derive logically from data furnished by the 'positive' sciences, though they may not have been formulated by any one of those sciences.

Engels's writings alternately support the more ruthless and the more tolerant view of philosophy; but even the latter was in tune with contemporary positivism, which did not wish to abandon philosophy altogether but only to reduce it to what could be deduced from the natural sciences. Either way, materialism is not an ontology but a method prohibiting the addition of speculation to positive knowledge. None the less, Engels uses the term 'matter' to denote either the totality of physical beings or what is left of things when they are stripped of qualitative differentiation. 'The real unity of the world consists in its materiality' (*Anti-Dühring*, I, 4): that is to say, all that *is* is the physical world perceptible by the senses; there is no invisible Nature or behind-the-scenes world different in kind from that observed by the scientist. Engels does not discuss whether materialism in the purely methodical or phenomenalistic sense is identical with the view that the world is a material unity, or whether this view is the same as that of the primacy of matter over mind. He oscillates between scientistic phenomenalism unburdened by metaphysical categories, and a substantialist materialism which holds that there is one true original form of Being whose different manifestations constitute the events of the empirical world. Matter, which is this original Being, is permanently and essentially characterized by 'motion', including change of all kinds; for otherwise the source of change would have to be looked for outside matter, in something like the 'first impulse' (*primum mobile*) of the deists. Motion is the form of matter, uncreated and indestructible as itself.

3. Space and time

In addition to change, matter possesses the inseparable attributes of space and time. The theories which, in Engels's day, set out to explain space and time jointly, i.e. apart from the psychological aspect of time, may be reduced, very broadly, to three types: (1) Space and time are autonomous and independent of physical bodies. Space is the container of bodies, but there could be empty space with the same properties as physical space; time is the container of events, but there could be time in which nothing happens. This is Newton's doctrine. (2) Space and time are subjective, *a priori* forms (Kant): they originate in cognition, but are not derived from experience; they are transcendental conditions of experience, prior to any possible factual knowledge. (3) Space and time are subjective and empirical (Berkeley, Hume); they are ways of ordering experience *ex post*, i.e. combining empirical data so that the mind may operate on them more efficiently. Engels does not share any of these views: he holds that space and time are 'basic forms of being' and are therefore objective (contrary to Hume and Kant), but (contrary to Newton) they are inseparable properties of material bodies and events. Strictly speaking, this implies that there is no such thing as time in itself but only relations of succession (before and after), 'time' being a secondary abstraction from these; similarly, there is no space as such but only relations of distance, direction, and extent. Engels does not say this explicitly, but it appears to be his thought. 'The two forms of the existence of matter are naturally nothing without matter, empty concepts, abstractions which exist only in our minds' (*Dialectics of Nature*, 'Dialectical Logic and the Theory of Knowledge'). The temporal and spatial infinity of the universe are natural consequences of the doctrine that matter is uncreated and indestructible.

4. *The variability of nature*

There exists, therefore, nothing except material bodies in a state of constant change and differentiation. Engels writes, it is true, that 'the world is not to be comprehended as a complex of ready-made things, but as a complex of processes in which things, though apparently no less stable than the concepts which are our images of them, incessantly change, come into being and pass away' (*Ludwig Feuerbach*, IV). But we must not take this literally, on the lines of some modern theories where events are primary and things are momentary condensations of events: for Engels elsewhere defines matter as 'the totality of material things, from which this concept [matter] is abstracted' (*Dialectics of Nature*, 'Dialectical Knowledge and the Theory of Knowledge'). His purpose in describing nature as a complex of processes and not of things is rather to emphasize the eternal changefulness and instability of the material world.

The principle of constant change is a keynote of dialectical thinking. In Engels's view it was Hegel's greatest achievement to point out that every form of Being turns into another, and that only the universe as a whole is exempt from the law of birth, change, and passing-away. Early modern scientists such as Copernicus, Kepler, Descartes, Newton, and Linnaeus were dominated by faith in the immutability of basic natural processes and classifications, whether in the heavens or in the structure of the earth and of organic beings. This view was revolutionized by their successors, starting with Kant's astronomical theory developed by Laplace, Lyell's discoveries in geology, J. R. Mayer and Joule in physics, Dalton in chemistry, Lamarck and Darwin in biology—all of whom demonstrated the perpetual variability of nature and the impossibility of hard-and-fast classification. Every observable fragment of reality proved to be only a phase in its unceasing development; all categories were approximate; more and more intermediate and transitional forms came into view. Man himself was seen to be the product of natural variations, and all his many-sided faculties were the continuation of forces to which Nature herself had given birth. It was labour that distinguished man from the rest of the animal creation and was the source of all his proper pride: manual effort was the cause of mental development. The observation of constant change assures us that man, like the earth and the whole solar system, is sentenced to destruction; but the law according to which matter evolves ever higher forms of existence assures us that these forms in which we participate—conscious reflection and social organization—will reappear somewhere in the universe, and will in due course again cease to be.

5. *Multiple forms of change*

But the dialectic of nature is not only a matter of constant change. The main difference between the mechanistic and the dialectical viewpoints is that the latter distinguishes change in multiple forms. The mechanicism of the seventeenth and eighteeth centuries, transmitted to the nineteenth by the German materialists Vogt, Büchner, and Moleschott, maintained that everything that happens in the world is nothing other than mechanical motion, i.e. the displacement of material particles, and that all qualitative differentiations in nature are subjective or merely apparent. They concluded from this that all branches of knowledge should pattern themselves on mechanics: the processes observed by them would prove to be particular cases of mechanical motion, obedient to the laws governing the movement of bodies in space. Engels was far from accepting this position, even as an ultimate ideal. He believed that the qualitative differentiation of forms of change was a real phenomenon, and that the higher or more complex forms could not be reduced to the lower. The higher, in fact, is defined as that which presupposes the lower but is not presupposed by it. Chemical phenomena, for this reason, are higher than mechanical ones, and those of the organic world are higher still; in the same way, there is an ascent from biology to mental phenomena and social processes. There is thus a multiplicity of forms of change or motion, and a corresponding natural hierarchy of sciences. The forms differ in quality: each of them presupposes all the lower ones, but these do not exhaust it.

However, this all-pervading hierarchy and irreducibility of higher forms is not explained by Engels in a manner free from ambiguity. When he distinguishes the various forms of motion (mechanical processes, molecular movement, chemical, biological, mental, and social phenomena, on the ascending scale indicated by Comte), he does not state clearly in what their irreducibility consists. Is it because the laws of the higher forms cannot be logically deduced from those of the lower (for example, the laws of social history from those of chemistry) or are not logically equivalent to them? Or is it an ontological irreducibility, in that there is something in the 'higher' processes which is not mechanical motion and cannot be causally explained by it? The first interpretation is weaker, for it does not exclude the hypothesis that the higher processes are no more than mechanical ones which occur statistically in a particular way; on the ontological level mechanical motion would then be the only form of change, but science would content itself for observational purposes with statistical laws concerning its manifestation in particular conditions. The second interpretation excludes this hypothesis, but it is not at once clear how the ontological irreducibility comes about, given the starting-point of a homogeneous material substratum of all processes without exception.

Whatever be the answer to this question, it is clear that Engels does not regard nature as uniform in all its changes, or reduce its multiplicity to a single pattern: the manifoldness is real, not merely subjective or due to the temporary insufficiency of our knowledge. Genetically, all higher forms are derivations of the lower (and the history of science to some extent reflects this order), and they are in some sense inherent in them; in other words, matter tends by its very nature to evolve higher forms of Being in the manner observable on earth. Engels does not explain, however, in what way the higher forms are potentially contained in the elementary attributes of matter.

6. *Causality and chance*

The view that the manifoldness of nature is real makes it possible to conceive the problem of causality in a different way from that of mechanistic materialism. The latter in its classical form reduced determinism to the principle that every event is conditioned in every detail by the totality of circumstances at the moment when it happens. If we call something an accident, we can only mean that we do not know its cause; the category of contingency is a subjective one. A perfect intellect, as Laplace suggested, could give a complete and accurate description of the universe at any time, past or future, if it knew the exact mechanical coefficients (position and momentum) of every particle at the present moment or any other. There can be no question of undetermined phenomena or, in particular, free will except as a purely subjective and erroneous sense of freedom. This form of determinism, represented in modern philosophy by Descartes (as regards the material world), Spinoza, and Hobbes, had many adherents among nineteenth-century mechanists.

Engels, however, took a different view. He believed in universal causality in the sense that he rejected the possibility of uncaused phenomena and also the existence of design in nature, conceived as the realization of a conscious intention: this would have been contrary to materialism, as it involved the primacy of mind over matter. But he regarded the general formula of universal determinism as completely sterile from the scientific point of view. If we say that there are five peas and not six in a certain pod, or that a particular dog's tail is five inches long and no more or less, or that a particular flower was fertilized by a particular bee at a certain moment, and so forth, and that all these facts were determined by the state of the particles in the original nebula from which the solar system developed, we are making a statement that is useless to science, and are not so much overcoming the contingency of nature as universalizing it. Explanations like these leave us exactly where we were; they do not enable us to predict anything or improve our knowledge in any way. The business of science is to formulate laws operating in particular spheres so that we can understand phenomena, foresee them, and affect them. Small differentiations are the effect of an infinite number of reactions and may be thought of as accidental; but science is concerned not with them but with the general laws that can be discerned amid the mass of deviations. 'In nature, where chance also seems to reign, there has long ago been demonstrated in each particular field the inherent necessity and regularity that asserts itself through this chance. What is true of nature holds good also for society' (*Origin of the Family*, IX). Engels did not formulate precisely his idea of chance, but his thought seems to be that it is neither an event of whose causes we are ignorant (as the mechanists hold) nor an event with no causes (the indeterminist view). If a phenomenon is contingent, it is so objectively but relatively. Phenomena that form part of a series of events subject to a certain regularity are inevitably disturbed by events belonging to a different type of regularity, i.e. a different form of motion. These disturbances are called accidents, not in themselves but from the point of view of the process to which the former events belong. A cosmic catastrophe which destroyed all life on earth would be accidental in relation to the laws of organic evolution, which do not, so to speak, provide for such an event; but it would not itself be unconditioned. An isolated fact such as the presence of five peas in a pod is the result of many detailed circumstances which we need not and cannot investigate, including the state of the wind, the dampness of the soil, etc. All these combine to produce a particular fact which is therefore not determined by purely botanical laws, for example that a certain seed grows into a pea and not a pine-tree. A general statement that every detail of every process is governed by strict necessity is a mere metaphysical phrase without explanatory value. Science is concerned with laws which operate, of course, in slightly different situations each time, the variations being the effect of chance, but which are nevertheless dependable in spite of deviations and disturbances; it is the laws that matter, not their precise functioning in every separate case.

This being Engels's view of regularity and causality, he approaches the question of freedom in a different manner from the usual one. Freedom does not mean the absence of causation, nor is it a permanent human attribute; it is not a question of suspending the laws of nature, or of enjoying a margin of free play around their edges. With one important modification Engels follows the conception of freedom that arose among the Stoics and reached Hegel through Spinoza: freedom is the understanding of necessity.

Freedom does not consist in the dream of independence of natural laws, but in the knowledge of those laws and in the possibility thus afforded of making them work systematically towards definite ends ... Freedom of the will means nothing but the capacity to make decisions with real knowledge. The freer a man's judgement is in relation to a particular question, so much the greater is the necessity with which the content of his judgement is determined. (*Anti-Dühring*, I, XI)

It appears from this that freedom as the understanding of necessity has a different meaning for Engels than for the Stoics, Spinoza, and Hegel. The free man is not he who understands that what happens must happen, and reconciles himself to it. A man is free to the extent that he understands the laws of the world he is living in and can therefore bring about the changes he desires. Freedom is the degree of power that an individual or a community are able to exercise over the conditions of their own life. It is therefore a state of affairs, not a permanent attribute of man. It presupposes an understanding of the environment and its laws; but it does not consist merely of such understanding, for in addition it requires the individual to affect his environment, or at all events it is only visible when he does so. A man or a community are not free or unfree in themselves, but relatively to their situation and their power over it. There can, of course, never be such a thing as absolute freedom, i.e. unlimited power over all aspects of every situation; but human freedom may increase indefinitely as the laws of nature and social phenomena are better known. In this sense socialism is a 'leap from the realm of necessity into the realm of freedom', wherein society takes control over the conditions of its being and the productive system, which have hitherto run riot and operated against the majority.

Engels thus puts the question of free will in a different way from his predecessors. He does not ask whether a conscious act of choice is always determined by circumstances independent of consciousness, but rather in what conditions human choices are most effective in relation to the end proposed, whether practical or cognitive. Freedom is the degree of effectiveness of conscious acts—not the degree of independence with regard to the laws which govern all phenomena, whether men are conscious of their operation or not; for, according to Engels, such independence does not exist.

7. *The dialectic in nature and in thought*

Dialectics, as understood by Engels, are the study of all forms of motion or activity in nature, in human history, and in thought. Thus there is an objective dialectic which governs nature, and a subjective dialectic which is the reflection of the same laws in the human mind. The term 'dialectic' is used in a double sense, either for the processes of nature and history or for the scientific study of those processes. If we are able to think dialectically, it is because our minds obey the same laws as nature does: 'the dialectic of the mind is only the reflection of the forms of motion of the real world, both of nature and of history' (*Dialectics of Nature*, 'Natural Science and Philosophy'). As this implies, Engels accepted the psychological view of logic in accordance with the naturalist doctrines of his time: i.e. he regarded its laws as facts, empirical regularities of the functioning of the nervous system. Only man, however, is able to think dialectically. Animals can perform operations involving 'reason' in the Hegelian sense, i.e. the elementary abstractions of induction, deduction, analysis, synthesis, and experiment—cracking nuts is the beginning of analysis, and performing animals demonstrate the power to synthesize; but dialectical thinking involves the ability to examine concepts, and this is peculiar to man.

Dialectics—in the sense of thought which perceives phenomena in their development, their internal contradictions, the interpenetration of opposites, and qualitative differentiation—came into being gradually during the ages. We find it in embryo in Greek and Oriental thought, and even in popular sayings like 'extremes meet'; but only German philosophy, and above all Hegel, gave it the form of a complete conceptual system. This, however, had to be reinterpreted in a materialist sense before it could be useful to science. Concepts had to be stripped of their self-generating power and recognized as the reflection of natural phenomena; the method which consisted in dividing ideas into contraries and synthesizing these in a higher unity was thus seen as an image of the laws governing the real world.

The laws of the dialectic may be reduced to three: the transition from quantity to quality and vice versa; the interpenetration of opposites; and the negation of the negation. These are the laws formulated by Hegel and apprehended as governing nature, history, and the human mind.

8. *Quantity and quality*

The law that quantity becomes quality, or, more precisely, that qualitative differences arise from the accumulation of quantitative ones, may be explained as follows. Quantitative differences are those which can be exhaustively characterized by the distance between points on a single scale—temperature, pressure, size, number of elements, etc. Differences that cannot be expressed merely in figures are qualitative. Now it is found throughout the natural world that the increasing or diminishing of the quantitative aspect of a thing leads at a certain point (usually clearly defined) to a qualitative change. The dialectic states, moreover, that qualitative changes are brought about *only* by quantitative increases or decreases. Changes of this kind occur in all fields of reality. A difference in the number of atoms of a given element in a molecule of a chemical compound produces a substance with quite different properties (for example, the series of hydrocarbons, alcohols, acids, etc.). Current of a certain strength causes a filament to become incandescent; bodies change their consistency according to temperature, and melt or freeze at a definite point. Light and sound waves are perceptible to human receptors within certain limits of frequency, and here again the threshold of perception represents a qualitative difference due to a quantitative change. The slowing-up of intracellular motion, and the consequent loss of heat, at a certain point causes a cell to die, which is a qualitative change. A sum of money has to be of a certain size in order to become capital, i.e. to produce surplus value; the co-operation of men at work is not merely a combination, but a multiplication of the strength of each. (Not all these examples come from Engels, but they are in accordance with his thought.) In general, qualitative changes resulting from a quantitative increase or decrease may be seen in all cases where we distinguish between an agglomeration and an integrated whole. Nature and society afford innumerable examples of situations in which the whole is not merely the sum of its parts, but the latter acquire new properties by being part of an integrated system, while the system creates new regularities that cannot be deduced from the laws governing its elements. This concept of the whole became, after Engels's day, an important subject of methodology and an essential category in such forms as *Gestalt* psychology, holism in biology, etc.; it can also be found in Greek thought, for instance Aristotle draws attention to the difference between an integral whole and a combination of elements. But the law of the transformation of quantity into quality sets out to generalize these simple observations into a universal principle. The fact that the structure of organisms depends partly on their size is also a particular case of the law: an animal with the structure of an ant could not be as large as a hippopotamus, and conversely. Even in mathematics, Engels argues, there are qualitative differences, for example roots and powers, the incommensurability of infinitely large or small magnitudes with finite ones, and so on.

The opposition of qualitative and quantitative differences throws a clear light on the contrast between Engels's materialism and that of the mechanists. The latter—for example, Descartes, Hobbes, Locke, and most of the French eighteenth-century materialists—endeavoured to show that qualitative differentiation is not inherent in the world itself but is a feature of our perception, and that the authentic or 'primary' attributes of things are 'geometrical' ones of size, shape and motion; everything else is an illusion caused by our subjective reaction to mechanical stimuli. Engels, on the other hand, reproduces to some extent, of course in a more exact form, the ideas of Francis Bacon, who believed that qualitative differences could not be reduced to quantitative co-ordinates. The law of the transformation of quantity into quality states merely, it would seem, that there are non-additive features in nature and society, or perhaps that there are no purely additive qualities, i.e. none that can be intensified indefinitely without causing new properties or the disappearance of existing ones.

9. Contradictions in the world

The second of Engels's dialectical laws is that of development through contradiction and the interpenetration of opposites. His remarks on this subject are in a more condensed form than the rest of his argument. He observes that 'the two poles of an antithesis, like positive and negative, are just as inseparable from each other as they are opposed, and despite all their opposition they mutually penetrate each other' (*Anti-Dühring*, Introduction). The phenomenon of polarity occurs in magnetism, electricity, mechanics, chemistry, the development of organisms (heredity and adaptation), and social life. It is not a question, however, of merely noting this fact but of arguing that nature contains in itself contradictions, the opposition and interpenetration of which is the source of all development. In Engels's opinion the existence of contradictions in nature is a refutation of formal logic, one of whose primary laws of thought, as they were called, was the principle of non-contradiction. As he writes, 'Motion itself is a contradiction: even simple mechanical change of place can only come about through a body at one and the same moment of time being both in one place and in another place, being in one and the same place and also not in it' (*Anti-Dühring*, I, XII). This is still more evident in more complex phenomena. 'Life consists precisely in this, that a living thing is at each moment itself and yet something else. Life itself, therefore, is a contradiction that is objectively present in things and processes, and is constantly asserting and resolving itself' (ibid.). Even the science of mathematics is full of contradictions. 'It is for example a contradiction that a root of A should be a power of A, and yet A to the power of one-half is the square root of A. It is a contradiction that a negative quantity should be the square of anything, for every negative multiplied by itself gives a positive. . . . And yet the square root of minus 1 is in many cases a necessary result of correct mathematical operations' (ibid.). In the same way, societies develop through the unceasing emergence of contradictions.

Engels was criticized for his view that contradictions are so present in nature that they cannot be described without violating logic, i.e. that logical contradictions are a feature of the universe. The great majority of contemporary Marxists hold that the principle of 'development through contradiction' does not involve rejecting the logical rule of non-contradiction; and they observe that when, following Hegel, Engels spoke of motion being a contradiction he was repeating the paradox of Zeno of Elea, the only difference being that Zeno declared motion to be impossible because contradictory, while Engels declared contradiction to be in the nature of things. Many Marxists now hold that it is possible to speak of contradiction in the sense of conflict or contrary tendencies in nature and society, and of these being the cause of development and the evolution of higher forms, without necessarily rejecting formal logic. There is nothing illogical in the fact that contrary tendencies exist in practice; we are not asked to believe that two propositions contradicting each other are true, but only that nature is a system of tensions and conflict.

10. *The negation of the negation*

Engels's law of the 'negation of the negation' is intended to give a more exact account of the stages of development through contradiction, and *mutatis mutandis* it agrees with Hegel's formulation. The law states that every system has a natural tendency to produce out of itself another system which is its contrary; this 'negation' is negated in its turn so as to produce a system that is in some important respects a repetition of the first, but on a higher level. There is thus an evolution in the form of a spiral: the opposition of the thesis and antithesis is resolved, and they are merged in a synthesis which preserves them in a more perfect form. A seed, for instance, develops into a plant, which is the negation of it; this plant produces not one seed but many, after which it dies; the seeds collectively are the negation of the negation. With insects we have the similar cycle of egg, larva, imago, and eggs in large number. Numbers are negated by the minus sign, which in turn is negated by squaring; it makes no difference that we can arrive at the same number by squaring the positive, 'for the negated negation is so securely entrenched in A^2 that the latter always has two square roots, A and minus A' (*Anti-Dühring*, I, XIII). History develops according to the same rule, from common ownership among primitive peoples to private property in class societies and public ownership under socialism. The negation of the negation consists in restoring the social character of property, not by returning to a primitive society but by creating a higher and more developed system of ownership. In the same way, the primitive materialism of ancient philosophy was negated by idealistic doctrines so as to return in the more perfect form of dialectical materialism. Negation in the dialectical sense is not simply the destruction of the old order, but its destruction in such a way as to preserve the value of what is destroyed and raise it to a higher level. This, however, does not apply to the phenomenon of physical death. Life contains the germ of destruction, but the death of an individual does not lead to his renewal in a higher form.

11. Critique of agnosticism

The basic problem of philosophy has also, as Engels said, its 'other side': the question whether the world is knowable, whether the human mind is capable of forming a true image of relations in independent nature. On this point the new materialism is firmly opposed to all agnostic doctrines such as, in particular, those of Hume and Kant. It rejects the idea that there is any absolute limit to knowledge, or that phenomena are radically different from unknowable 'things in themselves'. According to Engels, the agnostic viewpoint is easy to refute. Science is constantly transforming 'things in themselves' into 'things for us', as when it discovers new chemical substances that existed in nature but were not previously known. The difference is between reality known and unknown, not between the knowable and unknowable. If we are able to apply our hypotheses in practice and use them to foretell events, this confirms that the area under observation has been truly mastered by human knowledge. Practice, experiment, and industry are the best argument against agnostics. It has indeed happened that agnosticism played a useful part in the history of philosophy, as when French scientists of the Enlightenment sought to free their own studies from religious constraint by declaring that metaphysical problems were insoluble and that science was neutral *vis-à-vis* religion. But even this attitude smacked of evading real problems by pretending that they could never be solved.

12. Experience and theory

The prior condition of knowledge is experience. Engels, like J. S. Mill before him, adopts an empirical standpoint even in mathematics, at least as regards the origin of its fundamental notions:

The concepts of number and form have not been derived from any source other than the world of reality . . . Pure mathematics deals with the space forms and quantity relations of the real world—that is, with material which is very real indeed . . . But, as in every department of thought, at a certain stage of development the laws abstracted from the real world become divorced from that world and set over against it as something independent, as laws coming from outside, to which the world has to conform. (*Anti-Dühring*, I, III)

However, Engels's empiricism is a long way from that of most phenomenalists and positivists of his time. He does not hold that knowledge proceeds unidirectionally from raw fact to theory, nor does he regard theoretical generalizations as 'passive' constructions, i.e. as arising from accumulation and induction and exercising no reflex effect on the observation of new facts. Here as elsewhere, there is an interaction between facts and theories. Engels did not expatiate on the problems involved, but the main lines of his thought are clear. He is opposed to what he calls 'bare empiricism', i.e. uncritical belief in facts as, so to speak, interpreting themselves. In 'Natural Science in the Spirit World' (*Dialectics of Nature*) he points out that strict empiricism cannot provide an answer to the beliefs of spiritualists, who appeal to experiment and observation. Theory is essential to the interpretation of facts, and contempt for it is fatal to science. (For this reason Engels called Newton an 'inductive ass'.)

Facts do not interpret themselves, and to perceive their connections we need theoretical instruments which derive, it is true, from observation, but in time become independent elements of knowledge. In the progress of science there is a kind of mutual corroboration between experience and theory, though the former is always genetically prior to the latter. It appears that Engels does not regard the laws of science as merely the logical sum or economical formulation of individual statements of fact but as embodying something further, namely the necessity of the connection which they describe—a necessity which is not inherent in any one fact, nor in all together. There is in nature a 'form of universality'.

All real, exhaustive knowledge consists solely in raising the individual thing in thought from individuality into particularity and from this into universality, in seeking and establishing the infinite in the finite, the eternal in the transitory. The form of universality, however, is the form of self-completeness, hence of infinity; it is the comprehension of the many finites in the infinite. We know that chlorine and hydrogen, within certain limits of temperature and pressure and under the influence of light, combine, with an explosion, to form hydrochloric acid gas; and as soon as we know this, we know that it takes place wherever and whenever these conditions are present, no matter whether the phenomenon occurs once or is repeated a million times, or on however many heavenly bodies. The form of universality in nature is law. (*Dialectics of Nature*, 'Dialectical Logic and the Theory of Knowledge')

The necessity of a law concerning a particular causal connection is not, as Hume would argue, a mere habituation of the mind; it is inherent in the natural connection itself, and we recognize this by the fact that we not only observe the regular sequence of particular events but are able, as a result, to produce the events ourselves.

Engels's remarks on the empirical background of theoretical constructions are rather summary, but their general trend is clear enough. He is a radical empiricist as regards the genesis of knowledge (no valid knowledge is derivable elsewhere than from experience) and a moderate empiricist as far as method is concerned. The social process of knowledge leads to the forging of theoretical instruments, thanks to which we do not submit passively to facts but interpret and combine them. (The second law of thermodynamics, for instance, appeared to Engels an absurdity, as it posited an over-all diminution of energy in the universe.) Science is not merely the concise recording of facts, but the comprehension of something universal and necessary in the world of nature.

13. *The relativity of knowledge*

At the same time, Engels holds, it is impossible either for the whole of our knowledge or for elements of it such as natural laws to attain absolute validity. While accepting the traditional view that truth signifies agreement with reality, Engels follows Hegel in expounding the idea of truth as a process and as something essentially relative.

But in what does this relativity consist? Engels does not hold that the accuracy of a judgement is a matter of time or personality in the sense that it is true or false according to who pronounces it or in what circumstances. His belief in relativity is formulated in different ways. In the first place, knowledge is relative in the obvious sense that it is always incomplete, that man in his finitude cannot discover all the secrets of the universe. A second, more important aspect of relativity is one which applies especially to scientific laws. The way in which science usually advances is that theoretical explanations of observed facts are replaced in course of time by others which do not contradict the former ones but narrow the sphere of their validity. Thus Boyle's and Mariotte's law on the relation between the pressure, volume, and temperature of gases was corrected by Regnault's discovery that it does not apply outside certain limits of temperature and pressure. But we can never be certain that we have discovered once and for all the limits of applicability of a particular law, or that it may not have to be reformulated more precisely in the future. In this sense all scientific laws are relative, or are true only in a relative sense.

Thirdly, we can speak of the relativity of knowledge in the sense that the same collections of facts admit of different theoretical explanations, the scope of these explanations becoming narrower as science progresses, though it never disappears altogether. Fourthly, although there is a difference between a law of nature and a hypothesis (unless we deny the reality of causal connection, in which case every law is hypothetical), yet the basis of scientific generalizations can never be complete, since they comprise an infinity of possible individual facts. Hence any items of knowledge that lay claim to absolute validity must either be commonplaces like 'All men are mortal', or particular facts like 'Napoleon died on 5 May 1821'. Truly absolute knowledge, either in the sense of mentally reproducing the whole universe or of formulating a law of unalterable and final validity, is an unattainable goal to which we can only approximate indefinitely. In so doing, however, we come to possess an increasingly full and accurate picture of reality as a whole.

14. Practice as the criterion of truth

In Engels's view, the truest confirmation of the accuracy of our knowledge is the effectiveness of our actions. If, on the strength of certain information, we set about changing the world in some particular, and if we succeed in doing so, that is the best vindication of our knowledge. Practice, in this sense, is the criterion of truth, and we thus have a reason for eschewing any mental speculation which does not lend itself to practical confirmation. In some passages Engels interprets the notion of 'practice' so broadly as to include the verification of hypotheses in cases where there is no question of our acting upon the external world, for example in astronomical observation. But the importance of practice in cognitive activity is even wider than this. It is not only the best criterion but itself a source of knowledge, inasmuch as real, socially felt needs direct human beings to particular fields of inquiry and determine the range of the questions they seek to answer. In this way, practice supplies the true purpose and social motivation of the quest for knowledge. Thought, in this sense, is practically oriented, which does not mean that it is not 'objective'—i.e. capable of reflecting, subject to its historical and other limitations, the real, factual attributes and relationships of nature itself, independent of the human mind. On the other hand, Engels writes in 'Dialectical Logic and the Theory of Knowledge': 'Natural science, like philosophy, has hitherto entirely neglected the influence of men's activity on their thought; both know only nature on the one hand and thought on the other. But it is precisely the alteration of nature by men, not solely nature as such, which is the most essential and immediate basis of human thought, and it is in so far as man has learnt to change nature that his intelligence has increased.'

From this interesting remark we might suppose that Engels was inclined to regard the content of human knowledge as the result of interaction between man and nature, and not simply a reflection of nature in which practical action plays the part of a touchstone and a determinant of interests. That would mean, however, that the object of our knowledge is not reality itself but man's relations with nature. This is hard to reconcile with the belief that human thought is a more and more perfect reflection of the world as it exists independently of man's cognitive and practical activity. The passage, however, is not so unambiguously expressed as to justify far-reaching inferences, and Engels nowhere developed the idea, nor is it clear exactly what he meant by such terms as 'basis of human thought'. Nevertheless, we find here a hint of a concept significantly different from his opinion that thought is a copy of the real world.

15. The sources of religion

By this dialectical transformation of materialism Engels stands opposed to the whole of idealistic philosophy and to all his materialist predecessors, who did not advance beyond a mechanistic interpretation of the world. This applies to some extent even to Feuerbach: while crediting the latter with a large part in overcoming German idealism, Engels criticizes him for simply rejecting the Hegelian dialectic instead of discovering its rational content. In addition Feuerbach, like all previous materialists, was 'a materialist from below and an idealist from above': that is to say, he was unable to explain human history except in terms of

ideology, in particular the figments of religion, which he regarded
as the mainspring of historical change. Modern materialism is
consistent in this respect too, that it accounts for historical events
by regarding social consciousness as the product of material
conditions of life. It would even seem, though Engels does not
expressly say so, that he regarded historical materialism as the
logical consequence of philosophic materialism. As for religion,
which Feuerbach made the efficient cause of great historical
changes, Engels, in accordance with positivist evolutionism, saw it
as the fruit of human misconception and ignorance.

From the very early times when men, still completely ignorant of the
structure of their own bodies, under the stimulus of dream apparitions
came to believe that their thinking and sensation were not activities of
their bodies, but of a distinct soul which inhabits the body and leaves it at
death—from this time men have been driven to reflect about the relation
between this soul and the outside world. If upon death it took leave of the
body and lived on, there was no occasion to invent yet another distinct
death for it. Thus arose the idea of its immortality . . . Not religious desire
for consolation, but the quandary, arising from universal ignorance, of
what to do with this soul, once its existence had been accepted, after the
death of the body, led to the empty notion of personal immortality. In
exactly the same way the first gods arose through the personification of
natural forces and, as religion developed, assumed more and more an
extra-mundane form. (*Ludwig Feuerbach and the End of Classical German
Philosophy*, II)

Engels, after the fashion of the Enlightenment thinkers, saw
religion as the fruit of ignorance or want of understanding. He thus
abandoned the Marxian view of religion as a secondary alienation
due to the alienation of labour, in favour of an intellectualist
explanation. In this respect he also shared the ideas of nineteenth-
century evolutionism as to the origin and nature of religion.

CHAPTER XVI

Recapitulation and Philosophical Commentary

1. Marx's philosophy and that of Engels

Engels's viewpoint may be described in summary terms as both naturalistic and anti-mechanistic. He presents the universe as evolving dynamically towards higher forms, manifold in differentiation and enriching itself by inner conflict. His version of the dialectic is an anti-philosophical, anti-metaphysical one (though with some inconsistency on this point), which accepts the multiplicity of the universe as irreducible to a single pattern. It is allied to scientism and positivism by its confidence in natural science and mistrust of philosophy as anything but a set of intellectual rules; also by its general empiricist and determinist trend and by a (less definite) inclination to phenomenalism. On the other hand, it diverges from typical positivism by its critique of radical empiricism and its theory of the multiple forms of motion. (Yet even on this point Comte, for whom Marx and Engels expressed such profound contempt, anticipated Engels's views: he refused to reduce all phenomena to mechanistic models, and his classification of sciences was taken over by Engels with little modification.)

It should be added that Engels's evolutionism is apparently related to the separate parts of the universe and not to the universe as a whole. As we read at the end of the Introduction to *Dialectics of Nature*, the universe is infinite and eternal, reproducing the same forms in a never-ending cycle of birth and destruction. Particular fragments of the universe, particular astral systems, by force of internal necessity evolve higher forms of organic life and consciousness; but the universe as a whole does not evolve in this way. We may suppose that, as dwellers on the earth, we are now living in a part of the cosmos which is in a state of upward evolution; but from the point of view of nature as a whole this is a mere passing efflorescence, bound to repeat itself without ceasing in some nook or cranny of the universe.

Engels's observations were of course made in the light of contemporary science and mathematics, and many of them are now out of date. But the general lines of his thought—naturalism, cognition as a reflection of reality, knowledge as relative, the dialectic of nature—were upheld by later Marxists and regarded especially by the Russians (Plekhanov, Lenin) as the Marxist philosophy *par excellence*. At the same time, the dialectic of nature was criticized by some Marxists. The first to attack Engels's philosophy as radically different from Marx's was probably Stanislaw Brzozowski, while Max Adler also spoke of important divergences between the two founding fathers. Subsequently Lukács attacked the dialectic of nature, arguing that the idea that nature itself behaved dialectically was incompatible with Marx's view that the dialectic was an interaction between subject and object, leading ultimately to their unification. According to Marx, nature was not something ready-made and assimilated by man in the process of cognition; it was the counterpart of a practical effort, and was 'given' only in the context of that effort. The evident fact that man transforms nature does not itself invalidate the contemplative theory of knowledge, if praxis is merely the exploitation of natural forces or a criterion for verifying hypotheses. The dialectic, which according to Marx is the unity of theory and practice, cannot be formulated so as to relate to nature in itself as it presupposes the activity of consciousness.

The issue here raised as to whether the founders of scientific socialism saw eye to eye in their epistemological views may, it appears to me, be analysed as follows.

Engels's dialectic was formulated under the influence of Darwin's discoveries and in the intellectual atmosphere of Darwinism. The main trend of opinion, shared by Engels, was to interpret life, knowledge, and social phenomena from the point of view of naturalism, which treats human history as a prolongation and a special case of natural history, and assumes that the general laws of nature apply, in specific forms, to the destiny of mankind. Engels, of course, does not question that human history has special features, nor does he assert that the laws of the animal kingdom suffice to explain human society or can be applied to it without modification; indeed, he expressly rejects such a procedure, holding that evolving nature creates new qualities and that human society is an instance of this differentiation. None the less, writing in *Ludwig Feuerbach* of the difference between the history of the organic world in general and that of mankind, he observes that men, unlike animals, act according to conscious intentions, but that their intentions and acts as a whole conform to the 'objective' regularities of history, which are independent of whether men realize them or not. This last thought is in harmony with many of Marx's statements, but the passage itself is not in accordance with Marx if it means that the conscious character of individual acts, which does not affect the laws governing history as a whole, is the only feature that distinguishes the history of mankind. For it does not appear that the philosophical bases of Marx's Marxism are compatible with belief in general laws of nature having, as particular applications, the history of mankind and also the rules of thought, identified with psychological or physiological regularities of the brain. Whereas Engels, broadly speaking, believed that man could be explained in terms of natural history and the laws of evolution to which he was subject, and which he was capable of knowing in themselves, Marx's view was that nature as we know it is an extension of man, an organ of practical activity. Man, of course, did not create nature and it is not a subjective imagination; but the object of our knowledge is not nature in itself but our contact with it. In other words, when Marx spoke of knowledge having a practical character he did not simply mean that interest is determined by practical needs and hypotheses are confirmed by practical action. Human praxis is the true object of our knowledge, which can never free itself from the practical, situational manner in which it is acquired. We cannot contemplate the subject in itself, free from historical involvement; the *cogito* is an impossibility. But equally the object cannot be purged of the fact that it presents itself to man in the practical context, as a purely human object. Practical contact with nature is the horizon that our knowledge cannot overstep, and in this sense there is no ready-made nature that we can contemplate and then act upon. Nature, as far as we are concerned, is known only in terms of our acts and needs; knowledge cannot be divested of the fact that it is human, social, and historical knowledge. Or again: there is no transcendental viewpoint from which the subject can apprehend natural forms as they are, in order then to duplicate them in his own mind. The materialist interpretation of consciousness according to Marx is that knowledge and everything else in the mind—feelings, desires, imaginations, and ideals—are the product of social life and history. Man, therefore, cannot adopt a cosmic or divine viewpoint, throwing aside his own humanity and comprehending reality as it exists in itself and not as an object of human praxis.

There is thus a clear difference between the latent transcendentalism of Engels's dialectic of nature and the dominant anthropocentrism of Marx's view. The difference is also seen in the significance they respectively attach to Hegel and the Hegelian dialectic. Engels, who extolled Hegel's part in elaborating the conceptual framework of the dialectic, and who regarded the German workers' movement as the sole legitimate heir of classic German philosophy, saw it as the great merit of Hegelianism to have emphasized the transience of all forms of social existence. He criticized Hegel, on the other hand, for the non-dialectical conception of nature as repeating its cycle of evolution without a break, and especially, following the radical Young Hegelians of the 1840s, for the 'contradiction between system and method'. By this they meant that the dialectic speaks of ceaseless development and negation, so that no form of Being or society can be final, and the Absolute is always out of reach; yet Hegel represents certain forms of religion, philosophy, and the state as final and unimprovable, and thus sins against his own method.

But the supposed conflict between method and system cannot be resolved by recognizing the transience of all forms and the impossibility of final ones. Hegel's thought is not comprehensible without its fulfilment in the Absolute, and negativity as the Young Hegelians understood it is no longer Hegelian. The whole burden of Hegel's criticism of Kant and Fichte, especially of the 'bad infinity' or the notion of unending growth, was that any phase of development can only be understood in relation to a final state, without which so-called 'progress' is merely eternal repetition. Only an Absolute which is actually attainable, not somewhere dimly visible on the horizon, can provide the reference system that gives meaning to any stage of spiritual evolution. The notion that it is possible to salvage from

the Hegelian dialectic the idea of eternal progress while jettisoning the conservative idea of an ultimate goal is analogous to a
philosophy which, confronted with the contradiction between
God's omnipotence and man's free will, should abolish God and
claim that it had preserved the genuine essence of Christianity,
namely atheism. The contradiction, or tension, is itself the essence
of Christianity, and to remove one of its terms is not to
accommodate Christianity to critical thought but simply to
destroy it. In the same way, the notion of infinite progress without
that of the final unification of Being is not a critical absorption of
Hegelianism but a denial of it, and the first term of the
'contradiction' is not even specifically Hegelian: it comes from
Kant and Fichte, and if it is to be the kernel of dialectical thought,
such a dialectic has no need of the Hegelian tradition.

Marx's assimilation of Hegel, however, is not based on keeping
the method while rejecting the system, but on 'standing Hegel on
his feet instead of his head', which is a different thing. Marx in his
own fashion took over from Kant and Hegel the idea of history
culminating in the complete unity of man, the identification of
existence with essence and the abolition of contingency in human
life. Man, according to Marx, is not doomed to contingency, as
Stirner maintained (and as do modern existentialists, at least of
the atheistic kind); on the contrary, what has hitherto been contingent, though miscalled freedom, derives from the power of
objectified forces over man. To remove these forces and subject
man's existence to his own freedom, abolishing the difference
between empirical Being and species-essence, is to destroy the
contingency of existence. Man is no longer at the mercy of
alienated forces of his own creation; the individual is not a victim
of anonymous society, nor the owner of its objectified labour in the
form of capital; in short, man's Absolute Being is fully realized in
actual being. The latter, in consequence, ceases to be accidental;
its individuality expresses the universal essence of humanity, and
its freedom is historical necessity. Man's fundamental disunity can
thus be overcome, but not in the way Hegel suggested. Hegel,
having reduced man and his works to self-consciousness and the
externalization thereof, and regarding humanity as a stage in the
evolution of spirit, could not, on the basis of his own method,
reconstruct man as an integral being. Man's contingency cannot
be healed by an Absolute outside himself, and accordingly Hegel

does not cure the contingency of individual life, or else he does so only in the context of that life: in effect, he condemns empirical human individuality to a state of contingency throughout its existence, as may be seen in the permanent dichotomy between the state and civil society in Hegel's philosophy of law. To do away with contingency it is necessary, firstly, to take man as a complete physical being, working and contending with nature, and secondly, to comprehend that man's only reality consists in his being an individual—any other form of existence is the effect of the alienation of labour, an aberration of fortune which, however, is historically inevitable and is the condition of his liberation. Only when Hegelianism has thus been transformed in the sense of materialism (consciousness as a component of the complete man and an effect of practical activity) and individualism (the individual as the only subject, all other modes of existence being predicates of the actual man)—only then is it possible to look forward to man's true unification as predicted in the Paris Manuscripts and in *Capital*. Hegel has been 'stood on his feet'—the individual is the subject and universal Being is the predicate, instead of the other way round, and the starting-point of historical development is not the externalization of consciousness but that of natural human forces in the form of labour.

Marx, therefore, does not take over Hegel's method without his system, but transforms both together. In the new schema we still have the prospect of a kind of ultimate goal, namely what Marx calls the end of the past and the beginning of true history. It is a consummation in the sense that it puts an end once and for all to the historic duality between the individual and reified social Being, between self-objectification in labour and the alienation of its products. The healing of disunity and return to full integration are no less essential to Marx's doctrine than to Hegel's, although the disunity and therefore the return are conceived in different terms. The finality of the socialist transformation, as we have seen, does not imply the cessation of development but the extinction of all conflict between man's empirical life and his nature—the removal of all obstacles that alienated labour and the contingency of life opposed to the true, creative objectification of man's natural powers.

In a similar way to that in which Marx differs from Hegel, his idea that 'philosophy will be abolished by being realized' is different from the scientistic belief that philosophy will be superseded by the positive sciences. In Marx's view the abolition of philosophy is a natural element in the reintegration of man, as it consists in depriving thought processes of their autonomy *vis-à-vis* life as a whole. Thought becomes a direct affirmation of life, aware that it is itself conscious life and nothing else; the division between physical and intellectual work is done away with; thought can no longer withdraw to an 'independent' realm of its own; philosophy, which is the mind's aspiration towards the integrality of man, will disappear when that aspiration is realized. This is quite different from the view that philosophy will no longer have the right to a separate existence, and that anything worth while in it will be taken over by the various positive sciences.

The difference between these two interpretations of the human condition is clear, as is the sharpness of the division, in Marx's view, between the present world and that which is to come. On his own philosophical premisses he could never have made any concession to reformist strategy; the new society had to make a complete break with the old, and a revolutionary upheaval was the only valid form of social criticism. By contrast, on the assumption that progress continues throughout history but never reaches an absolute goal, it is easier to understand the view that reform within the framework of capitalism may be of value in itself.

To sum up the difference in the attitudes of Marx and Engels we may say that they exhibit a contrast, firstly, between naturalistic evolutionism and anthropocentrism; secondly, between the technical interpretation of knowledge and the epistemology of praxis; thirdly, between the idea of the 'twilight of philosophy' and that of its merging into life as a whole; and, fourthly, between infinite progression and revolutionary eschatology. Many critics have taken the view that Marx never uses the term 'materialism' in the same sense as Engels, as he always means by it the dependence of consciousness on social conditions and not the metaphysical primacy of matter over mind. Some, such as Z. Jordan, even argue that Marx had a much better title that Engels to be called a positivist, as he rejected any kind of 'substantialist' metaphysics. To some extent this is a question of terminology: Marx was certainly not a positivist in the historical sense of the word, as he did not share the phenomenalist theory of knowledge or the prohibition against looking for an 'essence' behind phenomena, and indeed often expressed himself to the contrary. It is true, however, that, unlike Engels, he did not concern himself with metaphysical questions about the primal substance and the origin of the world. In his early writings he expressly rejected metaphysical questioning, and it is of course one thing to do this and another to answer the question negatively. Certainly Marx is a 'materialist' in the broad sense of one who does not believe in spirit existing prior to matter, or who rejects the question of such existence as meaningless. As a rule, however, the term is used to denote a 'substantialist' belief in 'matter' as the substratum of all that can meaningfully be said to exist; or, more precisely, the belief that all objects have the properties that scientific and everyday experience ascribe to physical bodies. It is hard to call Marx a materialist in this sense, and Engels himself, as we saw, varies between scientistic phenomenalism (which is not a metaphysical doctrine but an intellectual rule) and true materialism, which goes beyond the reach of scientific rigour and, according to how it is formulated, is either obscure or unprovable.

A point of view represented especially by Catholic critics of Marxism, and recently defended also by L. Coletti, is that materialism is incompatible with Engels's dialectic of nature because the latter predicates the existence in nature of qualities, such as creativity, that belong only to spiritual beings. This criticism, however, is open to objection. The idea that nature can evolve forms that are qualitatively new (in the sense we have considered), and that some parts of nature obey laws that cannot be deduced from the universal laws of physics, does not involve any logical contradiction with materialism in the above sense. At all events, the theory of a multiplicity of irreducible qualities does not in itself conflict with materialism. There is, however, another way in which, materialism may be found incompatible with the dialectic. Engels clearly expresses the view that logical contradiction is a property of certain natural phenomena. Now the statement that a certain logical relationship occurs in nature may be reconciled with the philosophy of Hegel, Leibniz, or Spinoza (in the latter's case, with the proposition that *cogitatio* is an attribute of the whole universe), but none of these is compatible with Engels's form of materialism. If we interpret 'contradiction' and 'negation' in the non-logical sense of conflict or destruction, this point falls to the ground. It would seem, however, that Engels's casual identification of logical relationships with physical ones is due to his inadequate training in philosophy rather than to a deliberate theory. With all his wide knowledge and agility of mind, Engels was an amateur in philosophy. His critique of Kant's 'agnosticism' is astonishingly naïve: he shows complete misunderstanding by arguing that according to Kant new chemical substances could never be discovered because, if they were, a 'thing in itself' would become an object of Knowledge. It is also not clear how Engels could reconcile his psychologistic interpretation of logic (which is expressed in summary form and does not go beyond the commonplace views of his time) with his belief that human knowledge is a reflection of nature as it 'really' and independently exists. For, if the laws of thought are not obligatory rules independent of experience and of the existence of things, but are merely the way in which the human brain works and are thus particular cases of some general law of nature, the question whether knowledge is 'true' in the traditional sense has no meaning: cognitive activity would be a form of biological reaction and nothing more, and could be evaluated only from the point of view of its usefulness.

Despite Engels's inconsistencies and reckless generalizations, can the 'dialectic of nature' in some sense preserve its validity? Those Marxist critics who dispute the possibility have pointed out that in Marx's usage 'dialectic' refers to an interplay between the mind and its social environment: this cannot be transferred to nature, or constitute a set of universal laws whereof the laws of social life are only a manifestation. If it were so, the development of society and above all its revolutionary transformation would be the effect of 'natural laws', which is the opposite of Marx's view. However, if this criticism be accepted, it does not follow that orthodox Marxism forbids consideration of the irreducibility of various natural processes to a single model; all that is shown is that the term 'dialectic' does not apply to such consideration in the way that it does to social phenomena. With this reservation there seems to be no reason why Engels's speculations should be condemned out of hand, though it is a further question how far and in what sense they are true in detail. His ideas of logical contradictions in nature, or the dialectic in arithmetic, are certainly naïve; but the question of the multiplicity of qualities is not, nor does it seem improper to speak of an accumulation of quantitative changes leading to qualitative ones (in the sense suggested above, i.e. that most or even all the parameters by which natural phenomena are described are not indefinitely additive).

Certainly Engels's 'dialectic of nature' is full of obsolete examples and unfounded speculation in the realm of philosophic cosmology. He maintains that the emergence, in the history of the earth, of higher forms from lower represents an immanent necessity and that nature 'must', by virtue of some unknown law, produce the same forms in similar conditions. Although Engels himself at other times condemned this kind of arbitrary speculation, at all events in general terms, it belongs to a traditional philosophy of nature that was fairly widespread in the nineteenth century. This is not to say, however, that Engels's philosophy made any contribution to the development of science. As historians of science point out, there have in the past been moments of crisis when philosophical ideas played an important part in this way, for example the influence of Platonism on Galilean physics or of empiriocriticism on the theory of relativity. But no such heuristic role can be ascribed to Marx's or Engels's philosophy of nature; and its effect in the Soviet Union has been to stifle sciences, not bring them to birth. It may even be said that Engels is not wholly innocent in this respect: on the one hand, he emphasized that philosophic generalizations are valueless if they are not based on scientific experience, but on the other, in his critique of empiricism, he ascribed to philosophy a supervisory role *vis-à-vis* 'plain experience'. He failed to explain clearly how these principles could be reconciled or on what basis philosophy was entitled to criticize experience. The idea that it is so entitled could easily furnish a pretext for the subjection of science to ideology, which has in fact happened—of course, in political circumstances that have nothing to do with this part of Engels's doctrine.

Questions connected with the dialectic of nature constitute a much-publicized part of what is now codified as 'dialectical materialism'. How far they are scientifically and philosophically fruitful at the present time is a topic to be considered later.

2. *Three motifs in Marxism*

As with all great thinkers we can perceive in Marx's own

doctrine, considered as a whole, a degree of tension between heterogeneous strains of thought, and also between the sources which he wrought into a synthesis. From this point of view we may distinguish three principal motifs.

(1) The Romantic motif. In the main lines of his criticism of capitalist society, Marx is an heir to the Romantic movement. The Romantics attacked industrial society from a conservative point of view, deploring the loss of 'organic' ties and loyalties and the fact that human beings confronted one another not as individuals but as representatives of impersonal forces and institutions or the money power. On the one hand, personality was lost in anonymity and men tended to treat one another as embodiments of their social function or the wealth they possessed. On the other hand, genuine collective life disappeared as well: there were no true communities of the traditional kind, moral entities united not only by interest but by spontaneous solidarity and direct contact between individuals. The opposition between such organic communities and 'society' as a mechanical aggregate held in balance by nothing but the negative bond of interest is a theme that runs through pre-Romantic and Romantic philosophy, from Rousseau and Fichte to Comte. The dream of a return to perfect harmony and to a state in which no middle term intervened between the individual and the community, or the individual and himself, was an attack, expressed or implied, on liberalism and its theoretical basis in the social contract. Liberal philosophy assumes that men's conduct is necessarily governed by selfish motives and that their conflicting interests can only be reconciled by a rational system of laws which safeguards the security of all by limiting the freedom of each. This implies that men are one another's natural enemies, each one's freedom being the limit of everyone else's. Unlimited freedom would be self-destructive, for if no one agreed to respect the rights of others, all would be exposed to aggression and none would be safe; the social contract, in Hobbes's sense, prevents this by organizing the community on the basis of men respecting one another's freedom. Society is thus an artificial creation, a system of legislation to restrain natural egoism and provide security for all at the price of a partial relinquishment of freedom. In the Romantics' view this was indeed a true picture of industrial society, but it did not answer to the requirements of human nature. Man's natural destiny was to live

in a community based not on the negative bond of interest but on the independent, spontaneous need to communicate with others. Coercion and control would not be necessary in a society in which each individual freely identified with the whole.

Marx adopted the destructive part of the Romantics' view of contemporary society: witness his theory of alienation and of the power of money, and his belief in a future unity in which the individual would treat his own forces directly as social forces. The aspects of society that he attacked were the same as those whose devastating consequences had been noted by the Romantics: men were dominated by their own energies and skills in the form of the anonymous laws of the market, the abstract tyranny of money, and the ruthless process of capitalist accumulation. To Marx, as to the Romantics, the freedom contained in the Declaration of the Rights of Man, which allowed the individual to do what he liked so long as he did no harm to others, was the hallmark of a society dominated by the negative bond of self-interest.

Not only this, but the main features of the communist Utopia are also borrowed from the Romantics. Marx's basic principle is that all mediation between the individual and mankind will cease to exist. This applies to all constructions, rational or irrational, that interpose themselves between the individual and his fellows, such as nationality, the state, and law. The individual will voluntarily identify himself with the community, coercion will become unnecessary, the sources of conflict will disappear. The removal of mediating forms does not mean the destruction of individuality—on the contrary. As in the Romantic view, the restoration of organic links will at the same time restore the authenticity of personal life. As things are now, the individual wrenched from the community and enslaved to anonymous institutions is robbed of his personal life and obliged to treat himself as a mere object. The worker sees his whole effort as a means of biological survival, while the creative part of his work becomes alien to him; his personal qualities and abilities take on the form of a commodity bought and sold on the market like any other. The capitalist loses his own personality in a different but equally pernicious way: as the personification of money he is not master of his behaviour but must do as the market tells him, irrespective of whether his intentions are good or bad. On both sides of the gulf personality is extinguished as individuals turn into servants of alienated forces. The abolition of capitalism does not mean exalting the community at the expense of the individual, but restoring both at once. Instead of freedom being conceived in the liberal fashion as the private sphere of non-interference with others, it becomes the voluntary unity of the individual with his fellow men.

But the agreement between Marxism and Romanticism is only partial. Romanticism in its classic form is a dream of attaining social unity by reviving some idealized feature of the past: the spiritual harmony of the Middle Ages, a rural Arcadia or the happy life of the savage, ignorant of laws and industry and contentedly identifying with the tribe. This kind of nostalgia is, of course, the reverse of Marx's viewpoint. Although he shows traces of the Romantic belief in the felicity of the savage, these are not numerous or important and there is no suggestion in his work that mankind could or should revert to a primitive life-style. Unity will be recovered not by destroying modern technology or invoking primitivism and rural idiocy, but by further technical progress and by obliging society to put forward its utmost efforts to perfect its control over natural forces. It is not by retreating into the past but by strengthening man's power over nature that we can salvage what was of value in primitive society: the process is a kind of spiral, involving the maximum negativity of the present system. The destructive effects of the machine cannot be cured by abolishing machines, but only by perfecting them. Technology itself, by its negative aspects as it were, makes it possible to revive what it destroyed.

Because future unity will be obtained not by jettisoning the achievements of social development but by continuing it, that unity will reside in the human species as a whole and not in traditional forms like the nation or the village. The national community, which so many Romantics regard as the paradigm of organic life, is already being dissolved by the progress of capitalism, which sweeps away everything that does not serve its own expansion. The workers have no fatherland, and neither has capital: on both sides of the great conflict of the age, patriotism has lost its relevance. Nationalism may be exploited for political or other short-term ends or to justify protectionist policies, but its strength is collapsing under the remorseless pressure of cosmopolitan capital and the internationalist consciousness of the proletariat. From this point of view also capital, the destroyer of tradition, is clearing the way for the new society.

(2) If Marx parted company with the Romantics in this important feature of his Utopia, it was because of what may be called the Faustian–Promethean motif—a strong influence and, in some ways at least, a rival to Romanticism. It is hard to refer this motif to any particular school of thought: it appears in a wide variety of philosophies, including some strands of neo-Platonism (man as the head of created Being) and in texts of Lucretius and Goethe that were well known to Marx. We find it in Giordano Bruno and other Renaissance writers whom Marx regarded as models of fulfilled humanity, universal giants who had overcome the penury of the division of labour and had not only assimilated the entire culture of their day but had raised it to a higher level by their own efforts. This strain in Marx's thinking appears clearly from the answers he gave to his daughters' 'questionnaire'—favourite poets, Shakespeare, Aeschylus, Goethe; favourite heroes, Spartacus, Kepler; idea of happiness, fighting; most hated quality, servility. The Promethean idea which recurs constantly in Marx's work is that of faith in man's unlimited powers as self-creator, contempt for tradition and worship of the past, history as man's self-realization through labour, and the belief that the man of tomorrow will derive his 'poetry' from the future.

Marx's Prometheanism is of course of a special kind, and above all it relates to the species and not the individual. Marx believed, as he made clear in his defence of Ricardo against the sentimental critique of Sismondi, that the idea of 'production for production's sake' meant developing the riches of human nature as an end in itself, and that the progress of the species must not be held up by considerations of individual happiness. Even if the development of the species took place at the expense of a majority of individuals, it was in the end synonymous with the development of every individual; the progress of the whole always involves detriment to some, and the callousness ascribed to Ricardo was a proof of his scientific honesty.

Marx was certain that the proletariat as the collective Prometheus would, in the universal revolution, sweep away the age-long contradiction between the interest of the individual and that of the species. In this way, too, capitalism was the harbinger of socialism. By smashing the power of tradition, brutally rousing nations from their slumbers, revolutionizing production, and liberating fresh human forces, capitalism had made a civilization in which man for the first time was able to show what he could do, although as yet his prowess took non-human and anti-human forms. It was pitifully sentimental to upbraid capitalism in the hope of stopping or diverting its victorious advance. The conquest of nature must go forward; in the next stage, men would achieve mastery over the social conditions of progress.

A typical feature of Marx's Prometheanism is his lack of interest in the natural (as opposed to economic) conditions of human existence, the absence of corporal human existence in his vision of the world. Man is wholly defined in purely social terms; the physical limitations of his being are scarcely noticed. Marxism takes little or no account of the fact that people are born and die, that they are men or women, young or old, healthy or sick; that they are genetically unequal, and that all these circumstances affect social development irrespective of the class division, and set bounds to human plans for perfecting the world. Marx did not believe in the essential finitude and limitation of man, or the obstacles to his creativity. Evil and suffering, in his eyes, had no meaning except as instruments of liberation; they were purely social facts, not an essential part of the human condition.

In the 1844 Manuscripts, it is true, Marx refers to sexual relations, i.e. presumably a biological tie, as the paradigm of the truly human links which are apparently to be dominant in communist society. But the parallel is at once explained in a contrary sense from what we expect. It is not that the biological tie is a model for the social one, but that it has taken on a social character: man discovers in sexual relations to what extent his nature has been 'humanized', i.e. socialized—in what way his biology has become human, and his biological needs have become social needs. Contrary to social Darwinism and to liberal philosophy, Marx not only does not derive the social tie from biological needs, but represents the latter, and the biological conditions of human existence, as elements of the social tie. 'Socialized nature' is not a metaphor. Everything in man's being is social: all his natural qualities, functions, and behaviour have become virtually divorced from their animal origins.

For this reason Marx can scarcely admit that man is limited

either by his body or by geographical conditions. As his argument with Malthus showed, he refused to believe in the possibility of absolute overpopulation, as determined by the earth's area and its natural resources. Overpopulation was a purely social fact relating to the conditions of capitalist production, as technical progress and exploitation caused relative overpopulation in the shape of a reserve army of workers. Demography was not an independent force but an element in the social structure, to be evaluated accordingly.

Marx's ignoring of the body and physical death, sex and aggression, geography and human fertility—all of which he turns into purely social realities—is one of the most characteristic yet most neglected features of his Utopia. Among other things, it means that the popular analogy between Marx's soteriology and that of Christianity (the proletariat as redeemer, total salvation, the chosen people, the Church, etc.) is erroneous in one crucial respect. Salvation, for Marx, is man's salvation of himself; not the work of God or Nature, but that of a collective Prometheus who, in principle, is capable of achieving absolute command over the world he lives in. In this sense man's freedom is his creativity, the march of a conqueror overcoming both nature and himself.

(3) But Prometheanism too has its limits, at all events in the interpretation of the past: it is rivalled by the third motif, that of the rationalist, determinist Enlightenment. Marx often speaks of the laws of social life, operating in the same way as the laws of nature. By this he does not mean that they are a continuation of the laws of physics or biology, but that they impose themselves on human individuals with the same inexorable necessity as an avalanche or a typhoon. It is for objective scientific thought to study these laws as a naturalist does, without preconceived dogma, sentiment, or value-judgements, as Marx considered himself to have done in *Capital*. The normative concepts of alienation and dehumanization thus present themselves as the neutralized, non-evaluating concepts of exchange-value, surplus value, abstract labour, and the sale of labour-power. In the questionnaire we have already quoted, Marx's rationalism and philosophical scepticism appear in his favourite maxim, *De omnibus dubitandum*.

In this scientist approach we recognize the third conception of freedom, as formulated by Engels: freedom is the understanding of necessity, the extent to which men are able to turn the laws of nature to their own use, the level of material and social technique.

Here too, however, there is a reservation to be made. Belief in the 'laws' governing society is grounded in the interpretation of history up to the present, the 'prehistory' of mankind. Up to now men have been governed by forces they have created but do not control—currency, the market, religious myths. The gulf between the tyranny of economic laws and the mind which impotently observes them is closed by the appearance of the proletariat, conscious of its mission. From then on necessity is not imposed from without, and does not consist in the technical utilization of existing laws by enlightened social engineers. The very difference between necessity and freedom ceases to exist, and so presumably do 'social laws' as hitherto understood, i.e. as something like the law of gravitation. The latter, however, while it can be known and put to use, cannot be abolished, and operates whether we know about it or not. The term 'law' in this sense cannot apply to a social process that occurs only on condition of being understood, and this is precisely the case with revolutionary praxis. There is a crucial distinction here: the laws that have governed society up to now were independent of human knowledge; the fact that they are now known does not mean that they cease to operate. But the revolutionary movement of the proletariat is not the exemplification of a law in this sense, for although it is caused by history it is also the awareness of history.

While, then, the Romantic side of Marxism applies equally to the past and to the future (criticizing the dehumanization of man by capitalism, and looking forward to a state of unity), and while Prometheanism looks primarily to the future (for, although man has at all times been his own creator, he was not and could not have been aware of this fact), the deterministic aspect of Marxism relates to the past which still weighs upon mankind, though it is destined soon to be thrown off altogether.

The whole of Marx's thought can be interpreted in terms of these three motifs and their interrelation. They do not coincide, however, with the conventional 'sources' of Marxism. The Romantic strain derives variously from Saint-Simon, Hess, and Hegel; the Promethean from Goethe, Hegel, and the Young

Hegelian philosophy of praxis and self-knowledge (man as creator of himself); the determinist and rationalist from Ricardo, Comte (whom Marx derided), and again from Hegel. All three motifs are influenced by Hegelian thought, but in all of them it is transformed from Hegel's intention.

The three motifs are present uninterruptedly in Marx's work, but their respective strength varies at different times. There can be no doubt that Marx laid more weight on the purely scientific, objective, deterministic aspect of his observations in the sixties than in the forties. The other two strains did not lose their force, however, but continued to affect the direction of his work, the concepts he used, the questions he put and the answers he gave to them—even though, as often happens in such cases, he was not fully aware of their constant influence.

Marx was convinced that he had synthesized all the intellectual values available to him in a single picture. In the light of his own conception of his work, such questions as whether he was a determinist or a voluntarist, or whether he believed in historical laws or the power of human initiative, have no meaning. Ever since, as a student at Berlin, he came to the conclusion that with Hegel's help he had overcome the Kantian dualism between what is and what ought to be, he was in an intellectual position which enabled him to reject such questions.

3. *Marxism as a source of Leninism*

All these considerations, however, belong to the domain of social philosophy, and it was difficult to derive a precise political strategy from them at a time when there was already a strong movement in being which professed Marxism as its ideology. The philosophy required interpretation and specification, and this brought to light tensions and contradictions within Marxism which had not been noticeable on the plane of general soteriology and eschatology. The debate between necessity and freedom could be resolved in theory, but at a certain point it had to be decided whether the revolutionary movement must wait for capitalism to mature economically or whether it should seize power as soon as the political situation permitted. General principles were of little use in resolving this question. Marxism promised that society would become one and that all barriers between the individual and society would be removed; the next step was to draw practical conclusions and translate the promise into the language of a political programme. It was also necessary to define more clearly the idea of civilization as conditioned by class and at the same time universal. What exactly was meant by the state 'withering away', and how was this to be brought about in practice? Those who relied on the gradual and automatic development of capitalism into communism, and those who stressed the creative historical role of revolutionary initiative, could both find support in Marxist writings. The former accused the latter of seeking to violate the laws of history as laid down by Marx; the latter retorted that the former expected the impersonal process of history to make their revolution for them, which might mean waiting till the end of the world. Marx was quoted on either side of the argument, but, taken together, the quotations did not prove much, and, as usually happens, they were used to buttress positions adopted for other reasons.

Still more troublesome was the practical interpretation of all Marx's prophecies concerning the nature of communism. It was possible to argue as follows: according to Marx, all social antagonisms were based on class conflicts. When private ownership of the means of production was abolished, there would be no more classes and no social conflict except that due to the lingering resistance of the possessing classes. Marx envisaged that there would be no 'mediacy' in socialist society: this meant, in practical terms, the abolition of the liberal bourgeois separation of powers and the unification of the legislature, executive, and judiciary. Marx also envisaged the disappearance of the 'national principle': so any tendency to cultivate national separateness and national culture must be a survival of capitalism. Marx had declared that the state and civil society would become identical. Since the existing civil society was a bourgeois one, the simplest way to interpret this was by the complete absorption of civil society into the new state, which was by definition a working-class state ruled by the party that professed Marxism, the proletarian ideology. Marx had said that the negative freedom of the liberal bourgeois tradition would have no place in socialist society, as it only expressed the antagonistic character of society. The building of the new world could thus begin by substituting for negative freedom a higher form of freedom based on the unity of the individual and society. As, by definition, the proletariat's aspirations were embodied in the proletarian state, those who failed in any way to conform to the new unity deserved destruction as survivals of bourgeois society. What else, after all, was the meaning of the principle that human progress is always at the expense of the individual, and that this cannot be otherwise until absolute communism is achieved?

By arguing on these lines, the whole Marxist–Romantic theory of unity, classes, and the class struggle could be used (which does not mean that this was historically inevitable) to justify the establishment of an extreme despotism which professedly embodied the maximum possible freedom. For if, as Engels taught, the freest society is that which has most control over the conditions of its life, it is not a gross distortion of the theory to infer that society will be free in proportion as it is governed more despotically and subjected to more numerous regulations. Since, according to Marx, socialism deposes objective economic laws and enables men to control the conditions of their lives, it is easy to infer that a socialist society can do anything it likes—i.e. that the people's will, or the will of the revolutionary party, can ignore economic laws and, by its own creative initiative, manipulate the elements of economic life in any way it pleases. Marx's dream of unity could thus take the form of a despotic party oligarchy, while his Prometheanism would appear in the attempt to organize economic life by police methods, as Lenin's party did at the outset of its rule. Economic voluntarism, which was only abandoned when the new society was on the brink of ruin, was an application, and not too much a caricature, of Marxian Prometheanism—Chinese communism went through a very similar period, inspired by the same ideology and no less catastrophic in its results. Under socialism economic failure can only be seen as due to the ill-will of the governed, which in turn must be an effect of resistance by the possessing classes. The rulers had no need to seek the reasons for failure in doctrinal errors: as true Marxists they could blame them on the bourgeoisie and intensify repressive measures against the latter, as in fact they did. In short, the Leninist–Stalinist version of socialism was a possible interpretation, though certainly not the only possible one, of Marx's doctrine. If freedom equals social unity, then the more unity there is, the more freedom; as the 'objective' conditions of unity have been achieved, namely the confiscation of bourgeois property, all manifestations of discontent are relics of the bourgeois past and should be treated accordingly. The Promethean principle of creative initiative divided the field with historical determinism: initiative was vested in the political machine, while the backward masses were expected to accept their lot as a historical necessity which, once understood, was identical with freedom. There is nothing easier than to find passages in Marx which support the view that the superstructure is an instrument of the base and that both must be described in class categories. If there are new relations of production reflecting the interests of the proletariat, the superstructure—politics, laws, literature, art, and science—must conform to the demands of these relations as interpreted by the conscious vanguard of the proletariat. Thus the abolition of law as a mediating institution between individuals and the state, and the principle of servility in every manifestation of culture, could be regarded as a perfect embodiment of Marxist theory.

It is easy to reply to objections such as these that Marx (except perhaps for a short time after the revolutions of 1848) not only did not question the principles of representative democracy but regarded them as a necessary part of popular rule, and that although on two occasions he used the term 'dictatorship of the proletariat' (without saying what he meant by it), he had in mind the class content of the power system and not, as Lenin did, the liquidation of democratic institutions. It follows that the despotic socialism of history is not socialism as Marx intended it; the question, however, is how far it represents the logical outcome of his doctrine. To this it may be answered that the doctrine is not wholly innocent, though it would be absurd to say that the despotic forms of socialism were a direct outcome of the ideology itself. Despotic socialism arose from many historical circumstances, the Marxist tradition among them. The Leninist–Stalinist version of Marxism was no more than a version, i.e. one attempt to put into practice the ideas that Marx expressed in a philosophical form without any clear principles of political interpretation. The view that freedom is measured in the last resort by the degree of unity of society, and that class interests are the only source of social conflict, is one component of the theory. If we consider that there can be a technique of establishing social unity, then despotism is a natural solution of the problem inasmuch as it is the only known technique for the purpose. Perfect unity takes the form of abolishing all institutions of social mediation, including representative democracy and the rule of law as an independent instrument for settling conflicts. The concept of negative freedom presupposes a society of conflict. If this is the same as a class society, and if a class society means a society based on private property, then there is nothing reprehensible in the idea that the act of violence which abolishes private property at the same time does away with the need for negative freedom, or freedom *tout court.*

And thus Prometheus awakens from his dream of power, as ignominiously as Gregor Samsa in Kafka's *Metamorphosis.*

SELECTIVE BIBLIOGRAPHY

THE relevant works by Marxist authors whose writings are discussed in Volumes II and III are listed in the bibliographies of those volumes. When an edition of collected works of an author is available, individual works are not mentioned.

1. GENERAL WORKS. GENERAL ANTHOLOGIES AND BIBLIOGRAPHIES

Bibliographie marxiste internationale, Paris: Centre d'études et de recherches marxistes, 1964–.

Cole, G. D. H., *A History of Socialist Thought*, 5 vols., London, 1953–60.

Drachkovitch, M. M., (ed.), *Marxism in the Modern World*, London, 1965.

Drahn, E., *Marx-Bibliographie*, Charlottenburg, 1920.

Fetscher, I., *Der Marxismus. Seine Geschichte in Dokumenten*, Munich, 1967 (an anthology with a copious bibliography).

— — *Karl Marx und der Marxismus*, Munich, 1967 (Eng. trans. 1971).

Howe, I., *Essential Works of Socialism*, 2nd edn., New Haven and London, 1976.

Jordan, Z. A., *The Evolution of Dialectical Materialism*, London and New York, 1967.

Lachs, J., *Marxist Philosophy, a Bibliographical Guide*, Chapel Hill, 1967.

Lichtheim, G., *Marxism. A Historical and Critical Study*, London, 1961.

Masaryk, T. G., *Die philosophischen und soziologischen Grundlagen des Marxismus*, Vienna, 1899.

Papaioannou, K., *Marx et les marxistes*, Paris, 1970 (an anthology).

Plamenatz, J., *German Marxism and Russian Communism*, London, 1954.

Vranicki, P., *Geschichte des Marxismus*, 2 vols., Frankfurt, 1972–4.

Wolfe, B., *Marxism: 100 years in the life of a doctrine*, New York, 1965.

2. WORKS BY MARX AND ENGELS

Marx–Engels, *Historisch-kritische Gesamtausgabe* (*M.E.G.A.*), Frankfurt, 1927–35 (incomplete).

— — *Werke*, 39 vols., Berlin, 1956 ff. (incomplete).

— — *Collected Works*, Moscow, New York, and London, 1975 ff. (planned to comprise 50 vols.).

Anthologies in English

Marx–Engels, *Selected Writings in Sociology and Social Philosophy*, ed. T. Bottomore and M. Rubel, London, 1956.

— — *Basic Writings on Politics and Philosophy*, ed. L. Feuer, New York, 1959.

Marx, Karl, *Early Writings*, ed. T. Bottomore, London, 1963.

— — *The Essential Writings*, ed. D. Caute, London and New York, 1967.

— — *The Early Texts*, ed. D. McLellan, Oxford, 1971.

— — *Economy, Class and Social Revolution*, ed. Z. A. Jordan, London, 1971.

3. GENERAL WORKS ON MARX AND ENGELS; BIOGRAPHIES

Berlin, I., *Karl Marx*, 3rd edn., Oxford, 1973.

Blumenberg, W., *Karl Marx*, London, 1971.

Calvez, J.-Y., *La Pensée de Karl Marx*, Paris, 1956.

Cornu, A., *Karl Marx et Friedrich Engels*, 3 vols., Paris, 1955–62.

Künzli, A., *Karl Marx, eine Psychographie*, Vienna, Frankfurt, and Zurich, 1966.

McLellan, D., *The Thought of Karl Marx. An Introduction*, London, 1971.

— — *Karl Marx. His Life and Thought*, London, 1973.

Mayer, G., *Friedrich Engels. Eine Biographie*, 2 vols., The Hague, 1934.

Mehring, Fr., *Karl Marx. Geschichte seines Lebens*, Leipzig, 1918 (Eng. trans. 1936).

Meyer, A. G., *Marxism: The Unity of Theory and Practice*, Ann Arbor, 1963.

Nikolaevsky, B. I. and Maenchen-Helfer, O., *Karl Marx, Man and Fighter*, London, 1936.

Payne, R., *Marx: A Biography*, London, 1968.

Raddatz, F. J., *Karl Marx, Eine politische Biographie*, Hamburg, 1975.

Rubel, M., *Bibliographie des œuvres de Karl Marx*, Paris, 1956.

— — *Karl Marx. Essai de biographie intellectuelle*, Paris, 1957.

4. THE HEGELIAN LEFT, FEUERBACH, THE YOUNG MARX, THE MARX–HEGEL RELATIONSHIP

Avron, H., *Feuerbach ou la transformation du sacré*, Paris, 1957.

Bauer, B., *Kritik der evangelischen Geschichte des Johannes*, Bremen, 1840.

— — *Kritik der evangelischen Geschichte der Synoptiker*, Leipzig, 1841.

— — *Das entdeckte Christenthum*, Zurich, 1843.

Berlin, I., *The Life and Opinions of Moses Hess*, Cambridge, 1959.

Cieszkowski, A. Graf von, *Prolegomena zur Historiosophie*, Berlin, 1838, 2nd edn. 1908.

— — *Gott und Palingenesie*, 1842.

Dupré, L., *The Philosophical Foundation of Marxism*, New York, 1966.

Feuerbach, L., *Sämtliche Werke*, 10 vols., ed., W. Bolin und F. Jodl, Leipzig, 1903–11; 2nd edn. 1959.

Friedrich, M., *Philosophie und Ökonomie beim jungen Marx*, Berlin, 1960.

Garewicz, J., *August Cieszkowski w oczach Niemców . . . (August Cieszkowski in the eyes of the Germans)*, in *Polskie Spory o Hegla (Polish Controversies about Hegel)*, Warsaw, 1966.

Grégoire, Fr., *Aux sources de la pensée de Marx: Hegel, Feuerbach*, Louvain, 1947.

Hess, Moses, *Philosophische und sozialistische Schriften*, 1837–1850, ed. A. Cornu und W. Monke, Berlin, 1961.

Hillmann, G., *Marx und Hegel*, Frankfurt, 1966.

Hook, S., *From Hegel to Marx*, Ann Arbor, 1962 (1st edn. 1936).

Hyppolite, J., *Études sur Marx et Hegel*, Paris, 1955 (Eng. trans. 1969).

Istituto G. Feltrinelli, *Annali*, vii, 1964–5 (contains a bibliography).

Jodl, F., *Ludwig Feuerbach*, Stuttgart, 1921.

Kühne, W., *August Graf von Cieszkowski*, 1938.

Löwith, K., *Die Hegelsche Linke*, Stuttgart, 1962.

— — *Von Hegel zu Nietzsche*, Stuttgart, 1959.

McLellan, D., *Marx before Marxism*, London, 1970.

Maguire, J., *Marx's Paris Writings: An Analysis*, Dublin, 1972.

Panasiuk, R., *Filozofia i Państwo. Studium myśli polityczno-społecznej lewicy heglowskiej i młodego Marksa 1838–1841 (Philosophy and State. A Study of the Social and Political Thought of the Hegelian Left and the Young Marx 1838–1841)*, Warsaw, 1967 (with a good bibliography).

Popitz, H., *Der entfremdete Mensch*, Basle, 1953.

Rosen, Z., *Bruno Bauer and Karl Marx. The Influence of Bruno Bauer on Marx's Thought*, The Hague, 1977.

Silberner, E., *Moses Hess. An annotated Bibliography*, New York, 1951.

Strauss, D. F., *Das Leben Jesu*, 2 vols., Tübingen, 1835–6.

— — *Streitschriften zur Verteidigung meiner Schrift über das Leben Jesu*, Tübingen, 1838.

— — *Die christliche Glaubenslehre*, 2 vols., 1840–1.

Thier, E., *Das Menschenbild des jungen Marx*, Göttingen, 1957.

Tschiżewskij, D., *Hegel bei den Slaven*, Reichenberg, 1934.

Tucker, R. C., *Philosophy and Myth in Karl Marx*, Cambridge, 1961.

Walicki, A., 'Cieszkowski—filozoficzna systematyzacja mesjanizmu' ('Cieszkowski—a Philosophical Systematization of Messianism'), in *Filozofia Polska (Polish Philosophy)*, Warsaw, 1967.

— — 'Two Polish Messianists: Adam Mickiewicz and August Ciesz-kowski', in *Oxford Slavonic Papers*, vol. ii, Oxford, 1969.

Złocisti, T., *Moses Hess, der Vorkämpfer des Sozialismus und Zionismus*, Berlin, 1921.

Zółtowski, A., *August Graf von Cieszkowski's Philosophie der Tat*, Posen, 1904.

5. PRE-MARXIAN SOCIALISM

Abendroth, Wolfgang, *Sozialgeschichte der europäischen Arbeiterbewegung*, 1965.

Advielle, V., *Histoire de Gracchus Babeuf et du babouvisme*, Paris, 1884.

Angrand, P., *Étienne Cabet et la République de 1848*, Paris, 1948.

Babeuf, G., *Textes choisis*, Préf. et Comment. par G. et I. Willard, Paris, 1950.

Beer, M., *A History of British Socialism*, London, 1953.

Bernstein, S., *The Beginning of Marxian Socialism in France*, New York, 1965.

Blanc, L., *Organisation du travail*, Paris, 1839.

— — *Le Socialisme. Droit au travail*, Paris, 1848.

Blanqui, A., *Critique sociale*, 2 vols., Paris, 1885.

Bouglé, C., *La Sociologie de Proudhon*, Paris, 1911.

Bravo, G. M., *Wilhelm Weitling e il communismo tedesco prima del quarantotto (Wilhelm Weitling and German Communism before '48)*, Turin, 1963.

— — *Les Socialistes avant Marx*, 3 vols., Paris, 1970 (an anthology, with a copious bibliography).

Charléty, S., *Histoire du saint-simonisme*, Paris, 1931.

Cole, G. D. H., *The Life of Robert Owen*, London, 1930.

Cole, M. I., *Robert Owen of New Lanark*, London, 1953.

Del Bo, G., *Charles Fourier e la scuola societaria. Saggio bibliografico (Charles Fourier and the 'societaria' School. A Bibliographical Essay)*, Milan, 1957.

Dommanget, M., *Les Idées politiques et sociales d'Auguste Blanqui*, Paris, 1957.

— — *Victor Considérant. Sa Vie, son œuvre*, Paris, 1929.

Fourier, Ch., *Œuvres complètes*, Paris, 1841–5.

Fournières, E., *Les Théories socialistes au XIXe siècle de Babeuf à Proudhon*, Paris, 1904.

Garewicz, J., *Między marzeniem a wiedzą. Początki myśli socjalistycznej w Niemczech (Between Dream and Knowledge. Beginnings of Socialist Thought in Germany)*, Warsaw, 1975.

Gray, A., *The Socialist Tradition*, London, 1963.

Gurvitch, G., *Proudhon, sociologue*, Paris, 1955.

— — *Les Fondateurs français de la sociologie contemporaine: Saint-Simon et P. J. Proudhon*, Paris, 1955.

— — *Proudhon et Marx: une confrontation*, Paris, 1964.

Halévy, E., *Histoire du socialisme européen*, Paris, 1948.

Harris, D., *Socialist Origins in the United States. American Forerunners of Marx 1817–1832*, 1966.

Harrison, J. F. C., *Robert Owen and the Owenites in Britain and America*, London, 1969.

Harvey, R. H., *Robert Owen, Social Idealist*, Berkeley, 1949.

Jackson, J. H., *Marx, Proudhon and European Socialism*, New York, 1962.

Labrousse, C. E., *Le Movement ouvrier et les théories sociales en France de 1815 à 1848*, Paris, 1861.

Landauer, C., *European Socialism*, Berkeley, 1959.

Leroy, M., *Le Socialisme des producteurs. Henri de Saint-Simon*, Paris, 1924.

— — *Histoire des idées sociales en France*, 2nd edn., Paris, 1962.

Lichtheim, G., *The Origins of Socialism*, London, 1969.

Loubère, L. A., *Louis Blanc. His Life and his Contribution to the Rise of French Jacobin-Socialism*, Evanston, 1961.

Manuel, F. E., *The New World of Henri Saint-Simon*, Cambridge, Mass., 1956.

— — *The Prophets of Paris*, Cambridge, Mass., 1962.

Mazauric, C., *Babeuf et la conspiration pour l'égalité*, Paris, 1962.

Morton, A. L., *The Life and Ideas of Robert Owen*, London, 1962.

Owen, R., *A New View of Society, and Other Writings*, Intro. by G. D. H. Cole, London, 1927.

— — *The Book of the New Moral Worlds*, London, 1836–1844.

— — *The Future of the Human Race*, London, 1853.

Pollard, S. and Salt, J., (eds.), *Robert Owen. Prophet of the Poor*, London, 1971.

Proudhon, P. J., *Œuvres complètes*, Paris, 1923, ff. (incomplete).

Renard, E., *Bibliographie relative à Louis Blanc*, Toulouse, 1922.

— — *Louis Blanc, sa vie, son œuvre*, Paris, 1928.

Spitzer, A. B., *The Revolutionary Theories of L. A. Blanqui*, New York, 1957.

Saint-Simon, H., *Œuvres de Saint-Simon et d'Enfantin*, 47 vols., Paris, 1865–78.

— — *Selected Writings*, trans. F. M. H. Markham, Oxford, 1952.

Talmon, J. L., *The Origins of Totalitarian Democracy*, London, 1952.

Vidalenc, J., *Louis Blanc*, Paris, 1948.

Walch, J., *Bibliographie du saint-simonisme*, Paris, 1967.

Weitling, W., *Das Evangelium eines armen Sünders. Die Menschheit wie sie ist und wie sie sein sollte*, ed. W. Schäfer, Reinbek bei Hamburg, 1971.

Wittke, C., *The Utopian Communist. A Biography of W. Weitling, Nineteenth Century Reformer*, Louisiana State U.P., 1950.

6. THE FIRST INTERNATIONAL, BAKUNIN, LASSALLE

Archives Bakounine, Textes établis et annotés par Arthur Lehning, The Hague, 1961 ff.

Bakunin, M., *Œuvres*, 6 vols., Paris, 1895–1913.

—— —— *Gesammelte Werke*, 3 vols., Berlin, 1921–4.

—— —— *Sobranie sochineniy i pisem* (*Collected Works and Letters*), ed. Y. Steklov, 4 vols., Moscow, 1934–5.

Braunthal, J., *History of the International*, vol. i, 1864–1914, London, 1966.

Carr, E. H., *Michael Bakunin*, London, 1937.

Collins, M. and Abramsky, C., *Karl Marx and the British Labour Movement*, London, 1965.

Footman, D., *Ferdinand Lassalle*, New York, 1969.

Freymond, J. (ed.), *La Première Internationale*, 2 vols., Geneva, 1962.

Guillaume, J., *Karl Marx, pangermaniste, et l'Association Internationale des Travailleurs de 1864 à 1870*, Paris, 1915.

Lassalle, Ferdinand, *Gesammelte Reden und Schriften*, 12 vols., hg. und eingeleitet von E. Bernstein, Berlin, 1919–20.

—— —— *Reden und Schriften* ... *im Auswahl*, hg. von H. Feigl, Vienna, 1920.

Lehning, A., *From Buonarroti to Bakunin, Studies in International Socialism*, London, 1970.

Molnar, M., *Le Déclin de la Première Internationale*, Geneva, 1963.

Oncken, H., *Lassalle. Zwischen Marx und Bismarck*, Stuttgart, 1966 (1st edn. 1904).

Pyziur, E., *The Doctrine of Anarchism of M. A. Bakunin*, Milwaukee, 1955.

Ramm, T. (ed.), *Der Frühsozialismus*, Stuttgart, 1955.

Temkin, H., *Bakunin i antynomie wolności* (*Bakunin and the Antinomies of Freedom*), Warsaw, 1964.

7. VARIOUS ASPECTS OF MARX'S AND ENGELS'S DOCTRINES

Avineri, S., *The Social and Political Thought of Karl Marx*, Cambridge, 1968.

Amsterdamski, S., *Engels*, Warsaw, 1965.

Becker, W., *Kritik der Marxschen Wertlehre*, Hamburg, 1972.

Bober, M. M., *Karl Marx's Interpretation of History*, Cambridge, Mass., 1950.

Bottomore, T., *The Sociological Theory of Marxism*, London, 1972.

Croce, B., *Historical Materialism and the Economics of Karl Marx*, London, 1922.

Dahrendorf, R., *Marx in Perspektive. Die Idee des Gerechten im Denken von Karl Marx*, Hanover, 1952.

Eilstein, H. (ed.), *Jedność materialna świata* (*The Material Unity of the World*), Warsaw, 1961.

Feuer, L. S., *Marx and the Intellectuals*, New York, 1969.

Fleischer, H., *Marxismus und Geschichte*, Frankfurt, 1969 (Eng. trans. 1973).

Gamble, A. and Walton, P., *From Alienation to Surplus Value*, London, 1972.

Heller, A., *The Theory of Need in Marx*, London, 1976.

Hommes, J., *Der technische Eros: das Wesen der materialistischen Geschichtsauffassung*, Freiburg, 1955.

Horowitz, D. (ed), *Marx and Modern Economics*, London, 1968.

Jaroslawski, J., *Theorie der sozialistischen Revolution von Marx bis Lenin*, Hamburg, 1973.

Kamenka, E., *The Ethical Foundation of Marxism*, London, 1963.

Kedrov, B. M., *Engels i dialektika yestestvoznania* (*Engels and the Dialectics of Natural Science*), Moscow, 1970.

Klages, H., *Technischer Humanismus*, Stuttgart, 1964.

Krajewski, W., *Engels o ruchu materii i jego prawidłowości* (*Engels on the Movement of Matter and its Laws*), Warsaw, 1973.

Lifshitz, M., *Karl Marx und die Aesthetik*, Dresden, 1967.

Lobkowicz, N., *Theory and Practice. The History of a Marxist Concept*, Notre Dame, 1967.

Mandel, E., *The Formation of the Economic Thought of Karl Marx*, London, 1971.

Morawski, S., *Il Marxismo e l'estetica* (*Marxism and Aesthetics*, Rome, 1973.

Ollman, B., *Alienation: Marx's Critique of Man in Capitalist Society*, Cambridge, 1971.

Plamenatz, J., *Karl Marx's Philosophy of Man*, Oxford, 1975.

Post, W., *Kritik der Religion bei Karl Marx*, Munich, 1969.

Prawer, S. S., *Karl Marx and World Literature*, Oxford, 1976.

Reiprich, K., *Die philosophisch-naturwissenschaftlichen Arbeiten von Karl Marx und Friedrich Engels*, Berlin, 1969.

Robinson, J., *An Essay in Marxian Economics*, 2nd edn., London, 1967.

Rosdolsky, R., *Zur Entstehungsgeschichte des Marxschen 'Kapitals'*, Frankfurt, 1968.

Schmidt, A., *Der Begriff der Natur in der Lehre von Karl Marx*, Frankfurt, 1962 (Eng. trans. 1971).

Schwann, G., *Die Gesellschaftskritik von Karl Marx*, Stuttgart, 1974.

Tucker, R., *The Marxian Revolutionary Idea*, London, 1970.

Witt-Hansen, J., *Historical Materialism: the Method, the Theories*, New York, 1960.

Index